Redefining Airmanship

Other McGraw-Hill titles of interest

Piloting for Maximum Performance *by Lewis Bjork*

Optimizing Jet Transport Efficiency: Performance, Operations & Economics *by Carlos E. Padilla*

Aircraft Safety: Accident Investigations, Analyses and Applications *by Shari Stamford Krause, Ph.D.*

Airport Planning & Management-3rd Edition *by Alexander T. Wells, Ed.D.*

Airport Operations-2nd Edition *by Norman Ashford, H.P. Martin, and Clifton A. Moore*

Denver International Airport: Lessons Learned *by Paul S. Dempsey, Andrew R. Goetz and Joseph S. Szyliowicz*

Flying Jets *by Linda O. Pendleton*

Kitplane Construction-2nd Edition *by Ronald J. Wanttaja*

They Called It Pilot Error *by Robert Cohn*

Becoming a Better Pilot *by Paul A. Craig*

Redefining Airmanship

Tony Kern

McGraw-Hill

New York San Francisco Washington, D.C. Auckland Bogotá
Caracas Lisbon London Madrid Mexico City Milan
Montreal New Delhi San Juan Singapore
Sydney Tokyo Toronto

Library of Congress Cataloging-in-Publication Data

Kern, Tony
    Redefining airmanship / by Tony Kern.
    p. cm.
    ISBN 0-07-034284-9
    1. Aeronautics. 2. Airplanes — Piloting. I. Title.
  TL545.K4 1996
  629.132'52 — dc20   96-41788
                                                 CIP

McGraw-Hill

A Division of The McGraw-Hill Companies

11 12 13 14 15 QFR/QFR 1 5 4 3 2 1

ISBN   0-07-034284-9

The sponsoring editor of this book was Shelley IC. Chevalier, the editing supervisor was Sally Glover, and the production supervisor was Claire B. Stanley. This book was set in Garamond by McGraw-Hill's desktop publishing department in Hightstown, N.J.

Printed and bound by R.R. Donnelly & Sons Company.

 This book is printed on recycled, acid-free paper containing a minimum of 50% recycled, de-inked fiber.

McGraw-Hill books are available at special quantity discounts to use as premiums and sales promotions, or for use in corporate training programs. For more information, please write to the Director of Special Sales, McGraw-Hill, 11 West 19th Street, New York, NY 10011. Or contact your local bookstore.

This book is dedicated to the memory of Major General Glenn A. Profitt II. While serving as the director of plans and programs at the USAF Air Education and Training Command, General Profitt had the vision and perseverance to establish the U.S. Air Force's first career-spanning and integrated human factors training system. Ironically, he perished less than a year later in a human-factors-related mishap while flying as a passenger aboard a C-21 aircraft.

Contents

7 Know your environment   *157*

8 Know your risk   *197*

9 Situational awareness   *229*

10 Judgment and decision making   *253*

13 Instructing and evaluating airmanship 343

14 Understanding airmanship error *373*

15 The marks of an airman *405*

Acknowledgments

I have received guidance from so many people in writing this book that I cannot possibly list them all here. However, I would like to thank the major players that made this book possible. Special gratitude to Dave Wilson and Hughes Training for providing several of the case studies that make up the heart of this work. Also, special thanks to Ted Mallory at Northwest Airlines, Dr. John Lauber at Delta Airlines, and Al Mullen at Crew Training International for contributing materials and insights for the research effort. Dr. Bob Helmreich, Dr. John Wilhelm, Pete Connelly, and all the folks at the NASA/University of Texas Aircrew Research Laboratory have provided fantastic support throughout all my research efforts spanning many years. I also wish to acknowledge the assistance of the USAF Academy Library Special Collections, USAF Safety Center, the Air Force Historical Research Agency, Colonel Mark Wells, Dr. Alan Diehl, Dr. Richard Reinhart, Dr. Robert Alkov, Jim Quick, Bernie Hollenbeck, Brigadier General Chuck Yeager, Jim Simon, Ben Drew, Lieutenant Colonel Eileen Collins and NASA, Stephen Coonts, and Major General John Huston. I would like to also thank the University Aviation Association for helping me to validate the airmanship model. A very special thanks to General John Shaud and my friend John Nance for their contributions of the foreword and preface.

I would like to thank my beautiful wife, Shari, for her continuous editing, typing, moral support, and encouragement, and my sons, Jacob and Trent, for allowing me time between wrestling matches and living-room football to complete this book. Most importantly, I would like to thank the Lord for giving me the creative energy and perseverance to make this contribution to aviation safety and professionalism.

List of Contributors

Robert A. Alkov, Ph.D.
Southern California Safety Institute

Jack Barker, Ph.D.
Major, USAF

Eileen Collins
Lieutenant Colonel, USAF

Pete Connelly
NASA/University of Texas Air Crew Research Project

Stephen Coonts
Author

Alan Diehl, Ph.D.
Former NTSB and USAF accident investigator

Urban Drew
Major, USAF (Ret.)

J. D. Garvin, Ed.D.
Captain, USAF

Bernie Hollenbeck
Colonel, USAF (Ret.)

John Huston
Major General, USAF (Ret.)

Glen Hover
Major, USAF

Richard Jensen, Ph.D.
Director, Ohio State University Aviation Psychology Laboratory

John Lauber, Ph.D.
Delta Airlines, Corporate Safety and Compliance

Al Mullen
President, Crew Training International

John Nance
Author and airline pilot

Jim Quick
USAF Safety Center

Richard Reinhart, M.D.
Colonel, USAFR

Steve Ritchie
Brigadier General, USAFR

John Shaud
General, USAF (Ret.)
Executive Director, Air Force Association

Jim Simon
Major, USAF

Chuck Yeager
Brigadier General, USAF (Ret.)

Foreword

Throughout my involvement with the United States Air Force and aviation in general, I have held a strong conviction about developing complete aviators who are capable of using all the resources available to them as they go about flying in many and varied conditions. While I was commander of the USAF Air Training Command, we directed the integration of cockpit resource management (CRM) training into the T-1A "Jayhawk" program, making it the first military system to fully integrate behavioral factors into the crew coordination training process. We did this so that all of our pilots going through specialized undergraduate pilot training (SUPT) en route to crewed aircraft would have a basic understanding of crew coordination before they flew their first operational mission. I am equally convinced that this need extends beyond crewed aircraft and into the classrooms, briefing rooms, and cockpits of our fighter and reconnaissance aircraft as well. I suspect the same applies in commercial and general aviation.

But schoolhouse approaches can only do so much. We continue to lose far too many men and women to mishaps that are avoidable, in spite of the many organizational initiatives and regulations designed to prevent these occurrences. Make no mistake, our training programs are working well, and our safety record continues to improve. But there remains a need for a cultural change within aviation to put an end to undisciplined practices and restore basic airmanship as the signpost of excellence.

A cultural change of this type must come from within. In the military, it must come from the hundreds of soldiers, sailors, and airmen who strap on their aircraft every day and take off to do the nation's work. Top-level leadership must support the effort, but for the necessary change to take place, it needs to be championed from inside. Peer pressure—not the threat of punishment—is what will make flight discipline violations disappear and reinvigorate the pursuit of excellence. When lack of skill, currency, or knowledge becomes unacceptable, and when total competence is seen as the standard of airmanship, the cultural change will have occurred.

Outstanding airmanship has been called by many names over the years, from "Sierra Hotel" to "the right stuff." However, the complexity of modern flight operations has added new dimensions to airmanship and a shared meaning of airmanship may have been lost. Perhaps we are all a little at fault. In our zeal to improve operations and safety, we have studied the flight crew, aircraft, and mission from literally hundreds of angles. In the process, we have dissected flying operations almost beyond recognition. While the tacticians, social scientists, and engineers have added greatly to the sum of our knowledge on a variety of important and relevant subjects, the flyer's ability to merge all of this knowledge into a complete picture of airmanship has greatly suffered. Meanwhile, high technology has increased the air machine's capabilities by an order of magnitude. This factor might have created apathy in our younger flyers, who gradually became more interested passengers than committed pilots, and, because of technology, they get away with it—most of the time. This lax attitude is inappropriate for our high-risk operation.

While we should continue to debate such things as tactics and techniques, the fundamental essence of *who we are* and *what we stand for* should be universally understood. As the introduction to this book points out so astutely, "Airmanship is far too important for relativistic interpretation." There is a very real need in aviation to put it all back together, and *Redefining Airmanship* does an excellent job of doing just that.

The airmanship model outlined in Chapter 1 is right on target. There can be no question that flight discipline is the foundation of all that we become as airmen. Mishap data continues to show that we have a small percentage of aviators who, for one reason or another, fail to develop this cornerstone of professionalism. The remainder of the model illustrates a holistic picture of what real airmanship is, with skill, currency, and multiple knowledge bases supporting the capstones of airmanship—situational awareness and judgment. This unique synthesis is a welcome step towards a unified understanding of what is important to individual airmen. This type of synthesis can only be accomplished by someone with deep roots in aviation operations, training, and human behavioral factors. The author has all three. He presents a new way of looking at an old subject, reuniting airmanship in theory with the way it has always been in practice. Self-improvement and personal responsibility help define airmanship, and this approach is useful for new and experienced aviators alike as they sift through myriad aviation topics to find and develop what is needed on a personal level. A tip of the hat to *Redefining Airmanship* and especially the author's use of operational examples to provide the best reality check of all—the actual world of flying.

John A Shaud

General John Shaud, USAF (Retired)
Executive Director, Air Force Association

Preface

profession *(n.) A calling requiring specialized knowledge and often long and intensive academic preparation.*
Webster's Seventh New College Dictionary

carbon-based units *(n.) "Humans, Ensign Perez. Us."*
Dr. Leonard McCoy, chief medical officer, U.S.S.
Enterprise **(From** *Star Trek: The Motion Picture,*
Paramount Pictures, Inc.)

Aviation is an error-intolerant profession, the safe and orderly practice of which is continually threatened by the error-prone nature of its most indispensable component: humans. Yet the concept of aviation without human involvement is an absurdity. That's quite a catch-22, a creation compromised by its creator's tendency to be less than universally professional.

But what does being a professional have to do with airmanship? Well, the recent human revolution in aviation has led us to realize that the more we approach flying with the same solemnity and dedication required of the traditional professions, the safer and more effective we become—whether we're flying for hire, for fun, or for God and country. Modern airmanship, in other words, requires the stuff of professionals, not gimlet-eyed risk-takers. The stakes are simply too high to tolerate the alternative.

But can airmen really be referred to as professionals? Given the cost of today's air machines—and today's accidents—in human and monetary terms, there can be no reasonable tolerance of performance at less than a professional level. We are professionals, in other words, regardless of our niche in aviation, and we must act the part.

Don't cringe, but another definition is in order. "Professionals," you see, were long thought in Western culture to include only those who practiced the so-called "learned professions," principally law and medicine. But the English language has always been sufficiently elastic to admit new candidates under its definitional umbrella as civilization and technology evolved, and new, previously unheard-of pursuits, such as flying, were invented.

Aviation, for instance—the "practice" of flying—is a classic case. In fact, in giving training seminars for medical professionals, the definition I have long used to describe a surgical team ("highly trained professionals working in a time-critical and highly structured environment utilizing sophisticated implements and devices to achieve specific goals, in which the penalties for failure or inadequacy are potentially great in both human and monetary terms") also precisely describes the responsibilities and demands of aviating. By the nature and demands of their duties, good aviators are, indeed, professionals.

Airmanship, as comprehensively defined for the first time by Dr. Kern in this work, truly does refer to a special form of professionalism, because it presupposes the personal acceptance of neverending responsibility and the individual determination to strive for perfection in the operation of aircraft of all types. Airmanship, in other words, requires that the practitioner wholeheartedly accept the ethical standard that less-than-professional discipline and conduct never be tolerated, even in amateur (i.e., purely for pleasure) operations.

The pursuit of profit, by the way, is normally part of the test of what is and what isn't a profession, but aviation is a necessary exception to that rule. In fact, a professional aviator is simply a person who approaches aviating in a professional manner, regardless of whether he or she has ever earned a cent through flying. Why? Because the stakes are far too high to tolerate the idea of amateur airmanship, and because nonpiloting persons will not tolerate amateurish, unprofessional conduct even from a weekend flyer in a '49 Cub. You have only to search any local newspaper over a year's time to find examples of righteous indignation at the less-than-professional act of some pilot who lost track of the aircraft's fuel, decided to test the ice-carrying characteristics of an aircraft, or otherwise used poor judgment in what was obviously a not-for-hire operation. Is that an excessive burden, to require all of us all the time to be professional?

Not when you consider the license the public gives the military, the commercial operators, and the private operator to fly large, heavy machines over their heads and possessions. In fact, each time an aircraft lifts from the ground (by whatever means), its operator accepts a public trust, and that's just as true whether the pilot is flying a Navy jet off a carrier in the watery middle of nowhere, cruising a multibillion-dollar B-2 over the nation's capital, or launching a loaded civilian 747-400 over a the heart of an American city. The responsibility to use all available resources and intellect to safeguard that craft, its occupants, the investment of the owners — as well as those trusting souls below — is open-ended.

If this is beginning to sound like a lecture on personal responsibility, good! It is. Personal responsibility in aviation is more than just signing a form and taking command; it is accepting a lifetime commitment to use maximum effort and intellect in the pursuit of aviation excellence, regardless of what seat you occupy and regardless of your position or rank within any aviating organization, military or otherwise. The responsibilities to act, to speak up, and to communicate when necessary are also perpetual and indispensible — as well as a lesson we've too often paid for in blood. The responsibility for safety cannot be delegated, regardless of how low on the totem pole an airman perceives himself or herself to be.

These are all components of the revolution in thinking that began in the late 1970s — the human revolution in aviation. Chief among its lessons is the truth that carbon-based units (us, Ensign Perez) are an integral part of all aviating operations — even remotely piloted ones. Why is that a revolution? Because for the first time we are systemically admitting and cataloging our human limitations and building our aviation systems to tolerate human failure without compromising safety. It's a revolution, though, that has just begun, and to the extent that airmen and their institutions refuse to embrace its principles, its benefits are still not fully realized.

Too often we still send exhausted pilots to their deaths because it's against the prevailing machismo attitude to admit that pilots can get tired. Too often we still try to deny the human nature of flying and the human nature of airmanship. Too often we force pilots to defer to overly complex instruments and computers without considering the predictable results (such as mode confusion). Too often we still lose aircraft because someone aboard refused to speak up about a particular problem, a personal limitation, an operational decision, or some other challenge that was eventually answered by default, with disastrous consequences. And too often an entire aviation organization embraces dangerous, dysfunctional human conduct because no one will take the individual responsibility to point out a naked emperor.

In fact, unrestricted communication within all components of the aviation organization — the timely transfer of information — is the heart and soul of aviation safety. Today we know for certain that to achieve acceptable levels of risk in any aviation environment (including military combat), all available human intellect and knowledge must be fully incorporated and involved in the decision-making process, whether in the headquarters building, the corporate offices, the maintenance organization, or the cockpit. That's more than a theory. That's an axiom.

Translation: We've dismissed Captain James T. Kirk (who knew all, saw all, and flew solo) and hired Captain Jean-Luc Picard, who knew how to use all the fine, highly trained, highly paid human talent around him to wring the best decisions from even the worst situations.

What follows is a unique work that, perhaps for the first time, begins to define the professional ethos of being an airman, a professional engaged in committing periodic acts of aviation. Its author is a friend and colleague with whom I've worked long hours in battles against the slavering dragons of bureaucracy and the forces of inertia, trying to bring the enlightened disciplines of CRM (crew resource management) to the military world. His work is a breakthrough and a benchmark.

Lt. Col. John J. Nance, USAFR
Pilot and author

Introduction

Standards set by precedent are based on something less than average performance, and for that reason, one should not submit to them.
Field Marshall Erwin Rommel

The goal of this book is to provide structure for a lifetime of learning about airmanship, in the hopes of establishing higher personal standards related to your flying activities. This step is a very necessary first one towards personal excellence, because modern airmanship has taken on an ever-increasing and sometimes bewildering complexity, leaving many airmen to wonder what it means to be an expert. Aviators need to understand how all the various factors of airmanship fit together and, perhaps more importantly, how they interact with each other. This understanding is key to improving airmanship, whether you are just entering training or attempting to upgrade your aviation skills. The text seeks to demystify the many complex psychological and physiological aspects of airmanship in an attempt to return the information to the aviator, where it truly belongs. In academic jargon, this book is definitely an "applied" text.

Modern aviators face a tough task. We are required to combine physical, cognitive, team-building, and communication skills, while simultaneously monitoring, managing, and updating a dynamic situation in a relatively hostile environment. Because of the multidisciplinary nature of flight operations, flyers from all realms of the aviation universe — commercial, military, and general aviation — must draw from multiple bases of knowledge to make assessments and split-second decisions and then have the skill and proficiency to execute them safely and effectively. The stakes are high — literally life and death. It is not child's play.

Education and training programs abound to ensure that we meet the minimum requirements for safe operations. But what about the aviators who seek to reach their maximum potential? Where do they begin their quest for personal achievement? Clearly, an understanding of what good airmanship is must be the first step.

While this might not sound like a daunting task, it is often very difficult to get any two aviators to agree on a definition of modern airmanship. The fundamental meaning of airmanship has changed, evolving from a meaning of basic stick-and-rudder competence to something much broader — a complex mix of human, machine, and environmental elements. In the process, many flyers have become confused about what constitutes good airmanship, and this problem goes beyond mere semantics. Operational errors and aviation mishaps — roughly 80 percent of which still involve human error — are frequently blamed on "poor airmanship." Myriad approaches have been implemented to remedy the problem, from ergonomics that address human-machine interface to training for better crew coordination and situational awareness. While these initiatives have achieved various levels of success, *the individual flyer* remains the key to meeting the last great challenge in aviation — human error. Aviators seeking self-improvement continue to ask themselves, "Where do I focus my efforts?" Because all aviators are not alike, the answer to this question must come from within each of us, based on a valid and shared understanding of airmanship. The goal of this book is to provide a structure for such an understanding.

Airmanship is clearly too important for relativistic interpretation. Failures often result in tragedy and unnecessary deaths, not only for those who make the errors, but for innocent victims as well. All aviators who share the sky should have an in-depth understanding of airmanship — a common ideal for discussion, assessment, and improvement.

Those of us who are lucky enough to earn our living as flyers have a *professional responsibility* to seek continuous improvement. This group includes military, commercial, and corporate aviators of all types: pilots, nonpilot flight personnel, students, instructors, and check airmen. As professional aviators, we are obligated to seek the highest standards of airmanship.

General aviation enthusiasts may benefit even more from a comprehensive and integrated approach to airmanship than their military, commercial, and corporate colleagues, because most recreational flyers lack the assets or time to attend the formal training programs that are the military and industry standard. Although many "how-to" books are on the market for general aviation pilots, few (if any) offer a comprehensive picture or an integrated systems approach to understanding and improving airmanship.

Regardless of the niche of aviation in which we employ our aircraft, we all share a *moral responsibility*—to each other and to the public at large—to operate in a safe and efficient manner. The responsibilities of flight are far too great to rely on anything less than a shared interpretation of airmanship standards.

A systems approach to understanding and improving airmanship

This book proposes a new way of thinking about the nebulous concept of airmanship. Based on extensive historical research, it suggests a systems thinking approach, in which each element of airmanship is seen as making an impact on the whole, in a dynamic and complex human equation. Peter Senge, the director for the Center for Organizational Learning at MIT and the author of *The Fifth Discipline,* points out the need for a comprehensive learning approach to complex phenomena:

> *From a very early age we are taught to break apart problems, to fragment the world. This apparently makes complex tasks and subjects more manageable, but we pay a hidden, enormous price. We can no longer see the consequences of our actions; we lose our intrinsic sense of connection to a larger whole . . . we try to reassemble the fragments in our minds, to list and organize all the pieces, but . . . the task is futile— similar to trying to reassemble the fragments of a broken mirror to see a true reflection (Senge 1990).*

Similarly, Clay Foushee, a world-renowned aviation human-factors expert, has found that breakdowns of airmanship are most often caused by failures of integration and not by any lack of skill or proficiency. If this is the case, the solution to many of these errors must lie, at least in part, in understanding and internalizing the concept of airmanship, which is what this book is about.

The goal of this book is to create a composite picture of a successful aviator, in whom no facet of airmanship exists in isolation. The picture that emerges from this effort merges training, operational, and human factors into a single entity—airmanship. This approach is both necessary and appropriate. The airman is still the single largest variable on any aircraft, and no institutional training or evaluation system can ever approach the capability of the internal barometer that lies within each of us for assessing our personal state of competency. But that barometer is only as good as the internal model of airmanship we possess. Without a clear and valid picture of the ideal, the internal guide is useless, and in some cases in which the individual's picture of airmanship is skewed, it can even be detrimental to improvement.

The value of self-appraisal

The great mathematician Archimedes said "Give me a long enough lever, and I can move the world." This book provides each aviator with a mental structure for applying leverage at the most appropriate point. By evaluating their own performance in terms of an established ideal of airmanship, aviators are able to develop an accurate self-analysis—perhaps the most valuable and rare tool in aviation today.

Self-appraisal is not a natural task for many flyers, who tend to stay within their comfort zones and avoid areas in which they are less than skilled or proficient. Many flyers overcompensate in one area to make up for weaknesses in others. We have all seen the type; for example, a systems or regulations expert who has significant problems making a crosswind landing, or the golden-hands type who is smugly certain that his or her skill and proficiency can make up for any lack of regulatory knowledge. The airmanship model suggests that these compensatory approaches are inappropriate and perhaps even dangerous. Flyers should strive for balance across the areas of airmanship as an umbrella against the unknown situation that lurks in Murphy's closet.

Target audience and goals

This book is written and designed for aviators from all fields who seek to meet the professional responsibilities and moral obligations inherent in aviation. It is a book to facilitate personal achievement, competence, and expertise. The airmanship model provides a basis for improvement in four primary ways:

1. It provides a relevant structure for integrating a lifetime of learning across the various disciplines of aviation training and education, merging physical skill development with cognitive education and human factors. By identifying the common traits of successful aviators from the past and present, it provides a historically valid definition of airmanship. A conceptualization, or "big picture," of airmanship is presented as a model for self-assessment and improvement.

2. Through case-study analyses from military, commercial, and general aviation, the book provides a means to see the integrated effects of the various elements of airmanship in a variety of scenarios. Analyzing case studies allows individual aviators to apply airmanship lessons to their own individual flying environments.

3. The airmanship model creates a framework for continuing discussion and development by training and operational personnel. By allowing the various disciplines to see how the integration between airmanship factors occurs at the individual level, increased cooperation and dialogue might be possible between disciplines. In addition, trainers from all corners of professional and general aviation might also find this book useful for case-study and training-program curriculum analysis.

4. The model provides a guide for continuous personal improvement to guard against complacency in those who believe themselves "past the training stage." Traditionally, many flyers view training as something done at the beginning of a career, during an upgrade to a new position or aircraft, or a recurring annual annoyance to be endured. They tire of redundant training in areas in which they already excel.

This book offers a structure for relevant self-improvement based on an individual diagnosis of personal airmanship. Aviators who have been content with single areas of specialization can broaden their airmanship base and increase their professional competence. With some minor exceptions, the fundamental elements of good airmanship apply across the traditional divisions of aviation. The relationship between proficiency, systems knowledge, and situational awareness is just as applicable and important to a weekend flyer in a Cessna 172 as it is to military pilot flying a B-1 bomber at 500 knots and 200 feet, albeit at different levels of demand.

Overview of the sections and chapters

This book is organized in five parts. The first four deal with the definition and details of the airmanship model, and the final section discusses other subjects of special interest. The book is organized around case studies, which are designed to assist the reader in grasping the integration of factors in a real-world setting.

Section one: Origins

Section one outlines the historical origins of airmanship, illustrating the emergence of successful traits in airmen from the earliest mythological origins through the inventors, air racers, and military and commercial operations. Chapter 1 outlines the results of historical research, demonstrating that certain common themes appear throughout the history of human flight. The central thesis of this chapter is that success leaves clues and that modern aviators should not ignore the path outlined by our aviation ancestors.

Section two: Foundations

Section two begins the description of the airmanship model, which uses the analogy of airmanship as a building composed of a foundation, pillars, and capstones. The chapters in this section describe the foundation stones of the airmanship model: *discipline, skill,* and *proficiency.*

Chapter 2 revolves around the theme that the foundation of all airmanship is personal discipline. It goes on to point out that a single failure of discipline can be the first step on a very slippery downhill path towards continuous compromise and noncompliance. A case study of failed discipline is presented that dramatically illustrates what lies at the far end of this dangerous path of poor discipline.

Chapter 3 builds on the idea of discipline to describe personal skill and proficiency as the second foundational block of airmanship. It overviews research findings to illustrate several key points about skill and proficiency essential to good airmanship, debunking several myths about experience along the way. It concludes with a step-by-step guide for developing a personal plan for skill and proficiency improvement that is tailored to your individual needs.

Section three: Pillars

Good airmanship requires us to draw on multiple knowledge bases. Section three outlines five critical bases of knowledge, described as "pillars of knowledge." They include a knowledge of *self, aircraft, team, environment,* and *risk.*

Chapter 4 discusses the importance of knowing one's self, which Socrates pointed out as the key to all wisdom. It begins with a discussion about the need to understand our physiological aspects, which have an impact on the next two areas of discussion, our mental and emotional processes. Tools for self-assessment are suggested, and the need to maintain self-monitoring throughout education, training, and flight operations is clearly pointed out. Case studies are used throughout the chapter to illustrate main points.

Chapter 5 discusses the importance of the second pillar of airmanship knowledge: knowledge of your aircraft. Although this knowledge is usually taken for granted as a prerequisite for safe flying, this chapter points out several subtleties about systems knowledge, including probing the history of the aircraft type for frequent ergonomics-related errors and the need to establish a personal relationship with maintenance personnel, as well as the aircraft itself. Once again case studies help develop the central themes in a real-world sense.

Chapter 6 points to the inherent importance of understanding your team. The team is defined as anyone you interact with or might have reason to interact with in the course of training and operations. Fundamentals of teamwork research are overviewed, including effective leadership and followership trends. Teamwork is discussed in congruence with the cockpit/crew resource management (CRM) initiatives that are so effective in the commercial and military aviation arenas. Special emphasis is given to two groups of aviators who sometimes view themselves as "CRM-exempt": fighter and general aviation pilots.

Chapter 7 describes the knowledge requirements relating to the three-part environment in which flyers must operate. The physical, regulatory, and organizational environments are seen as distinct, yet related, with each requiring an in-depth knowledge base for successful airmanship to develop. Short case studies are used to impart the relevance and importance of the information. Under the physical environment, the chapter details the need to understand basic weather and atmospheric phenomena. The need to understand the basic "rules of the road" for operation in the international, national, and local airspace systems are outlined and described as critical to understanding the regulatory environment. Several other regulatory concerns are also discussed and viewed as essential for good airmanship. Finally, organizational or corporate environment knowledge requirements are reviewed, and strategies for improvement in all areas are recommended.

Chapter 8 addresses the essential requirement for an aviator to understand multiple sources of risk, completing the discussion of the required pillars of airmanship knowledge. This chapter crosses into all areas of airmanship. Multiple case studies are used to detail common risk factors and root out hidden sources of danger that many aviators might be unaware of. Risk-management strategies are presented, and the individual's role and responsibility is established.

Section four: Capstones

Section four discusses the capstones of airmanship: *situational awareness* and *judgment*. This section concludes with two case studies of successful airmanship taken to the extreme in the commercial and military environments.

Chapter 9 shows all previously discussed aspects of airmanship as feeders into the complex and critical phenomenon known as situational awareness (SA). Through a detailed look at this multifaceted subject, we are able to see why an integrated systems approach is so vital to understanding airmanship. Some background theory is discussed to aid understanding, but the primary emphasis is on techniques for preventing loss of SA, recognizing this loss if and when it occurs, and recovering safely from a total loss of situational awareness. SA, in turn, feeds the judgment and decision-making process.

Chapter 10 takes a look at the ethereal concept of judgment. It quickly debunks the myth that judgment is somehow an innate possession of a select few. In fact, judgment is simply the culmination of solid airmanship development in the previously described elements of the model. Judgment is exercised through the effective management of these foundations and pillars and is known as decision-making. Techniques for maximizing these skills are discussed and strategies recommended for personal use.

Chapter 11 brings together all aspects of the airmanship model in two thrilling case studies. Captain Al Haynes described the keys to his success in recovering a stricken United Airlines DC-10 after the total loss of flight controls. The second example is from the military sector and details the heroics of an F-16 flight lead on a daring search-and-rescue mission during the Persian Gulf War. Both examples demonstrate that airmanship skills are developed over time and, perhaps more importantly, the time of need is not of your choosing. They demonstrate that the new "right stuff" requires the total airmanship package and that none of us go it alone up there anymore — we're part of a team.

Section five: Special topics

The final section deals with topics of current interest and importance, including inhibitors to effective airmanship, instruction, evaluation, and understanding and correcting human error. The book concludes with a short chapter detailing 10 common principles of airmanship — the marks of an airman.

Chapter 12 points out common obstacles to achieving one's airmanship potential and recommends strategies for dealing with them effectively. Conflicting demands on your time, poor role models, hazardous attitudes, and peer pressure are among several inhibitors discussed and analyzed. Once again, the focus is on individual solutions to these challenges.

Chapter 13 points out that airmanship needs to be taught and evaluated and recommends strategies for teaching and evaluating from a systems approach. Techniques for prebriefing, inflight instruction, evaluation, and critiquing are presented, emphasizing tools for near-term implementation.

Chapter 14 discusses the usefulness of error. By understanding what certain types of error mean to us, we are able to refine our training to correct them. Additionally, interpreting error is shown to be critical for successful instruction. Various taxonomies of error are discussed, along with implications for safety, effectiveness, and efficiency. Several case studies are used for illustration.

Chapter 15 concludes with a description of the marks of an airman. These are designed as a short reference list to keep airmanship development on track long after this book has been put on the shelf. Caution must be exercised here, because the characteristics themselves mean little if the model is not understood.

The appendix provides an excellent tool for evaluating airmanship in a crew environment and illustrates how much good research can help the aviator, if he or she knows where to find it.

Finally, this book seeks to motivate all flyers to seek higher standards of personal airmanship. In the final analysis, individual aviators are responsible for their own development. With the airmanship model to use as a guide, good airmanship becomes a matter of personal choice, plain and simple. Although it might sound like a trite cliché, a mentor of mine once explained that a person's future is determined by 10 letters and seven words; "If it is to be, it is up to me." These words certainly apply to airmanship development.

But good airmanship is more than just avoiding errors and preventing accidents. Flying is naturally exciting and fun. Improving airmanship is a way to get more out of your flying—to enjoy the exhilaration of a personal best. As you read this book, reflect on your own experiences, goals, and desires for achievement. Put yourself in the cockpits of military fighters, commercial airliners, and general aviation aircraft. Ask yourself what you would have done or, perhaps more importantly, how well you are prepared to face similar situations in your own flying future.

Suggestions for reading this book

This text is organized around case studies and is meant to be a "reader." Read each chapter reflectively and question what these concepts mean to your own flying environment. This book does not represent the final word on any aspect of airmanship but merely seeks to provide a fundamental structure for future education, training, and discussion. Therefore feel free to argue any of the points contained herein. Although this model is based on extensive study by highly qualified researchers, their assertions are not gospel, and a better understanding of airmanship is certain to be had by lively debate.

Finally, many things are said more than once in this book. This repetition is by design, to assist you in understanding the many interrelationships that make up an expert airman. Keep in mind, however, while perfect airmanship is the ideal, improvement is the goal. It is doubtful whether any of us can achieve perfection in all areas of the airmanship model, and, if we did, it would likely not last through the next airborne challenge. Seek modest and continuous improvement. Listen to yourself, refine your procedures and techniques, and take an active stance in making our skies a safe, fun, and profitable environment for all who fly.

Author's disclaimer

This book represents a collective viewpoint of what good airmanship is, defined by literally hundreds of flyers, living and dead. It was compiled through several years of historical research into successful acts of airmanship by some well-known—and some not so well-known—aviators. It is not designed to tell you how to fly your aircraft. Many others are more qualified to take on that task. Rather it is written to explain *what* and *why* you need to learn to become an expert airman and further to explain the interrelationships between the multitude of required skills and knowledges. This book is about responsibility to ourselves and each other.

This book is not the official position or policy of any part of the U.S. government, military or otherwise. Although I am an Air Force pilot and much of what I have learned and experienced in my military career has shaped the way in which I view airmanship, this book is the product of individual research. But neither is it one man's opinion. I would not be so arrogant or bold as to suggest that my personal views on airmanship should be adopted by all—or by any, for that matter.

This is not a "safety" book *per se*. Although improving airmanship will undoubtedly improve safety, this book is about personal responsibility and self-improvement—two worthy goals in their own right. I have lost seven close friends to human-error mishaps, all of whom for some reason or another were unable to come up with a critical piece of the airmanship puzzle when they needed it most. Perhaps you have experienced a similar experience, a lost feeling when someone you had a chance to influence makes a fatal error— a feeling that begs the question, "What could I have done?" In a very real sense, this book is my effort to "do" something to prevent these occurrences in the future. This book seeks to provide a structure for personal airmanship development as a guide to avoid pitfalls and a road map for success.

The airmanship model is presented here as a simple yet powerful tool for personal development. Use this model in any way you see fit, modify any or all of it, argue with it, improve it, or design your own model. Just don't ignore it. The underlying goal of this entire effort (and my life for the past several years) has been to facilitate and stimulate interest and discussion about airmanship on a personal level—the only level where it truly matters. Wishing you fair skies and favorable winds.

Redefining Airmanship

1

The roots and essence of airmanship

History is the discovering of the constant and universal principles of human nature.
David Hume

Late in the summer of 1916, Leutnant Hermann Göring, who later commanded the entire Luftwaffe in World War II, sat in a briefing room where his squadron commander asked for volunteers to fly a particularly hazardous reconnaissance mission deep into heavily defended French territory. Leutnant Wilhelm Hubener, a squadronmate of Göring, related what occurred:

> *Everyone groaned when the mission was announced—except Göring. He quickly got up, tapped his observer on the shoulder and the two of them went to their craft and took off. When they returned several hours later their plane was riddled with bullet holes, but they had the photographs. Our C.O. said to Göring, "Hermann, as an officer you are only a Leutnant, but as a Flieger, you are a General" (Wills 1968).*

The notion that flyers have a separate professional identity beyond their official rank or aeronautical rating is not unique to this example. Most aviators intuitively understand the existence of this unofficial hierarchy of airmanship but are unsure of how to advance within it. In the words of Tom Wolfe, the author of *The Right Stuff,* this so-called "pyramid of professionalism" defines our prowess as aviators. So what is it then that makes a great aviator? What are the elements of this invisible pyramid of airmanship, and how does one climb it? Does it even matter anymore in today's high-tech cockpit? What has *airmanship* come to mean in modern aviation?

1

No shared definition of airmanship

When asked to define good airmanship, most aviators have difficulty. The most common response, "I know it when I see it" (Kern 1995), doesn't provide much guidance for the new flyer seeking improvement. Expert flyers are said to have *good hands, judgment, discipline, common sense,* and *situational awareness,* but no one seems to agree with an all-encompassing picture of superior airmanship. Perhaps the inability to put our finger on a precise definition of airmanship illustrates a problem that goes beyond mere semantics. How can we train to become what we can not define and might not fully understand?

Failures of basic airmanship continue in the face of ever-improving technologies and training and have consequences that go well beyond the safety of the individual aviator or the aircraft. Although these critical breakdowns in airmanship are tragic in their own right, if they occur in a military setting, they can have far-reaching and often unexpected implications on mission objectives, interservice (joint) operations, and international trust. Consider the following incidents:

1. Two F-15 pilots under AWACS control misidentified, fired upon, and destroyed two U.S. Army helicopters, resulting in an international incident. During the accident investigation, one of the shooters lamented, "Human error did occur . . . It was a tragic and a fatal mistake which will never leave my thoughts, which will rob me of peace for time eternal. I can only pray the dead and the living find it in their hearts and their souls to forgive me." Further details are even more disturbing. Rules of engagement were not clearly understood, communicated, or followed (USAF 1994).

2. A B-52 bomber crashed while executing prohibited maneuvers at a U.S. Air Force base. The investigation revealed that a rogue aviator had been allowed to consistently violate U.S. Federal Aviation Regulations (FARs) and military regulations for at least three years. Even worse, this same aviator was the chief of standardization and evaluation of all aircrew members in the wing. A minimum of five wing and operations group commanders had the opportunity to intervene during this time period (USAF 1994).

3. An F-16 commanded by an experienced fighter pilot was on a routine ferry flight for military sale to a foreign country. The fully functional aircraft never made it. The pilot ran out of fuel, and the aircraft crashed en route to a divert base (Kern 1994b).

4. A tower air traffic controller called conflicting traffic "on short final" to an F-16 pilot conducting a simulated emergency approach well outside of prescribed operational guidelines. Although the pilot was unable to identify the traffic in question, he elected to continue the approach, resulting in a midair collision and the deaths of 24 army personnel who were struck by the burning wreckage as they waited to board a C-141 for training (Cross 1994).

Airmanship failures

Airmanship failures cross all aviation boundaries. The evidence suggests that while some aspects of military flying may be more demanding than commercial and general aviation, the types of errors remain relatively constant. Pilots suffer from inadequacies in discipline and knowledge, lose situational awareness, and make bad decisions.

Military airmanship failures

In the military, airmanship failures have become more than just a safety problem. A study of more than 800 critical incidents from Operation Desert Shield/Storm revealed that tactical aircrew error had significant operational, safety, and training implications (Kern 1994a). Although flyers tend to believe that the adrenaline and focus of combat improve their performance, many examples indicate that the opposite is at least as likely to occur. Consider the following airmanship errors from combat scenarios in Desert Storm:

1. Two A-10 pilots were flying a close air support (CAS) mission when they misidentified British Warrior armored vehicles as an Iraqi armored column. They fired Maverick missiles into the allied vehicles, killing 9 and wounding 11 British soldiers. A five-month British investigation into the incident attributed "no blame or responsibility to British forces." The British media splashed the incident across tabloid headlines for months afterward. A highly publicized

trial followed, and British families demanded compensation from the United States (Powell 1991).

2. On a ground attack mission, a fighter wing commander chose to disregard an established standard operating procedure (SOP) to stay above 8000 feet, an altitude chosen to avoid the high surface-to-air missile (SAM) threat below that altitude. After launching a Maverick missile, he saw enemy troops debarking an armored personnel carrier (APC), reversed, and descended to "gun" the new target. He pulled off the attack run at 6000 feet and took a SAM hit, which rendered his aircraft incapable of further combat in the war and set an extremely bad example for those under his command (Armstrong 1993).

3. In a similar example on the last day of the war, an F-16 descended below an established altitude restriction to attack a retreating Iraqi column. This decision was made in spite of the fact that the mission pilots had been briefed "not to expose yourselves unnecessarily, [because] the war will be over in a few days." Disregarding procedure and advice, the flight lead descended below a weather deck and was shot down. The consequences of the error did not end with the shootdown, since an Army rescue helicopter attempting to rescue the downed pilot was shot down, killing all five on board (Armstrong 1993).

4. A fatigued B-52 crew returning from a combat mission mishandled a minor malfunction and created a catastrophic self-induced emergency that culminated in the crash of the aircraft on final approach to its island destination, killing several crewmembers. An analysis of the accident uncovered poor crew coordination, flawed procedural knowledge, confusion, and a delayed ejection decision (USAF 1991).

It is perhaps illustrative that the only loss of a USAF bomber in the entire Gulf War was the result of a self-induced emergency caused by poor airmanship. United States military airpower has become so effective that we fly with relative impunity against distant enemies, but we cannot escape from what has become our most persistent and devastating adversary—our own mistakes. These military examples illustrate similar patterns in the commercial aviation.

Commercial airmanship failures

Airmanship failures in the commercial sector are perhaps more tragic. Military aviators understand the risks of combat flying and have made a conscious decision to participate. General aviation accidents, while tragic, usually result in only a few casualties, at most. But commercial errors often result in much greater loss of life, and the majority of the victims are innocent fare-paying passengers who are just trying to get somewhere on time. The following are but a few examples of commercial airmanship errors that have resulted in the devastating loss of innocent lives (Helmreich et al. 1995):

1. A Delta Airlines B-727 crashed on takeoff from Dallas/Fort Worth because the crew failed to extend the flaps.
2. An Air Florida Boeing 737 crashed while taking off from Washington due to icing on the wings. The first officer was ineffective in alerting the captain to abnormal instrument readings.
3. A British Midlands B-737 experienced an engine fire inflight. The crew shut down the wrong engine and crashed along a motorway.
4. An Eastern Airlines L-1011 crew became distracted by an abnormal landing gear indication and failed to note the autopilot disengagement as they orbited just outside of Miami. The aircraft crashed, killing all aboard.

General aviation airmanship failures

It should come as no surprise that general aviation pilots are not exempt from failures of airmanship. One might suspect that the majority of airmanship errors occur at the lower end of the experience scale in general aviation. While the vast majority of general aviation pilots are safe and well disciplined, the following incidents illustrate that poor airmanship is not confined to military or commercial flying.

1. A 37-year-old general aviation pilot with only 300 total flying hours was seen flying 5 to 7 feet off the runway surface to pick up speed to attempt an aerobatic maneuver at the end of the runway. The aircraft stalled and spun into the ground, killing the pilot (NTSB 1980).
2. An Idaho flight instructor required a new student to perform 60- to 70-degree bank turns while flying at extremely low

altitude along the Snake River. The result was predictable—two dead in a crash into terrain (NTSB 1994).

3. An experienced general aviation pilot went off supplemental oxygen while cruising at FL250 with a known pressurization problem in his Cessna 340 twin. He became hypoxic and could not resume control of his aircraft, which eventually ran out of fuel and crashed nearly four hours after takeoff (NTSB 1994).

Why do aviators continue to make these mission- and life-threatening errors? At least part of the problem may lie in the inability of aviators to consistently integrate skills and knowledge—to put it all together at the moment of truth. But the problem is not a lack of available information.

The need for a new approach: Integration vs. information

Professional journals and flying safety magazines abound with new findings on everything from pilot personality scales to situational awareness. Since the early 1980s, we have witnessed an explosion in the study of the psychological aspects of manned flight (Provenmire 1989). Unfortunately, this knowledge explosion has not translated into a significant decline in errors of basic airmanship. A landmark study by Dr. Clay Foushee, an internationally recognized expert on aviation human factors, found that failures of airmanship occur not because of a lack of proficiency or skill but because of an inability to coordinate skills into effective courses of action (Foushee 1985), indicating that one or more pieces of the airmanship puzzle might be missing in some aviators.

Everyone agrees that better airmanship is a worthy goal, but the method for best cracking this nut remains controversial. Efforts are made on many fronts, under such keywords as "realistic" training, judgment training, situational awareness, risk management, cockpit resource management (CRM), ergonomics, and stress awareness. While these studies have led to a general demystification on how the mind and body of an aviator works, they have not proven to be the complete or definitive answer to poor airmanship. In fact, this increasing specialization has caused a disaggregation, or splintering, of relevant and important information for young aviators, making it even more difficult for them to integrate attributes, skills, and knowledge into a personal and comprehensive whole. Flyers continue to ask the basic questions: "What do I need to know? Where do I go to get it?"

Historically, aviators have relied heavily on accident investigations and mishap analysis to identify areas for airmanship improvement. Although the study of aircraft and aircrew failures produces many valuable lessons, the problem is that these lessons come almost exclusively from negative examples. To date, the approach has been, "Pilot X did that and crashed, so don't do what pilot X did." Recent advances in the science and techniques of aircraft mishap investigation has created a system that can discover and recreate what went wrong with an aircraft or crewmember with incredible precision and detail. These are valuable and powerful lessons, but there is a more positive approach to improving airmanship.

Success also leaves clues in its wake, and these successes can be even more valuable tools than the continuous mantra of negative examples that flow from accident investigations. The study of airmanship successes should be as detailed and developed as the study of the negative examples, but, unfortunately, it is not. An analysis of the common traits of successful airmen can help to answer the two elusive questions that prompted me to write this book: "What is airmanship?" and "How can we develop it?" This discussion must naturally begin with aviation roots—the origins of airmanship.

The origins of airmanship

From the beginning of our existence, humankind has dreamed of flying. We were forever gazing skyward at the majestic flight of great birds, wondering what it must look like and feel like to have control of that endless blue. Before humans ever left the ground, the idea of flight was intoxicating. It remains so today. From the childlike awe of the uninitiated spectator watching a military aerial demonstration team for the first time to the seasoned operational pilot making a night crosswind landing at weather minimums, flying looks good and feels good when it is done right.

Unfortunately, humans have simply not yet mastered the art of flying. To be fair, birds have a 30-million-year head start, which may account for why sparrows never seem to make a mistake. When was the last time you saw a barn swallow misjudge a landing on a telephone wire, suffer a midair collision, or fail to pull out of a dive in time? Natural flyers integrate their instincts and actions into a level of aerial artistry that human pilots may never hope to achieve. But the ideal of *better integration* of both internal and external factors is the key to improvement for human flyers, who continue to struggle in their new environment. Although birds have natural instinct and a lifetime of practice to integrate their flying skills, humans come to flying relatively late in life and with a completely different learning style.

As adult learners, we tend to compartmentalize and fragment the information we are given. This style is the very nature of how we are taught to learn. We like to break things down and study the parts. But while the scientific method might help us understand complex concepts, it does little to help flyers integrate or apply these skills. For example, a kinesiologist might understand the firing of every neuron and the twitch of every muscle fiber in a golf swing, but unless he or she understands how each part relates to the others in actual practice, the knowledge is of little practical use. Knowledge alone does not lead to a lower handicap or to better airmanship. An airman must understand all parts of airmanship and how they interact to be able to effectively integrate these knowledges and skills in the dynamic environment of flight. Simply stated, a flyer must have the "big picture" of what airmanship is to become an expert airman.

Mythological origins of airmanship

The idea of a flyer's human frailty preceded flight itself. Ancient Greek mythology tells the story of the sculptor and inventor Daedalus, who was imprisoned with his son, Icarus (Fig. 1-1), in a tower on the island of Crete (Bulfinch 1934). In an attempt to flee the wrath of King Minos, Daedalus secretly constructed two pairs of wings from feathers and wax. Daedalus counseled his young son, "I charge you to keep at a moderate height, for if you fly too low the damp will clog your wings, and if too high the heat will melt them. Keep near me and you will be safe." With those words the two soared away towards the freedom of Sicily. Icarus was immediately taken with the breathtaking beauty and feelings of immortality that flight can bring. He flew higher and higher, not heeding the desperate warnings of his father, who understood what the solar heat would do to the waxed wings that held his son aloft. Icarus ignored his father, and the wings eventually gave way, sending Icarus spinning downward to his death. Daedalus never forgave himself and cursed his own inventive spirit for having caused his son's death. Once safe in Sicily, Daedalus hung up his wings as an offering to the god Apollo.

1-1 *From our earliest imaginings, humans understood that the intoxication of flight could lead to poor judgment. This is the oldest known rendering of the legendary Icarus, originally published in 1493.* USAF Academy Library Special Collections

This tale illustrates several common threads between humankind's earliest mythical notions of flight and the realities of today's high-tech flight environments. As in the story of Daedalus and Icarus, successful flying still requires an integrated understanding of your flying mechanism, your teammate(s), an immensely hostile environment, and yourself. As in the case of Icarus, a lack of self-discipline still causes unnecessary death and suffering. And Daedalus' self-damning condemnation sounds hauntingly familiar to any present-day instructor pilot who has lost a current or former student to a pilot-error accident.

But while our dreams of flying are as old as the human race itself, humans had to wait thousands of years after the mythical flight of

Icarus to begin to see these elements of airmanship played out within the reality of manned flight. (Although the balloon was developed by the Montgolfier brothers in France in 1783, for the purposes of this text, manned flight refers to the powered and controlled flight of aircraft.)

The inventors: More than mechanics

History is only partially correct when it records Orville and Wilbur Wright as inventors. Practically everyone knows that they built the first successful self-propelled, heavier-than-air aircraft—the Wright Flyer—and that they changed the world forever on December 17, 1903, near Kitty Hawk, on the Outer Banks of North Carolina. (There is historical controversy relating to the Wright Brothers' claim to be the first to fly a powered aircraft. For the purposes of this text, the documentation surrounding the Wright Brothers early flights, not their claim to be first, is what illustrates the earliest known reflections on airmanship.) What most do not know, however, is that Wilbur and Orville were airmen long before this historic date with destiny. Modern aviators can learn a great deal from the carefully planned and professional approach to airmanship taken by the brothers Wright.

The Wright Brothers sought first to understand the nature of flight. After learning of European attempts at sustaining powered flight, Wilbur and Orville began experimenting with kites and gliders in 1896—fully seven years before successfully completing their famous flight at Kitty Hawk (Hallion 1978). They quickly surmised that to sustain flight, they would need to control the climbing, descending, and turning of the aircraft. After consulting the U.S. Weather Bureau, they selected an isolated beach at Kill Devil Hills, North Carolina, and conducted more than 700 successful glider flights by the end of 1902 (Fig. 1-2). In addition, they constructed a crude (by today's standards) wind tunnel to test their various aerodynamic designs and hypotheses. They eventually developed a sophisticated "wing warping" design that, aided by a rudder, allowed the pilot to make coordinated turns by moving his body from side to side while laying prone over the wing in a cradle arrangement. Climbing and descending was accomplished through the use of a forward set of small winglets or canards. The aircraft was dynamically unstable and extremely difficult to fly. (In a traditional aircraft, the center of gravity is in front of the center of lift, lending itself to positive control. In a dynamically unstable aircraft, these positions are reversed, requiring constant control inputs to maintain control. It is interesting to note that modern combat aircraft, such as the F-16, require computer-assisted flight controls to aid the pilot. The Wright Brothers did not have this advantage, further attesting to their skill as pilots.)

1-2 *Prior to the historic flight in 1903, the Wright brothers sought to understand the nature of their aircraft and flight itself. Here Orville seems to be learning about ground effect on a glider flight in 1902.*
USAF Academy Library Special Collections

By any measure, Orville and Wilbur Wright were excellent airmen as well as inventors. In fact, the Wright brothers personify many of the ideal characteristics looked for in modern flyers. Both brothers clearly understood the aerodynamic nature of flight. As builders of their own aircraft, they were intimately familiar with its design, capabilities, and limitations. They had the ideal teamwork situation. Although they had contrasting personalities, they complemented each other in a synergistic fashion and had known and trusted each other their entire lives.

They also knew that the physical environment would play a major role in the success or failure of their efforts and sought out the expertise of the U.S. Weather Bureau to improve their chances. Finally, through more than 700 glider sorties, they became disciplined, skilled, and proficient at flying their creations. When the final piece of the puzzle, a powerful lightweight engine, was put into place, it came as no surprise to them that they succeeded. But even with all of these factors working in their favor, success was not automatic.

On the morning of December 14, Wilbur and Orville flipped a coin to see who would attempt the first powered flight. Wilbur won, but he stalled the aircraft on takeoff. In a letter to his father that day, Wilbur related his optimism in spite of the setback. "The machinery all worked in entirely satisfactory manner, and seems reliable. The power is ample, and but for a trifling error due to lack of experience with this machine . . . the machine would undoubtedly have flown beautifully" (Aymar 1990). Orville was slightly more curt about the "trifling error" but no less optimistic in his telegram home the next day. "Misjudgment at start reduced flight . . . Power and control ample. Rudder only injured. Success assured. Keep quiet." (Aymar 1990). Three days later, with Orville at the controls, the Wright Flyer flew into the air and into history. The aircraft flew not just once, but four times that day, proving that their optimism was well founded.

Unlike the mythical father and son team of Daedalus and Icarus, the Wright brothers had spent years integrating the various aspects of airmanship—knowledge of aircraft, each other, the environment, and themselves. Although primarily regarded as mere inventors, it is clear to aviators and historians alike that Orville and Wilbur Wright glued together more than just aircraft parts. Their comprehensive and integrated approach to flying established the first standards of airmanship.

The racers: Quest for high performance

It was not long until the ranks of aviators swelled to include competitors, adventurers, and daredevils, as well as the inventors. Unfortunately, the lessons learned from early airmen often perished with the flyers in these early competitions. The survivors, however, took lessons from the Wright brothers and other early pioneers of flight and flew to the limits of the new technologies. In doing so, this new breed of flyers began to test their own limits as well as the limits of their aircraft. Many aspects of modern airmanship, such as planning, knowledge, and discipline, can be traced to the experiences and lessons learned from these courageous men and women.

In 1909, the legendary aviation pioneer Glenn Curtiss was informed that he would be America's competitor for the Gordon Bennett Aviation Cup in the first international aeroplane contest to be held at Rheims, France, in August of that year (Aymar 1990). This event was to be the first of hundreds of such contests, where aircraft of all types competed against one another, limited only by the imagination of their creators and the laws of physics. These speed races were usually short in duration—20 kilometers was typical—and extremely dangerous. In 1909, the fastest planes in the world reached top end speeds of between 40 and 60 miles per hour, but it was still not uncommon to see 20 to 30 crashes over the course of a race week. Although the European press had reported speeds of up to 60 miles per hour for the French aviators Blériot and Latham, Curtiss "still felt confident," citing the press account's tendency for exaggeration (Aymar 1990).

Curtiss's plan for victory illustrates several elements of early airmanship. In secrecy, Curtiss developed a new eight-cylinder, V-shaped, 50-horsepower engine with which to challenge his primary competition, the Frenchman Louis Blériot (Fig. 1-3). Blériot was expected to win the competition in the same style monoplane in which he had crossed the English Channel. Curtiss approached the race differently than his competition. Unlike Blériot, Curtiss possessed limited means so far from his homeland and "had not the reserve equipment to bring out a new machine as soon as one was smashed." (Aymar 1990). He crafted a plan that would protect his irreplaceable aircraft until the final, winner-take-all race. After arriving in France, having transported his aircraft and engine separately by ship and rail, the American refused to take part in any preliminary events, much to the dismay of fellow Americans at the site. Curtiss was determined to protect his aircraft in spite of intense peer pressure to do otherwise. He explained:

It is hard enough for anyone to map out a course of action and stick to it, particularly in the face of the desires of one's friends; but it is doubly hard for an aviator to stay on the ground waiting for just the right moment to go into the air. It was particularly hard for me to stay out of many events at Rheims held from day to day, especially as there were many patriotic Americans there . . . . These good friends did not realize the situation. America's chances could not be imperiled for the sake of gratifying one's curiosity (Aymar 1990).

Glenn Curtiss' intuitive understanding of his machine, his team (the Americans), and himself was reinforced by an iron discipline to stick to the plan, regardless of outside pressures to perform. These qualities of airmanship are as critical to flyers today as they were on the eve of the great race. On August 29, 1909, at 10 a.m., the American

1-3a *Glenn Curtiss protected his aircraft in spite of considerable pressure to fly in demonstration events. This iron discipline paid off when the American won the first Gordon Bennett Aviation Cup. His caution is evidenced here as he prepares for engine start.* USAF Academy Library Special Collections

1-3b *Blériot, Glenn Curtiss' nemesis and primary competition, was far better equipped to fly multiple events during the Rheims Air Races, a fact Curtiss clearly understood.* USAF Academy Library Special Collections

fans had to wait no longer. Glenn Curtiss took his new engine and aircraft into the French sky and claimed the Gordon Bennett Aviation Cup victory for the United States, defeating Monsieur Blériot by a mere six seconds (Aymar 1990). In doing so, he became the first international champion in aviation history—and he did it with principles of airmanship that remain viable and essential to aviators today.

First combat: Ultimate competition

Scarcely two years after Glenn Curtiss's victory in France, airmanship took on a vital new importance. Italy and Turkey were at war in Libya, and both sides possessed military aircraft. In less than six months nearly every mode of employment now used in military aviation had surfaced, including reconnaissance, air defense, air superiority, transport, ground attack, and bombardment (D'Orlandi 1961). The brave airmen who flew these first combat missions had no experience, doctrine, or regulations to guide them. For this very reason, their experiences shed a special light on the nature of good airmanship. Unspoiled by previous training or indoctrination on how tactical airpower *should* be employed, they merely responded to what worked.

The first air combat took place against a backdrop of desert ground warfare on Libyan soil, where the Italians and Turks were fighting over economic interests in an ebb and flow campaign in and around Tripoli. The Italians, under the command of naval Captain Umberto Cagni, had occupied Tripoli on October 5, 1911, and were expecting an enemy counterattack of unknown size from an unknown location at any moment (D'Orlandi 1961). After 17 days, the Turk counterattack came and, to a large degree, succeeded in pushing the Italians back from their defensive positions. The Italian high command was in desperate need of reliable intelligence as to the locations and strength of the enemy. Native sources of information had become extremely unreliable, even hostile to the Italian troops, who were forced to disarm the population in the occupied areas. (D'Orlandi 1961). Faced with uncertainty, the Italians turned to Captain Carlo Piazza, the commander of the newly formed *air flotilla* for assistance in obtaining aerial observation of enemy troop locations, strength, and movements. The air flotilla was composed of five fully qualified pilots, six reserve pilots with "lower qualifications," nine aircraft—two Blériots, three Nieuports, two Farmans, and two Etrichs, all equipped with 50-horsepower engines and furnished with their own hangar—one sergeant, and 30 men (D'Orlandi 1961). Within the confines of this small group of combatants, the roots of tactical combat airmanship took shape.

Captain Piazza flew the first operational military mission on October 23, 1911, lifting off at 0619 a.m. in a French Blériot aircraft. Exactly one hour and one minute later, he returned to the field and reported "several enemy encampments, each with 150 to 200 men" (D'Orlandi 1961). With this simple report, Captain Piazza changed the nature of warfare forever by eliminating the element of surprise obtained by hiding behind terrain. War had become three-dimensional. Two days later, the Italian airmen discovered that the third dimension did not offer immunity from bullets, however, when Captain Ricardo Moizo returned to the airstrip with three bullet holes in the wing of his Nieuport (D'Orlandi 1961).

The increasingly important role of the aircraft in the military plans of the Italians and the small size of the flotilla made it apparent that *tactics* and *regulations* needed to be developed to protect the aircraft and pilots from enemy fire to the maximum feasible extent. Captain Piazza "laid down the rules for his young pilots without experience in warfare" (D'Orlandi 1961). These rules established the first tactics, risk-management procedures, and regulatory environment for combat aircraft. They were designed to guide the actions of the pilots by balancing the need to accomplish the military mission against the value of the aircraft and crewman. From the earliest moments of combat aviation, it was understood that it made more sense to make "what if?" decisions on the ground than in the complex and dynamic environment of flight. In the modern combat environment, where airspeeds have increased by a factor of 10, only the time to decide has significantly changed.

The aces: Icons of airmanship

Nothing touches the secret ambitions of modern aviators more than stories and exploits of the legendary military aces, but many modern flyers are surprised to learn that the very icons of individualism they secretly long to emulate were often the ultimate team players. This fact not withstanding, most honest flyers would admit to occasional daydreams of diving out of the sun behind an unsuspecting enemy and engaging in an individual duel to the death. After surviving several imaginary close calls, the would-be ace destroys the adversary with a high deflection shot and utters those immortal—and very cool—words: "Splash one!" The USAF chief of staff put these feelings into words at a gathering of the American Fighter Aces Association on September 23, 1960. "As a young boy dreaming of becoming an airman, if I had a choice between becoming chief of staff of the Air Force or becoming a fighter ace, I would have chosen to become a fighter ace" (White 1973).

We can learn much about airmanship from these legendary figures if we delve beneath the surface of mere numbers and types of aircraft destroyed. In September 1918, Eddie Rickenbacker became commander of the famous "Hat in the Ring" squadron. The same day he assumed command, he downed two enemy aircraft against seven-to-one odds, a feat that earned him the Medal of Honor (White 1973). But it is interesting to note that as a commander, he placed heavy emphasis on teamwork as opposed to individualism, and he further stressed constant and calculated risk management. According to General Thomas White, "It has been said that in no other Allied squadron in France . . . was there so much fraternalism, and such subordination of the individual to the organization" (White 1973). As commander, Rickenbacker instituted a mandatory "buddy system" and demanded that his pilots look after each other. His ideas on risk management and training sound more like an outtake from a current cockpit/crew resource management (CRM) course than the words of one of the most feared men in the skies over World War I France and Germany.

> *The experienced fighting pilot does not take unnecessary risks. His business is to shoot down enemy planes, not to get shot down. His trained hand and eye and judgment are as much a part of his armament as his machine-gun, and a fifty-fifty chance is the worst he will take—or should take— except where the show is of the kind that . . . justifies the sacrifice of plane or pilot (White 1973).*

Aces from subsequent wars continued to sound the same message of teamwork, tactics, and discipline. Major Richard Bong recorded 40 confirmed victories in World War II, making him the greatest American ace in history. "He was an expert at teamwork and a firm believer in having a strong, aggressive wingman," wrote General White. "He was a master tactician and an outstanding shot" (White 1973). In the European theater, Lt. Col. Robert Johnson amassed 28 victories in only 13 months and usually gave most of the credit to his wingman. "Anytime you lose your wingman," he said, "you've lost 75 percent of your eyes and fighting strength."

Aces from Korea and Vietnam add new verses to the same song of discipline and teamwork. Captain Joseph McConnell, Jr., was the leading ace in Korea and recorded 16 kills. When asked about the secret of his success, he responded, "It's the teamwork out here that counts, the lone wolf stuff is out. Your life always depends on your wingman, and his life on you" (White 1973). Colonel (later Brigadier General) Robin Olds, flying with the 7th Tactical Fighter Wing in Vietnam, tied his multiple MiG kills to the concept of discipline. "Every aspect of a fighter pilot's life demands the strictest discipline. Flying itself takes discipline. It is, in fact, both the end result of discipline and the constant application, through self-discipline, of the lessons of training . . . (that makes a great flyer)" (White 1973).

The great combat aces expanded and extended the early concepts, adding to the picture of airmanship. Complete airmen are seen as disciplined, skilled, and proficient, possessing an understanding of their aircraft, themselves, their teammates, their environment, and their risk. When each of these attributes are developed, flyers maintain a high degree of situational awareness and make consistently good decisions. They have good judgment. But the fighter pilots were not the only combat veterans who exemplified superior airmanship. Bomber crews developed the ideal of teamwork beyond anything yet seen in aviation.

1-4 *Early combat was unfettered by doctrine or excessive guidance. Flyers from this era merely responded to what worked or, in this case, what didn't.* USAF Academy Library Special Collections

Bombers: Courage and teamwork

The bomber crews of World War II flew against incredible odds and developed teamwork into an art form that may never again be matched. They had to, because their very lives depended on the synergistic effects of the bomber formation and the cohesion of individual crews. Ordered to strike targets deep in enemy territory without fighter support, these crews fought in the big leagues of aerial combat, against some of the most lethal fighter pilots the world has ever seen—the German Luftwaffe. The combat losses of the day were staggering and unimaginable in today's combat environment. According to airpower historian Mark Wells:

> *Bomber Command records indicate that between 1942 and 1945 the RAF Bomber Command flew more than 300,000 operational sorties. During that period about 8,000 aircraft were lost and a further 1,500 written off for battle damage. The human toll was just as high. In six years of conflict*

*almost 56,000—or about half [emphasis added] of the
125,000 aircrew who served with the command were killed,
8,400 injured, and a further 11,000 held prisoner or missing until the war's end. The US forces did not get off lightly
either. About 26,000 Eighth Air Force airmen were killed
. . . another 20,000 made prisoner (Wells 1995).*

The sacrifices made by these brave men are inspirational. The following excerpt was among the personal effects of Captain Arthur H.
Allen of the 94th Bomb Group (Heavy), discovered after he perished
on July 26, 1943, on a mission over Hannover, Germany. It illustrates
the essence of teamwork as seen by a B-17 aircraft commander and
combat veteran just days before his death.

*I suppose every pilot figures he has the best crew in the world,
but I am sure that I do. And that, coupled with a lot of luck,
is the main reasons for returning from our missions safely.
It's not right to say that any one man on the crew is any more
important than the other, because we are a team, and as a
team every one has a vital job to do. Ken is my copilot, and to
me more important than any other man on the crew. For
seven months we've flown together, and as pilot and copilot
we've learned to think the same and simultaneously, a combination that really counts in combat. When the going is
toughest, and the fighters are buzzing all over the place, the
sky black with flak and the formation whiplashing all over
the sky, I don't have to tell Ken what to do—he knows and he
does it. We've become such a coordinated team that we even
duck in rhythm. It is my honest opinion that Ken is the
coolest pilot in the business. And, if God sees fit to let him get
it one of these days, I just pray that I get it with him, because
I don't care to face death with any other man in the seat beside me.*

*Down in the nose are Greg and Carl. You have to speak of
them together, because just as Ken and I are a team, so are
they . . . Greg is Irish, and as a navigator, there isn't a better
one in the Group. He makes mistakes, as we all do, but he has
one fine characteristic that most navigators don't have. He
will admit that he doesn't know where he is while there is still
time for us to work the problem out together. I think I would
follow him into hell if he should plot a course that way. But*

even more important to the crew is the fact that he is really the nucleus of the crew. To the officers he is a pal and a companion, to the enlisted men he is a big brother. He's the type of guy that if we were to lose him, it would be very difficult for the crew to function properly (Allen 1983).

Captain Allen went on to detail the other members of his crew in the same manner, pointing out their strengths and weaknesses, personalities, and tendencies under stress. He was by no means the exception. Wells summarized the thoughts of many veterans related to the World War II experience.

Self-esteem, cohesiveness, and teamwork [were] the underpinnings of aircrew morale. Of these elements, it was principally the spirit of teamwork that allowed these men to participate daily in combat operations. This was true whether they flew single-seat fighters or multiplace bomber aircraft. The cohesive unit not only set standards for behavior but also rewarded good performance with acceptance and security (Wells 1995).

1-5 *Modern aviators have difficulty understanding the level of risk associated with early World War II daylight bombing missions. Here a crew stationed in the United Kingdom is briefed in front of a "flak map" to show the likely areas of greatest risk.* USAF Academy Library Special Collections

It is safe to say that the bomber crews of the second World War possessed an appreciation for teamwork that remains unsurpassed—and perhaps even unapproached today. It is interesting to note that the survivors of what was arguably the most vicious air combat environment the world has ever known share a common belief that teamwork was the single most essential ingredient in their success and survival. We should not underestimate its role in the development of modern airmanship.

The essence of airmanship: A historical model and definition

The point of these historical discussions is that good airmen share common traits and, further, that the successes of the past have left clues to the nature of good airmanship. This chapter began by asking the questions "What is airmanship?" and "How do we develop it?" The answer to these questions can be found by looking back at the history of manned flight in and out of combat, as well as forward at the likely demands of future technologies and conflicts. Interestingly, both lenses produce similar views of airmanship. Historically, great aviators tend to possess certain common qualities and characteristics, and a glimpse into the crystal ball of future technology or potential enemies suggests little change. The changes that *have* occurred over time appear to be changes of degree only, not fundamental shifts in the nature of what constitutes superior airmanship. This analysis reveals three fundamental principles of expert airmanship, regardless of the time frame analyzed: *skill, proficiency,* and the *discipline* to apply them in a safe and efficient manner. Beyond these basic principles, five areas of expertise were identified as common among expert airmen. Expert airmen have a thorough understanding of their aircraft, their team, their environment, the risks or enemy, and themselves. When all of these elements are in place, the superior aviator exercises consistently good judgment and maintains a high state of situational awareness. Each of these areas are described in detail in following chapters, but they are illustrated together in Fig. 1-6 to illustrate the conceptual relationships.

An expert aviator combines many factors into a comprehensive whole and further understands that all of these factors are dynamic, requiring constant and calculated attention. Flyers with the right stuff understand the capabilities and limitations of themselves; their team;

Airmanship

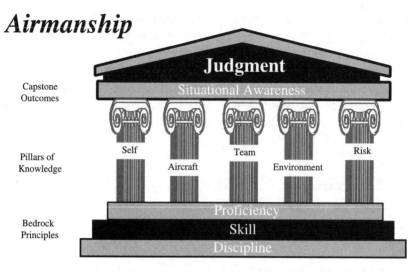

1-6 *The airmanship model. Airmanship is the consistent use of good judgment and well-developed skills to accomplish flight objectives. This consistency is founded on a cornerstone of uncompromising flight discipline and developed through systematic skill acquisition and proficiency. A high state of situational awareness completes the airmanship picture and is obtained through knowledge of one's self, aircraft, team, environment, and risk.*

their aircraft; the physical, regulatory, and organizational environment; and the multiple risks associated with the flight. An expert flyer builds on a bedrock of flight discipline, skills, and proficiency. No single-focus flyer approaches excellence. Experts in military tactics who can't fly an aircraft effectively due to lack of proficiency don't add much to the combat power of their country. Conversely, golden-hands pilots who don't understand the rules of engagement or who misidentify friendlies as foes can do tremendous damage to their country's cause with a single error.

But the combination of tactical and technical expertise is still not enough. Even if airmen understand the risks associated with enemy systems and tactics and can outfly everybody in their squadrons, they are ineffective on the air/land battlefield if they cannot integrate with their wingmen, crew, or the joint and combined team. This integration requires a special set of skills that have come to be known as *human factors*. Total airmanship blends technical and tactical expertise, proficiency, and a variety of human factors to smoothly and effectively integrate the capabilities of the pilot and the machine into the joint/combined team. Total airmanship leads to improved situational awareness, fewer mistakes, increased operational effectiveness, improved training, and safer flying operations. By eliminating gaps in airmanship, a flyer is better able to handle the rapidly changing and dynamic environment of flight.

To be certain, other factors are involved in airmanship. Opportunity, external situational factors, and just plain luck figure into the success or failure of nearly every decision that an aviator makes. This model does not attempt to explain all success in flight with a simple graphic representation. Nor does it suggest that good airmanship has always been a "balanced" approach, as suggested by the model. Many extremely successful flyers have compensated for a lack in one area with an abundance in another. History shows that this "single-focus" approach might work on occasion, but more often than not it results in failure rather than success. Murphy is patient, and if a weakness can be found, he will find it sooner or later.

While these factors brought out in the airmanship model might not be possessed by each and every aviator, in general terms, most successful airmen possess those qualities illustrated. The ideal of a solid foundation of discipline, skill, and proficiency, combined with multiple bases of knowledge from which to make decisions, is sound guidance for today's flyers.

But developing total airmanship is not a simple learning task. Reaching this level of expertise starts from within and begins with a motivation to improve, to develop an understanding of the skills and knowledge that are required to consistently master the extremely complex, and often hostile, environment of flight. Consistent good judgment and high situational awareness have always been the benchmarks of good airmanship, but methods for achieving these desirable outcomes have not been made clear. Many have suggested that judgment and situational awareness are innate qualities, something one either has or does not have. That is *not* the approach taken by this book. History suggests that there are specific underpinnings to good airmanship that can be systematically developed through study, training, and self-discipline. The development of these skills are the focus of following chapters, which detail each part of the airmanship model, illustrating the principles of total airmanship with factual case studies from aviators past and present.

References

Allen, Arthur, H. 1983. Ramblings of a bomber pilot. *Stories of the Eighth: An Anthology of the Eighth Air Force in World War Two.* John Woolnough, ed.: Hollywood, Fla.: The 8th Air Force News.

Armstrong Laboratory. 1993. Collection of critical incidents (#00144) from Operation Desert Shield/Storm. Brooks AFB (HRMA), Texas.

Aymar, Brandt. 1990. *Men in the Air: The Best Flight Stories of All Time from Greek Mythology to the Space Age.* New York: Crown Publishers.

Bulfinch, Thomas. New York: The Modern Library. *Bulfinch's Mythology.*

Cross, Christopher. 1994. Telephone interview. September 8.

Curtiss, Glenn, and August Post. 1990. The first international aeroplane contest. *Men in the Air.* Brandt Aymar, ed.: New York: Crown Publishers, pp. 220–225.

D'Orlandi, Renato (translator). 1961. *The Origin of Air Warfare.* 2d. ed. Rome: Historical Office of the Italian Air Force.

Foushee, H. C. 1985. Realistic training for effective crew performance. Proceedings, 4th Aerospace Behavioral Engineering Technology Conference. pp. 177–181.

Hallion, Richard, and Bildstein, Roger. 1978. *The Wright Brothers: Heirs of Prometheus.* Smithsonian Press.

Helmreich, Robert. L., Roy E. Butler, William R. Taggart, and John A. Wilhelm. 1995. Behavioral markers in accidents and incidents: Reference list. NASA/UT FAA Aerospace Crew Research Project Technical Report, 95-1.

Kern, Anthony. 1994a. A historical analysis of tactical aircrew error in Operations Desert Shield/Storm. U.S. Army Command and General Staff College (CGSC) Monograph. June 2.

————. 1994b. Ignoring the pinch. *Torch Magazine.* September.

————. 1995. What is airmanship? A survey of military aviators. Unpublished research in progress. United States Air Force Academy.

National Transportation Safety Board. ASRS. 1994. Aeroknowledge (CD-ROM)

————. 1980. Report brief. Accident ID #LAX89DUJO4. File no. 140.

Powell, Stewart. 1991. Friendly fire. *Air Force Magazine.* December. 59.

Provenmire, H. 1989. *Cockpit Resource Management: An Annotated Bibliography.* October.

United States Air Force (USAF). 1994a. Air Force Regulation (AFR) Report on B-52 mishap. June 24. pp. 110–14.

————. 1994b. Report on Blackhawk shootdown. Vol. 12:13. July. AFR pp. 110–14.

Wells, Mark K. 1995. *Courage and Air Warfare*. London: Cass Publishing.

White, Gen. Thomas. 1973. Air Force Fact Sheet 73-7, AF Aces. Secretary of the Air Force Office of Information.

Wills, Kelly, Jr. 1968. *Feldfliegerabteilung 25—The Eyes of Kronprinz* from the Recollections of Dr. Wilhelm Hubener. 9 (1): *The Cross and Cockade*.

2

Flight discipline: The cornerstone of airmanship

There is only one kind of discipline—perfect discipline.
General George Patton

A personal view from Eileen Collins, Lt. Col, USAF

Over 4000 flight hours in 30 different type aircraft.

The need for a disciplined approach to airmanship exists at all levels of flying. Flying in any environment is a job that involves a relatively large amount of risk. As an aviator, one way that I attempt to minimize potential problems is to study systems and procedures. In-depth systems knowledge and memorized procedures are required for safe operation of any aerospace vehicle. In addition to traditional types of study, I review procedures during my normal daily activities. For example, while jogging or swimming, I mentally review critical action procedures I would need to perform in an emergency. One can imagine a variety of emergency situations, and then mentally rehearse what steps to take to minimize damage or loss of life.

My approach to airmanship and preparation has always been disciplined. The techniques that I use for studying the complex systems I employ today are the same ones I have used throughout my career, beginning with my first flights in undergraduate pilot training.

If those of us involved in flight operations at any level do our jobs to the utmost of our ability, not only will we fly safe missions, but we will preserve the forward momentun of aviation at all levels, from the single-engine general avaiation pilot, to those who fly in space. This is a duty that is important to us as individuals and as team players. Each of us should know how our aircraft operates and how *we* operate in both normal and emergency situations. Not only could we save lives and aircraft, but we can promote and preserve interest in aviation for generations to come.

Lt. Colonel Eileen Collins has logged over 4000 hours in 30 different types of aircraft and was the first female pilot of a Space Shuttle. She became an astronaut in 1991 and has logged nearly three million miles in space. She served as the pilot on STS-63, the first flight of the Russian-American Space Program. (She also gave the author his first instrument checkride nearly two decades ago while serving as a T-38 Talon instructor pilot and flight examiner at Vance Air Force Base, Oklahoma.

Superior airmanship cannot exist without uncompromising flight discipline. A flyer can have the best hands around, possess more technical knowledge than the engineers who designed the aircraft, know more about military tactics than a weapons school instructor, and have more experience than Chuck Yeager—and still be a walking time bomb. This chapter looks at the phenomenon of flight discipline from both the individual and organizational perspectives and seeks to drive home three essential lessons.

1. Violations of flight discipline have an insidious creeping effect on an aviator's good judgment.

2. Flight discipline violations are contagious.

3. The best defense is a personal standard of zero tolerance for violations of flight discipline in any form.

This chapter defines flight discipline and discusses common rationales for departure from regulatory compliance. A case study of poor flight discipline is presented, which illustrates the contagious nature of flight discipline violations and shows the consequences of a tolerant organization, a state that can lead to a cultural norm of noncompliance. Although the example is taken from the military arena, it is illustrative of other examples from the commercial and general aviation sectors. Guidelines for combating flight discipline violations are recommended at the conclusion of the case study.

A sacred trust

At a gut level, most aviators can determine reasonable from unreasonable courses of action, regardless of the nature of the flight they are undertaking. From the beginning of an aviator's training, a flyer is taught that many options are available to complete flight objectives. Military flyers are taught that "flexibility is the key to airpower" and are given considerable latitude in employing tactical methods for accomplishing a mission. This tactical flexibility is one of the major strengths of military airpower and should not be changed. Similarly, commercial and general aviation pilots enjoy considerable freedom of action with which to address inflight events. The ability to operate safely and efficiently within this flexible environment is an essential prerequisite of good airmanship. Airmen are given a sacred trust not to take advantage of a system designed to allow flexibility and creativity. Because of this sacred trust, flight discipline must be the foundation of an aviator's airmanship package.

Flight discipline defined

Flight discipline is the ability and willpower to safely employ an aircraft within operational, regulatory, organizational, and common sense guidelines—unless emergency or combat mission demands dictate otherwise. Let's look more closely at this definition, beginning with the concepts of "ability" and "willpower." An actual violation of flight discipline must be a conscious and willful act. If one inadvertently flies below a specified altitude due to lack of skill or knowledge, it is not a breech of flight discipline. It is certainly poor airmanship, but the failures fall into other categories, such as skill, proficiency, or environmental knowledge. This distinction is important. Breeches of flight discipline are too important to be discussed without precision. As a community of aviators, we must be able to identify violators and either reform them or remove them from our midst. This cannot be done without a precise and shared understanding of what is meant by flight discipline.

The second part of the definition speaks to compliance with "operational, regulatory, organizational, and common sense guidelines." Operational guidelines deal with the limitations of the aircraft itself, usually defined in associated flight manuals and technical orders. One fact that is often forgotten is that flight manual ops limits can and often do change. These changes can be a result of new engineering data or because someone else paid for the lesson in blood. In either case, a flyer must stay current with technical changes to the operational limitations of his or her flying machine.

Regulatory compliance also means knowing and complying with any and all international, federal, state, and local flying directives. Depending on the nature of the flying, regulations can include a dizzying amount of information, including guidance from the International Civil Aeronautical Organization (ICAO), Federal Aviation Regulations (FARs), and other regulations and special directives that exist within your flying environment. Once again, these rules can and do change often, and it is the responsibility of the individual to keep abreast of the changes.

Organizational rules and instructions are the next level of compliance. Unfortunately, they are often seen by many aviators as "soft rules," or ones that are the easiest to violate with impunity. An airline company or military command may establish certain rules or "operating instructions" (OIs) to further delineate desired guidelines for their flyers. An example would be an additional fuel reserve during winter months or reduced crosswind limits for less-experienced pilots.

A wise man once remarked that "common sense isn't all that common," and he was certainly on target when it comes to actions taken by undisciplined aviators. Most aviators understand that written rules and regulations cannot apply to every situation, but a particular breed go out of their way to find loopholes in existing regulations. One example of such an approach was illustrated by a group of pilots who found a way to get around a new flight manual restriction against accomplishing 360-degree rolls. They rationalized that two 180-degree rolls were still "legal" as long as they rolled out of the maneuver in the same direction in which they rolled into it. In doing so they had made a "legal" maneuver more difficult and dangerous than the prohibited one. Clearly, these pilots understood the intent of the new prohibited maneuver, but they rationalized their way out of compliance in a classic example of a violation of common-sense flight discipline.

The final portion of the flight discipline definition addresses special circumstances in which it might be necessary and logical to depart from the first three areas of compliance (operational, regulatory, organizational), but certainly not from the fourth (common sense). Rules and regulations cannot foresee all of the eventualities and requirements of combat or emergencies. In these instances, a flyer must use what assets are available to get the job done. But even under extreme circumstances, caution must be exercised. Once the immediate concern is met, an aviator should quickly return to the safe environment of compliance.

Flight discipline seems like a simple enough mandate, and, in fact, most aviators do follow procedures and regulations to the letter. There are those, however, who find various excuses to depart from the proven path of compliance.

Five common excuses

Occasionally the temptation to violate flight discipline gets the best of an aviator. It can take the form of an intentional procedural short-cut or temporarily exceeding an operational limitation. The very nature of airborne operations means that there is usually little organizational oversight of an aviator's actions in flight. Aviators are simply trusted to do the right thing. There is often little or no obvious impact from a small violation of flight discipline, which can lead an unwary aviator down a path of rationalization for foolish actions. Five common rationalizations for poor flight discipline follow.

1. If no one knows about the infraction and nobody gets hurt, what's the problem?
2. Everyone knows that there are safety margins built into all the regulations.
3. Rules are simply to protect inept flyers from themselves.
4. This business is overregulated. Pilots did this for decades before the government stepped in.
5. I can't push the envelope and really improve if I am bound by all these silly rules.

It is easy to see how pilots could buy into any or all of these rationalizations if they were looking for a reason to do so. Can a "superior" airman violate a rule now and then and still be OK? The short answer to this unprofessionalism is an unqualified "No!"

Policies, procedures, and regulations exist for a variety of reasons, and the implications of poor discipline are often unseen by the violators themselves. In the military environment, the implications can be subtle and extended. Military aviators are often called upon to plan, train, and fight with members of other services and countries in joint and combined operations. Procedures, policies, standard operating procedures (SOP), and rules of engagement (ROE) are all developed for specific purposes, which might not be obvious or necessarily agreeable to the individual crewmember who must execute them with unquestioned precision. Small breeches of discipline that seem safe or acceptable in the sterile training environment of day-to-day operations around the "home drome" can have catastrophic implications in other more complex environments.

In commercial aviation, the consequences of poor flight discipline can be even greater than in the military. With today's congested airspace and jumbo jet airliners, the casualties from even a single accident can be so severe that margins of safety must be built in to protect innocent lives.

In general aviation, the less-structured environment must rely heavily on self-regulation. This environment, coupled with lower experience levels, mandate a conservative approach. But before we move on, let's analyze each one of the "five excuses" and see if they hold water in light of the consequences of failed discipline just discussed.

If no one knows about the infraction and nobody gets hurt, what's the problem? The real problem is that you can never have an intentional infraction in which "no one knows." The most significant one who knows is the perpetrator. Psychologists tell us that getting away with something once is very likely to lead to subsequent attempts at the same or similar activities. Noncompliance can be a slippery downhill path. In short, you are never really alone, and an intentional deviation will likely preprogram you to try it again.

Everyone knows that there are safety margins built into all the regulations. Yes, there usually are, and for good reasons. Safety margins are designed to account for a combination of error tolerances in instruments, navigation equipment, and some human error. To disregard a regulation based on an assumed built-in margin of safety presupposes that everything else is working perfectly. When a flyer makes this decision, he or she is betting his or her life (and often the lives of others) on a flawed assumption.

Rules are simply to protect inept flyers from themselves. The logic here is that regulations are designed for the lowest common denominator, so a superior airman can violate them and still remain safe. This rationalization is partially true, in that regulations must be designed to protect all flyers, the skilled as well as the marginally proficient. That fact makes this excuse one of the most persuasive. The flaw in the logic is that aviators must operate under a mindset of assumed compliance if the system is to work. In our automobile traffic system, if someone consistently rolls through stop signs or dashes under yellow lights at high speed, it is only a matter of time before someone who is following the rules gets hurt by the unsafe drive. Because no single aviator owns the sky, we share a moral responsibility to follow established guidelines, regardless of our skill or proficiency.

This business is overregulated. Pilots did this for decades before the government stepped in. The first half of this statement is sure to incite heated debate in many sectors of the aviation industry, but the performance capabilities, complexity, and sheer number of the aircraft operating today mandates a certain level of regulation. The depth and rationale for all of the existing regulations might appear unnecessary to individual operators from the point of view of their own aircraft and flying patterns. However, when you step back and look at the big picture of aviation, in which crop dusters, pleasure flyers, commercial jets, and military aircraft share the same sky, the need for regulatory oversight becomes readily apparent.

I can't push the envelope and really improve if I stay within the rules. Many flyers feel that they are developmentally constrained by the existing regulatory or organizational environment. Somehow they feel that they must practice outside of existing guidelines to become their best. Most high-risk maneuvers require instructor supervision or special-use airspace to perform. Nearly all safety regulations are based on lessons that have been paid for in blood by those who attempted what you might be contemplating. Aviators who feel the need to "push the envelope" to improve, are well advised to adhere to regulatory guidelines for doing so. Pushing a personal or an aircraft envelope can rapidly lead to exiting it, which is frightening at best—and potentially fatal if the dice roll the wrong way.

The "one better" trap

The competitive nature of aviators can cause even a small infraction of discipline to get out of hand and rapidly escalate to dangerous proportions. In many circumstances, colocated flyers try to outdo each other in a high-stakes game of "I can do better than that." Several decades ago at a European fighter base, the commander established a policy of having the last plane airborne each Friday afternoon accomplish a high-speed "flyby" of the flight line to boost troop morale. What began as an innocent initiative to fire up the maintenance troops quickly became a "hot dog" competition for the pilots. Initially, the flybys were safe and by the book. The pilots maintained prescribed airspeeds and altitudes. But gradually, the passes got lower and faster, even provoking complaints from local officials and nearby residents. But the practice continued, and eventually a pilot lost control at low altitude and crashed near a residential neighborhood.

A similar pattern of poor flight discipline developed during a military exercise on Guam, where Air Force and Navy aircrews were practicing joint attack operations. At the end of each day's flying, a simulated airfield attack took place, in which attack and bomber aircraft attempted to penetrate the air defense perimeter of the island, overfly Andersen Air Force Base, and then recover to the field for landing. The crews typically flew every other day, debriefing and mission planning on down days. Every afternoon, the aircrews who were on the ground would find their way to the rooftops with video cameras in hand, in time for the airfield attack, which was better than many air shows. Evenings were spent with liquid refreshments, watching the videotapes and debating who had the best performance of the day. By the third day of the exercise, jets were screaming by at rooftop level and spectators would drop flat on the rooftops, legitimately fearing for their lives. If this trend had not been stopped, there was only one way that this would have ended—with another disaster. But the on-site commander intervened and issued an order for a mandatory "Officers' Call" formation. Once all of the pilots, navigators, and weapons systems officers (WSO) were assembled, the commander made it perfectly clear that the next pilot who flew below the minimum altitude or above the maximum airspeed would receive a one-way ticket home and suffer severe administrative punishment. The commander's actions averted the likely disaster. These scenarios are not all that rare, and they demonstrate that once the downward slide away from disciplined flying begins, aviators are often incapable of stopping it themselves. Breaking rules is addicting to many flyers. Once they start, they need help to stop. The competitive nature of flyers, coupled with a system that intentionally provides minimal guidance and little oversight, sets the stage for temptation. The best safeguard against weak moments is to remember the words of General Patton that opened this chapter and to keep in mind the two most dangerous words any aviator can utter: "Watch this!"

Rogue aviators

The patterns of poor air discipline are not always of the "mass hysteria" type just described. There are also aviators, usually of high experience and skill, who see the built-in flexibility of aviation as a chaotic environment that can be manipulated to satisfy their own egos—often with tragic results. These rogue aviators are often popular and possess considerable social skills. They have learned what rules they can break, when, and with whom. They are typically perceived much differently by superiors than by peers or subordinates. Their brand of poor airmanship can have far-reaching implications. Because of their perceived skill and expertise, many younger flyers see these rogues as role models and begin to copy their brand of so-called "airmanship." The case study that follows illustrates the dangers posed by a rogue aviator. Note the escalating nature of the violations as the situation progresses. It appears that flight discipline violations follow similar patterns of escalation whether occurring in an individual or a group. Also pay attention to the tolerant posture of the organization, which allowed these antics to continue until disaster stopped the show.

A case study of failed airmanship and leadership

On a sunny afternoon in June 1994, Czar 52, a B-52H assigned to an Air Force base in the Pacific Northwest, launched at approximately 1358 hours Pacific Daylight Time (PDT) to practice maneuvers for an upcoming airshow. The aircrew, under the command of a highly experienced instructor pilot (IP), had planned and briefed a profile that grossly exceeded aircraft and regulatory limitations. On preparing to land at the end of the practice airshow profile, the crew was required to execute a "go-around," or missed approach, because another aircraft was on the runway. At midfield, Czar 52 began a tight 360-degree left turn around the control tower at only 250 feet altitude above ground level (agl). Approximately three quarters of the way through the turn, the aircraft banked past 90 degrees, stalled, clipped a power line with the left wing and crashed. Impact occurred at approximately 1416 hours PDT. There were no survivors out of a crew of four field-grade officers (McConnell 1994).

The accident investigation revealed a disturbing pattern of poor airmanship on the part of the pilot in command that spanned a period of at least three years. This case study illustrates how a lack of flight discipline can become contagious and, if left unchecked, can lead to the development of a rogue aviator who blatantly disregards and even revels in defying operational limits and regulations. As one crewmember said about the accident, "You could see it, hear it, feel it, and smell it coming. We were all just trying to be somewhere else when it happened." (Anonymous 1995).

Lt. Col. Bob Hammond (a pseudonym) was the aircraft commander of Czar 52 and was undoubtedly flying the aircraft at the time of the accident (93rd Medical Group 1994). He was also the chief of the wing standardization and evaluation (stan-eval) branch, ironically making him responsible for enforcing airmanship standards for all the aviators on the base. The squadron commander from the local B-52 squadron, who was an IP flying as the copilot, was also killed in the accident. A great deal of evidence suggests that considerable animosity existed between the two pilots who were at the controls of Czar 52. The squadron commander had unsuccessfully tried to have Hammond grounded for what he perceived as numerous and flagrant violations of air discipline while flying with his aircrews. The vice wing commander of the base was also on board, added to the flying schedule at the last minute on the morning of the mishap. It was to be his *finis flight*, an Air Force tradition in which an aviator is hosed down following his last flight in an aircraft. On landing, he was to be met on the flight line by his wife and friends for a champagne toast to a successful flying career. The radar navigator position was filled by the bomb squadron operations officer. It is interesting to note that all of the crewmembers aboard Czar 52 were senior field-grade officers. It was discovered during the accident investigation that many junior officers had openly refused to fly with Lt. Col. Hammond because of his reputation for poor flight discipline.

While all aircraft accidents that result in loss of life are tragic, those that could have been easily prevented are especially so. The crash of Czar 52 was the result of actions taken by a singularly outstanding "stick-and-rudder" pilot, but one who, ironically, practiced incredibly poor airmanship. This analysis shows in dramatic detail how an aviator lacking in flight discipline is literally an accident waiting to happen. This is true even when the individual is experienced, knowledgeable, and skilled. Good airmanship cannot coexist with poor flight discipline.

By nearly any measure, Bob Hammond was a gifted stick-and-rudder pilot. With more than 5200 hours of flying time and a perfect 31–0 record on check rides, Lt. Col. Hammond had flown the B-52G and H models since the beginning of his flying career in March of 1971 (Aeronautical Order 1989). He was regarded by many as an outstanding pilot, perhaps the best in the entire B-52 fleet. He was an experienced instructor pilot and had served with the Strategic Air Command's 1st Combat Evaluation Group (CEVG), considered by many aviators to be the "top of the pyramid" for heavy drivers. But between 1991 and June of 1994, a pattern of poor airmanship began to surface. Perhaps his reputation as a gifted pilot influenced his commanders, who were unable to interrupt or stop this destructive pattern of activity. Peers and subordinates of the mishap pilot had raised desperate warning signals about Lt. Col. Hammond's lack of flight discipline over a period of three years prior to the accident. But Hammond's superiors perceived him differently. The following comments were typical from Lt. Col. Hammond's superiors:

- "Bob is as good as a B-52 aviator as I have seen." (McConnell Tab V-3.3.)

- "Bob was . . . very at ease in the airplane . . . a situational awareness type of guy . . . among the most knowledgeable guys I've flown with in the B-52." (McConnell Tab V-2.8.)

- "Bob was probably the best B-52 pilot that I know in the wing and probably one of the best, if not the best within the command. He also has a lot of experience in the CEVG, which was the Command Stan Eval . . . and he was very well aware of the regulations and the capabilities of the airplane." (McConnell Tab V-6.3.)

A far different perspective on Lt. Col. Hammond's flying is seen in statements by junior crewmembers who were required to fly with him on a regular basis.

- "There was already some talk of maybe trying some other ridiculous maneuvers . . . his lifetime goal was to roll the B-52." (McConnell Tab V-21.4.)

- "I was thinking that he was going to try something again, ridiculous maybe, at this airshow and possibly kill thousands of people." (McConnell Tab V-21.7.)

- "I'm not going to fly with him; I think he's dangerous. He's going to kill somebody some day and it's not going to be me." (McConnell Tab V-27.10.)

- "[Lt.] Col. Hammond made a joke out of it when I said I would not fly with him. He came to me repeatedly after that and said 'Hey, we're going flying Mikie, you want to come with us?', And every time I would just smile and say, 'No. I'm not going to fly with you.'" (McConnell Tab V-32.10.)

- "Lt. Col. Hammond broke the regulations or exceeded the limits . . . virtually every time he flew." (McConnell Tab V-32.3.)

The reasons for these conflicting views may never be entirely known, but they hint at a sophisticated approach to breaking the rules that is often a pattern with aviators who begin to stray from the proven path of regulatory compliance. The following situations detail Hammond's descent toward disaster:

Situation one: Base airshow, May 19, 1991

Lt. Col. Hammond was the pilot and aircraft commander for the B-52 exhibition in the 1991 airshow. During this exhibition, he publicly violated several regulations and technical order (T.O.) limits of the B-52 by exceeding bank and pitch limits and flying directly over the airshow crowd in violation of Federal Aviation Regulation (FAR) Part 91. In addition, a review of a videotape of the show leaves the distinct impression that the aircraft may have violated FAR altitude restrictions as well. No disciplinary actions were taken as a result of the 1991 B-52 airshow exhibition. However, some aircrew members had already begun to lose faith in a system that would allow one of its senior instructor pilots to so blatantly disregard regulations. One pilot, when asked why more crewmembers didn't speak up about the violations, said, "The entire wing staff sat by and watched him do it (violate regulations) in the '91 airshow. What was the sense in saying anything? They had already given him a license to steal." (Anon. 1995)

Situation two: Change of command flyover, July 12, 1991

Less than two months after the 1991 airshow, Lt. Col. Hammond was the aircraft commander and pilot for a flyover for the bomb squadron change-of-command ceremony. During the practice and actual flyover, Lt. Col. Hammond accomplished passes that were estimated to be "as low as 100 to 200 feet." (McConnell Tab V-3.5.) Additionally, he flew steep bank turns of greater than 45 degrees (the limit is 30 degrees) and extremely high pitch angles in violation of flight manual restrictions, as well as a wingover—a maneuver in which the pilot rolls the aircraft onto its side and allows the nose of the aircraft to fall through the horizon to regain airspeed. The flight manual recommends against wingover type maneuvers because the sideslip can cause damage to the aircraft.

The cultural climate surrounding Lt. Col. Hammond's actions might have influenced his decision to continue these unauthorized maneuvers. As with the previous situation, the flyover mission plan was developed, briefed, and executed without intervention. Even though a change-of-command flyover required approval by the USAF vice chief of staff, no such approval was requested or granted. (AFR 110-14, AA-2.7). Although the senior staff was spurred to action by the magnitude of Hammond's violations, the response, a verbal reprimand, appeared to be little more than a slap on the wrist. This point was certainly not missed by other flyers in the wing, some of whom were beginning to see Lt. Col. Hammond as a role model.

Situation three: Base air show, May 17, 1992

Lt. Col. Hammond once again was selected to fly the B-52 exhibition at the base air show. The profile flown included several low-altitude steep turns in excess of 45 degrees of bank and a high-speed pass down the runway. At the completion of the high-speed pass, Lt. Col. Hammond accomplished a high pitch angle climb, estimated at more than 60 degrees nose high. At the top of the climb, the B-52 leveled off using a wingover maneuver (AFR 110-14, "Executive Summary," 1:5). This time the Deputy Commander for Operations (DO) reprimanded Lt. Col. Hammond, stating "If you go out and do a violation and I become aware of it, I will ground you permanently." (McConnell Tab V-3.10.) Apparently the message fell on deaf ears, because the pattern of poor flight discipline continued unabated. Further deterioration of airmanship should not have come as a surprise.

Situation four: Global power training mission, April 14–15, 1993

This time the breach of discipline occurred on a training mission. Lt. Col. Hammond was the mission commander of a two-ship mission to the bombing range in the Medina de Farallons, a small island chain off the coast of Guam in the Pacific Ocean. While in command of this mission, Lt. Col. Hammond flew a close visual formation with another B-52 to take close-up pictures, in violation of Air Combat Command (ACC) regulations. Later in the mission, Hammond permitted a member of his crew to leave the main crew compartment and work his way back to the open bomb bay to take a video of live munitions being released from the aircraft, also in violation of current regulations. (McConnell Tab V-26.20.) This episode demonstrates that his lack of air discipline was not limited to showing off in front of crowds; this aviator did not hesitate to violate any regulation he felt did not apply to his personal view

Situation five: Base air show, August 8, 1993

In spite of (or perhaps because of) his previous performances, Lt. Col. Hammond was again selected to fly the B-52 exhibition for the 1993 base air show. The profile was much the same as the previous air shows, but the bank and pitch angles were noticeably steeper. Following the low altitude, high-speed pass, he executed a high-pitch maneuver that one crewmember estimated to be 80 degrees nose high—10 degrees shy of completely vertical. It appeared as if Bob Hammond was caught up in a personal version of the "one better" trap. He might have felt that each performance had to be better than the last one.

By now, other crewmembers at the base had grown accustomed to Lt. Col. Hammond's air show routine. But his ability to consistently break the rules with apparent impunity was having a more insidious effect on younger, less-skilled crewmembers. In one example, a B-52 aircraft commander who had seen several of Lt. Col. Hammond's performances attempted to copy the "pitch-up" maneuver at an airshow in Canada—with near disastrous results. (McConnell Tab V-26.12.) The navigator on this flight said "we got down to seventy knots and . . . felt buffeting" during the recovery from the pitch up. (McConnell Tab V-26.12.) This airspeed is approximately 80 knots below minimum inflight airspeed for flaps-up maneuvering in the B-52. (If the 70-knot figure is accurate, the aircraft had already stopped flying and the resultant "recovery" was merely a fortunate pitch down into the recovery cone. The aircraft could just as easily departed controlled flight.) At 70 knots, the B-52 is in an aerodynamically stalled condition and is no longer flying. Only good fortune or divine intervention prevented a catastrophic occurrence in front of the Canadian audience.

A second example occurred at Roswell, New Mexico, when a new aircraft commander was administratively grounded for accomplishing a maneuver he had seen Bob Hammond do at an air show. It was a flaps-down, turning maneuver in excess of 60 degrees of bank, close to the ground. His former instructor said of the event, "I was appalled to hear that somebody I otherwise respected would attempt that." (McConnell Tab V-32.7.) The site commander was also appalled and sat the young flyer down and administered corrective training. The bad examples set by a rogue aviator had begun to be emulated by junior and impressionable officers and had resulted in one near disaster and another administrative action against a junior officer, both of whom were just trying to be like Bob.

Situation six: Yakima bombing range, March 10, 1994

Lt. Col. Hammond was the aircraft commander on a single-ship mission to the Yakima Bombing Range to drop practice munitions and provide an authorized photographer an opportunity to shoot pictures of the B-52 from the ground as it conducted its bomb runs. Lt. Col. Hammond flew the aircraft well below the established 500-foot minimum altitude for the low-level training route. In fact, one crossover was photographed at less than 30 feet, and another crewmember estimated that the final ridge line crossover was "somewhere in the neighborhood of about three feet" above the ground and that the aircraft would have impacted the ridge if he had not intervened and pulled back on the yoke to increase the aircraft's altitude. The photographers stopped filming because "they thought we were going to impact . . . and they were ducking out of the way." (McConnell Tab V-28.8.) Lt. Col. Hammond then joined an unbriefed formation of A-10 fighter aircraft to accomplish an impromptu flyby over the photographer. This mission violated ACC Regulations regarding minimum altitudes, FAR Part 91, and Air Force Regulation (AFR) 60-16 regarding altitude restrictions during overflight of people on the ground.

During the flight, crewmembers strongly verbalized their concerns about the violations of air discipline and regulations. At one point, Lt. Col. Hammond reportedly questioned the manhood of the radar navigator (RN) when the RN would not violate regulations and open the bomb doors for a photograph with weapons on board. On another occasion, following a low crossover, the navigator told Lt. Col. Hammond that the altitudes he was flying were "senseless." (McConnell Tab V-28.9.) But the real hero on this flight was Capt.

Eric Jones, a B-52 instructor pilot who found himself in the copilot seat with Lt. Col. Hammond during the low-level portion of the flight. On this day, it took all of his considerable skills, wits, and guile to bring the aircraft and crew safely back home. After realizing that merely telling Lt. Col. Hammond he was violating regulations was not going to work, Capt. Jones feigned illness to get a momentary climb to a higher altitude. But in the end it was once again Lt. Col. Hammond at the controls. The following is Capt. Jones's recollection of the events that took place then:

> We came around and [Lt.] Col. Hammond took us down to 50 feet. I told him that this was well below the clearance plane and that we needed to climb. He ignored me. I told him (again) as we approached the ridge line. I told him in three quick bursts 'climb-climb-climb.' . . . I didn't see any clearance that we were going to clear the top of that mountain . . . It appeared to me that he had target fixation. I said 'climb-climb-climb' again; he did not do it. I grabbed ahold of the yoke and I pulled it back pretty abruptly . . . I'd estimate we had a crossover around 15 feet . . . The radar navigator and the navigator were verbally yelling or screaming, reprimanding [Lt.] Col. Hammond and saying that there was no need to fly that low . . . his reaction to that input was he was laughing— I mean a good belly laugh. (McConnell Tab V-28.9.)

On returning from the mission, the crewmembers discussed the events among themselves and came to the conclusion that they would not fly with Lt. Col. Hammond again. Capt. Jones "vowed to them that never again would they or myself be subjected to fly with him. That if it required it, I would be willing to fall on my sword to ensure that didn't happen." The next day, Eric reported the events to the squadron operations officer, stating, "I did not ever want to fly with Lt. Col. Hammond again, even if it meant that I couldn't fly anymore as an Air Force pilot." (McConnell Tab V-28.13.) Captain Jones was told that it wouldn't come to that, because he "was joining a group of pilots in the squadron who had also made the same statement." (McConnell Tab V-28.13.)

This episode illustrates that individual crewmembers can combat flight discipline violations by others in several ways, and the crew on this day took assertive and proactive steps in the correct order. First, they verbalized their concerns to the offender, specifically pointing out the regulatory violations and their personal unwillingness to participate. When this did not work, they took whatever actions were necessary to get the aircraft and crew out of danger, including assuming temporary control of the aircraft to avoid disaster. Finally, when they had returned safely, they reported the incident and emphasized the severity of the violation by refusing to fly with the perpetrator again. Unfortunately, others were less assertive or proactive. One B-52 instructor pilot summed up the many feelings:

> *Everybody had a Hammond scare story. (He) was kind of like a crazy aunt . . . the parents say 'Ignore her' . . . 'He's about to retire' . . . 'That's Bob Hammond, he has more hours in the B-52 than you do sleeping.' Yeah, he might have that many hours, but he became complacent, reckless, and willfully violated regulations.* (McConnell Tab V-39.7.)

By June 1994, the organizational climate at the base was one of distrust and hostility. In spite of these facts, Lt. Col. Hammond was selected to perform the 1994 air show. "It was a nonissue," one of the commanders said. "Bob was Mr. Air Show."

Situation seven: Air show practice, June 17, 1994

Lt. Col. Hammond flew the first of two scheduled practice missions for the 1994 airshow. The profile was exactly the same as the accident mission except that two profiles were flown. Once again they included large bank angles and high-pitch climbs in violation of regulations and technical order guidance. This was in spite of the fact that the wing commander had directed that the bank angles be limited to 45 degrees and the pitch to 25 degrees. (These were still in excess of regulations and technical order guidance.) Both profiles flown during this practice exceeded the wing commander's stated guidance. However, at the end of the practice session, the DO told the wing commander that "the profile looks good to him; looks very safe, well within parameters." (McConnell Tab V-2.23.)

The base crewmembers were not comfortable with the situation. In fact, one of the squadron navigators refused to fly the air show if Lt. Col. Hammond was going to be flying. The squadron operations officer was also uneasy. "I had this fear that he was again going to get into the air show . . . that he was going to try something again, ridiculous maybe and kill thousands of people." (McConnell Tab V-21.7.)

It wasn't just the flyers that were getting nervous. Major Theresa Cochran, the nurse manager in emergency services, attended an airshow planning session in which Lt. Col. Hammond briefed that he planned to fly 65-degree bank turns. The wing commander quickly told him that he would be limited to 45 degrees maximum. Major Cochran recalled Lt. Col. Hammond's response in a prophetic discussion between herself and a coworker who was also in attendance at the planning session.

> *Colonel Hammond's initial reaction was to brag that he could crank it pretty tight . . . he said he could crank it tight and pop up starting at 200 (knots). Bob and I looked at each other, and Bob is going, "He's (expletive deleted)," and I said "I just hope he crashes on Friday, not Sunday, so I will not have so many bodies to pick up." . . . those words did return to haunt me."* (McConnell Tab V-19.7.)

On June 24, at 1335 Pacific Daylight Time, Lt. Col. Bob Hammond taxied to runway 23 for his final airshow practice. At 1416 PDT, the aircraft impacted the ground, killing all aboard.

The role of the organization in flight discipline

Just as "up" has no meaning without the concept of "down," leadership must be defined in terms of followership. Aviators have a moral and professional obligation to follow existing regulatory guidance, but the organization has an equal, if not greater, responsibility to ensure compliance. Lt. Col. Hammond gave his organization plenty of ammunition with which to take action, by refusing to follow written regulations and flight manual limitations, as well as ignoring the verbal orders and guidance given by his commanders. Even when verbal reprimands and counseling sessions focused on the specific problem of airmanship, he steadfastly refused to follow orders. At one point, only weeks prior to the accident, he clearly stated his feelings on the issue of guidance from senior officers: "I'm going to fly the air show and yeah, I may have someone senior in rank flying with me, . . . he may be the boss on the ground, but I'm the boss in the air, and I'll do what I want to do." (McConnell Tab V-21.10.) Bob Hammond was clearly out of control and was in desperate need of a strong disciplinary hand to stop his downward spiral toward disaster.

Instead of intervening, the organization served the role of enabler by allowing the violations of flight discipline to continue unchecked.

Lessons learned

The lesson learned and implication for current and future aviators and organizations is that trust is built by congruence between word and deed. Other flyers are quick to pick up on any disconnect. Retired Air Force Gen. Perry Smith wrote, "Without trust and mutual respect among leaders and subordinate leaders, a large organization will suffer from a combination of poor performance and low morale" (Smith 1986). He was right on target in this case. If the organization is unable or unwilling to step in to stop a negative trend in airmanship, there is usually only one ending to the story.

How could this occur? How could an otherwise outstanding aviator slip so far off track? The airmanship lesson to be taken from this tragedy is that flight discipline is a cultural as well as an individual phenomenon. It must be uncompromising. The organization must have a stated policy that unwarranted violations of regulations, in any degree, are intolerable. Once a single event goes unchallenged—an irreversible process may be set in motion. While Bob Hammond was at the controls when the crash occurred, those who cheered him on or sat by complacently for over three years share responsibility for the loss. In the words of Army Lt. Gen. Calvin Waller, "Bad news doesn't improve with age" (Waller 1994). Airmen at all levels must act on information or evidence of noncompliance. Individual flyers must also learn to recognize the traits of the rogue aviator; while Lt. Col. Hammond stood out like a beacon in the night, many others still operate today to a lesser degree.

A final perspective

Violations of flight discipline are certainly not limited to rogue aviators or to military operations. The files of the National Transportation Safety Board are filled with tales of tragedy that began with a single violation of flight discipline. The crash of Czar 52, like most accidents, was part of a chain of events. These failures included an inability to recognize and correct the actions of a single rogue aviator, which eventually led to an unhealthy command climate and the disintegration of trust between flyers. However, in most aircraft mishaps, the crash is the final domino to drop in the cause-and-effect chain of events. In this case, however, scores of young and impressionable aviators watched a rogue aviator as their role model for over three years. They, in turn, pass along what they have learned. The final domino in this chain of events might not yet have fallen.

The antithesis to the rogue aviator is the flyer who can be creative and decisive to meet mission demands while remaining within established guidelines. With a solid and uncompromising foundation of flight discipline, this flyer develops skills and proficiencies, the next elements of airmanship.

References

93d Medical Group/SGP. 1994. Medical statement to the Accident Board. August 19.

Aeronautical Order (PA) Aviation Service. 1989. 92d Bombardment Wing, Combat Support Group. March 10.

Anonymous. Personal interview with captain pilot, 525th BMS.

McConnell, Col. Michael G. 1994. Executive summary. AFR 110-14 USAF Accident Investigation Board, vol. 1, ed.: 1. June 24.

Smith, Perry. 1986. *Taking Charge: A Practical Guide for Leaders.* Washington, D.C.: National Defense University Press.

Tab V-26.18. 1994. Air Combat Command message, DTG 281155Z. February.

Waller, Lt. Gen. (Ret.) Calvin. 1994. U. S. Army Command and General Staff College (CGSC) lecture slides. March.

3

Skill and proficiency: Two edges of the airmanship sword

There comes a time in every man's life when he is called upon to do something very special; something for which he and only he has the capabilities, has the skills, and has the necessary training. What a pity if the moment finds the man unprepared.
Winston Churchill

A personal view from Brig. Gen. Chuck Yeager, USAF (Retired)

After more than a half century of aggressive flying, I've formed a few impressions about good airmanship. If I could leave three impressions in the minds of tomorrow's pilots, it would be these:

Lesson 1. Complacency will kill the best of pilots. I've known some very senior test pilots who hurt themselves while pleasure flying because they got complacent. When you start to get that feeling that everything is going just right, you had better watch out, because that's when the airplane is about to jump up and bite you in the rear end. Teach yourself to avoid complacency; it will save your butt someday.

Lesson 2. Know your aircraft, especially the parts of it that can save your life. I always prided myself on knowing more about my egress systems than anybody else, and when I had to punch out of the NF-104, it saved my life. I had taken a pretty good lick on the head from the ejection seat rocket motor after ejection.

The burning motor cracked the visor on my pressure suit helmet and ignited the rubber seals. The suit was pressurized with 100 percent oxygen, and it created what was basically an acetylene torch a few inches from my face. Although I was still dazed from the hit to the head, I knew that if I got the visor up, it would stop the flow of oxygen. I did, and it did, and I lived to fly another day.

The point is, that simple book knowledge might not be good enough. What might seem clear at groundspeed zero can be pretty confusing when the stuff hits the fan. You need to know your systems well enough to use them under stress — and in some cases even in a semiconscious state. One of the best ways to learn is to get to know your maintainers. If you make it a habit to get out to the airplane 15 minutes early, you can use that time to get to talk to mechanics, crew chiefs, etc. This is time well spent. Not only will you get to know the mechanics, but you will get to know the aircraft on an individual basis — just like they do.

Lesson 3. If you are going to fly, do it right. What I really admire in a flyer is professionalism and consistency. I'm really impressed by a guy or gal who goes out there day after day and does it right — not fancy or flamboyant, but just consistently good performance. Lots of pilots talk a good game, and sometimes their stories get better with each telling. Don't measure yourself by the stories of others. Seek to improve yourself—that's the mark of a true pro.

An expert airman flies well. Your aviation heritage and flight discipline are of little use to you without flying skills and proficiency. Skill development and maintaining proficiency is serious business, because unfortunately, we never know when we will be called upon to execute at the very limits of our capabilities. This is true in the military case study that follows, as well as in commercial and general aviation. Murphy is watching and waiting for the unprepared.

Case study: Prepared and ready

On May 12, 1975, fully two years after the peace treaty was signed that officially ended U.S. participation in the war in Southeast Asia,

Cambodian military forces seized the U.S. naval vessel *SS Mayaguez* and her crew. General Louis L. Wilson recalled the national embarrassment that accompanied the unprovoked act of piracy against an American vessel:

> *At the heart of the matter . . . was the implication of the ship's seizure with respect to U.S. credibility and self-respect . . . the brazen, unprovoked seizure of the ship . . . was widely viewed as an arrogant affront . . . founded on the belief that the United States lacked the will or ability to act decisively.* (*Talon* 1979.)

But the United States did act, with decisiveness and heroism. By noon on May 15, nearly 230 Marines and USAF crewmen had been placed on Koh Tang, a small island in the Gulf of Thailand, where the crew of the *Mayaguez* was believed to be held. The national command authorities had ordered a rescue mission.

First Lieutenant Don Backlund was assigned to the 40th Air Rescue and Recovery Squadron at Nakhon Phanom, Thailand, when the *Mayaguez* rescue was ordered. Only four years out of the Air Force Academy, the 26-year-old was to fly one of four search and rescue (SAR) helicopters for the extraction of American soldiers who were now pinned down at Koh Tang by heavy concentrations of hostile Cambodian soldiers. Earlier that day, eight of nine helicopters depositing Marines on the island had been lost or severely damaged. The situation was grim and becoming more critical by the moment. As the day wore on into evening, American troops had been unable to quell the dug-in Cambodian troops, and the entire rescue mission appeared to be in jeopardy. Alarmed by the fact that daylight was nearly gone and frustrated by delays, Backlund radioed the forward air controller (FAC) and in no uncertain terms summed up the urgency of the situation. It was time, he insisted, to get the action going.

Lieutenant Backlund summed up all of his skill, proficiency, and knowledge of enemy placement and capabilities and made a risk assessment. American troops were in trouble, and he was their best—and possibly last—chance. He made a low-level, high-speed run into the beach and saw the survivors pop their marking smoke. Lieutenant Backlund swung the helicopter's tail toward the beach and set her down. Despite the cumulative efforts of daylong airstrikes and the fact that three other helicopters were spewing minigun fire up and down the length of the tree line, enemy resistance was almost fanatical. At one point, Cambodian soldiers, seeing the marines escaping their grasp, stormed the helicopter and attained hand-grenade range. Only when the crew was sure that all 25 marines and airmen were aboard did Lieutenant Backlund pull out.

For his efforts in the *Mayaguez* rescue, Don Backlund received the Air Force Cross for valor. He summed up his feelings about his role in the *Mayaguez* mission in a film prepared to celebrate the 25th anniversary of the Air Force Academy: "One day there was peace and the next day there was war, and there was no time to get in shape. There was no time to relearn your procedures. There was no time to study the things you should have been studying all along."

When destiny called, Don Backlund was prepared. His skill and proficiency was all that stood between life and death for more than a score of young Americans. When the opportunity presented itself, this skill and proficiency level led to an inner confidence that allowed him to coolly perform what others have called a suicide mission *(Talon 1979)*.[1]

Skill building

Flying requires multiple skills that are developed in and out of the formal training setting. In addition to the obvious physical flying skills, it also requires communication skills, decision-making skills, team skills, and, perhaps most importantly, self-assessment skills. The ability of a flyer to assess his or her current talents and proficiencies, as well as to seek appropriate training as required, allows for logical risk assessments. These skills must be built by all flyers who desire to achieve high levels of airmanship, regardless if they are military pilots like Lieutenant Backlund or commercial or general aviation pilots who find themselves needing to call on the sum total of all their skills to handle a severe emergency.

Traditionally, flying training programs build to a predetermined minimum skill level for safe operations, which usually is validated by some type of check ride or evaluation, and then turn you loose into your flying environment. Certain currencies and annual training

1. This story was partially reprinted and paraphrased from an article titled "In Memoriam," which appeared in the USAF Academy newspaper, *The Talon*, in November 1979, after the death of that year of Major Backlund in a tragic accident.

requirements continue to guarantee minimum levels of proficiency, but these requirements are designed to keep you safe, not necessarily to help you reach your potential as an aviator. Unless you are lucky enough to be selected for some advanced training program, individual excellence requires you to personally manage your own development. It requires you to work training into every minute of that precious flying hour, something that many aviators are hesitant to do. Flying a back course localizer instead of a visual overhead pattern or doing an extra set of traffic pattern stalls makes your flying time much more productive than cruising at altitude or accomplishing another visual pattern—unless of course, your proficiency in these areas needs the work. Finally, reaching your individual potential requires a personal plan of action, based on an honest and valid self-assessment.

This type of discipline is different than we spoke about in Chapter 2. It is a commitment to seek continuous improvement, based on a rigorous evaluation of strengths and weaknesses. Socrates said to "know thyself" is the first pillar of wisdom, and he was surely right when it comes to airmen. Self-awareness is central to skill-building and improvement in both the formal training environment, where you have a responsibility to help your instructors identify areas that require additional training, and after formal training is complete. But there is more to skill-building than self-awareness. Before we can begin to completely grasp the individual nature of improving airmanship skills, it is helpful to understand that there are different levels of skill that we progress through on our way to individual airmanship excellence.

Four levels of skill

In any flyer's career, he or she is required to meet certain minimum standards to pass evaluations and legally operate an aircraft. Once these standards are met, however, the drive to continue to improve skills often diminishes for many reasons. Some are simply satisfied to become flyers, their goal has been met when they pass the check ride, and the motivation to improve is gone. Others face competing demands on their time. The pursuit of advanced degrees, family, church, and other activities cause their study habits to be redirected into other interests. Whatever the reasons, it is an unfortunate fact that once a certain skill level has been achieved, many flyers stop improving. One way to describe and hopefully avoid this airmanship paralysis is through an understanding of four levels of airmanship skill.

Level one: Safety

The first level of skill is reached, at least temporarily, by all flyers who eventually check out in an aircraft. They are able to master the required events to a level that satisfies an instructor that they will not injure themselves or others if turned loose into the skies. At this point, flyers can say to themselves, "I'm good enough to be safe." This level of skill is usually marked by some formal event like a solo flight or an initial qualification check ride. We typically take great pride in this event, because we have demonstrated to the world that we have what it takes to fly airplanes, and rightfully so, for many never make the attempt, and others do not make the grade. But this level is the most rudimentary skill, and much improvement is ahead for the dedicated aviator.

Level two: Effectiveness

The second level of skill comes about when we achieve a basic mastery of the activities required to perform the duties of our flying. For a commercial airline pilot, it might be the line check; in the military, it is likely to be the mission qualification; and for the general aviation enthusiast, it means being able to handle the local and cross-country environments that you wish to operate in on your own. Unless one desires a further upgrade, such as moving from first officer to captain, or upgrading to instructor, this level is the last at which you *must* perform. In professional flying, it earns you a paycheck. In general aviation, it allows you to practice your hobby. Many flyers see little reason to develop additional skills beyond this effectiveness level. Motivation to develop beyond this level must come from within, unless you are in an "up or out" environment such as the military or you seek a higher paycheck through an upgrade to a higher-paying position. But achievers seek higher levels.

Level three: Efficiency

Some aviators seek to develop skills to go beyond the basics of safety and effectiveness. They seek to fly more efficiently than the standards require. It might mean flying a profile that maximizes training while saving fuel and flying time or minimizes unnecessary inflight communications. The approach taken by these flyers is "not only can I do my job safely and effectively, but I will do it with a minimum of resources to save time and money for myself and my organization." Efficiency requires that you learn techniques as well as procedures, and it means experimenting with new and more efficient methods of operation. A balance must be struck when pursuing efficiency, because it can quickly become an end unto itself, diminishing effectiveness and even safety on occasion.

Level four: Precision and continuous improvement

The final level of expertise is characterized by the few aviators who constantly seek greater precision. If such a pilot flies a 3-knot airspeed window on final approach on Tuesday, she or he will try to make it a 2-knot window on Wednesday and try for perfection on Saturday. These flyers seek perfection, not as an obsession—but as a continuing motivation for personal improvement. They have refined their self-assessment capabilities to a fine edge. They analyze every error, and when things go right, they ask "How could it be done better?" This level of skill requires maturity and consistent and honest self-appraisal. Few ever reach this level.

Understanding skill levels (Fig. 3-1) helps us to understand when we hit plateaus in our airmanship skill development, but a plan is required to work through them. As we progress through our developmental stages of airmanship, we may find ourselves in and out of formal training programs that are designed to develop and hone our skills for various purposes. But throughout our development, we should be continuously assessing and training ourselves. At the most fundamental level, we are responsible for our own training.

The big "C": Confidence as a skill multiplier

The big reward for high levels of skill and proficiency is confidence, and in flying, confidence is critical. Like in other fine motor skill tasks that require finesse and "feel," confidence is a prerequisite for success in flying. If you doubt the impact of lost confidence on physical capabilities, ask any baseball pitcher what happens when he begins to doubt his ability to throw a strike. Or ask a basketball player who has missed four consecutive free throws what the mind does to the hands. These same aspects apply to flight operations. There are two ways to approach a high crosswind landing. The first way is to say, "I'll never get this thing on the ground." The pilot's teeth clench, the hands grip the controls like they were the edge of a cliff, and the toes curl up on the rudder pedals. This pilot's lack of confidence strips him or her of the very sensitivity needed to feel what the wind

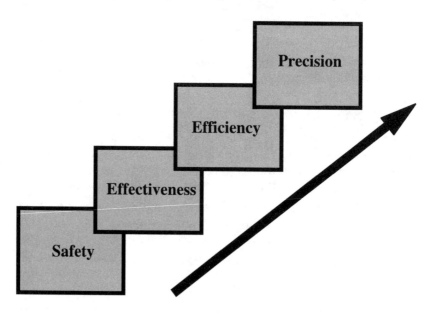

3-1 *Airmen should strive to build on successive layers of expertise. Safety comes first, followed by effectiveness, efficiency, and precision. These blocks are not made of concrete and require continuous attention and assessment to maintain.*

is doing to the aircraft, and inputs will be abrupt and poorly timed. The belief that he or she "can't do it" becomes a self-fulfilling prophecy.

On the other hand, a confident pilot approaches the situation much differently. At the first twinge of apprehension, he or she uses self-talk to quickly stabilize his mental state. "Yeah, these wind are stronger than any I've dealt with before, but it's just a matter of degree. A little more rudder, a little more power—no sweat. If it starts to look bad, I'll just take it around and I will have learned enough from the first effort to do it right on the second try." This pilot stays alert but relaxed, maintaining a watchful posture. The pilot feels what the aircraft is doing and is much more capable of applying smooth control inputs in a timely manner than his or her unconfident counterpart. The only difference between the two flyers is the mental state—confidence. Confident pilots have a much easier time staying in control of themselves and, therefore, the aircraft.

So where does this magical quality come from? Where do you get your dose of the big "C"? The answer lies in your personal approach to training, in the level of skill and proficiency you believe that you possess. If you have done your best, *really done your best*, to prepare yourself by maximizing your instruction and flying time, then you can legitimately assume you possess the skills that an aviator with your experience level should possess—and perhaps even more. You know that if the conditions are within legal limits, you should have the abilities to meet the challenge. In short, you are confident.

If, however, you have "skated" through your training, secretly hoping that the instructor or evaluators would not discover your hidden weaknesses, if you are the type who meets the minimum standards and is satisfied, then the nagging doubts of incompetence creep in to destroy your inner confidence at the very moments you need it the most.

A counterbalance to fear

Fear is a natural emotion for pilots to experience, especially while learning to fly, although none of us like to admit it. When we must speak of it, we tend to prefer words like anxiety, eagerness, or apprehension. Call it what you like, fear is a fact of life in the flying game. It can stem from unusual circumstances or lack of a required skill or proficiency to address a specific challenge of flight. Most flyers intuitively understand that fear is something that must be worked through, and that if it is not brought under control, even mild apprehension can develop into a mind-numbing and crippling experience that we all know as panic. In the words of David Bowie, to overcome fear we must "turn and face the strain." To do so requires a mature approach to self-assessment and training. Only you know what your weaknesses are, and if you want that magical skill multiplier we call confidence, you must seek the training to strengthen your weak areas.

Confidence acts as the counterbalance to fear. As the first signs of apprehension appear, the confident flyer is able to approach the upcoming event with a learning attitude instead of dread. Confidence born of hard work and discipline causes a flyer to approach a new situation with a mature mindset that says "Others have learned to do this, I can too." This approach is rational and sound if you have prepared. If you have not, however, it is a dangerous self-delusion.

The dangers of overconfidence

There is no question that confidence is necessary for an aviator. But how much is needed? How much confidence is justified? How much is too much? How can we be sure that we don't cross the line from confidence to overconfidence or cockiness? These are important questions that we all need to come to grips with on a personal level. Overconfidence occurs when our egos write checks that our knowledge and skill can't cash. Dirty Harry, the fictitious San Francisco police detective played by Clint Eastwood, used to mutter, "A man's got to know his limitations." This is sound advice and is discussed in greater detail in the next chapter. It is summarized best in this manner: Once you become aware of your demonstrated capabilities, that is, the flight events you have actually accomplished successfully on a regular basis in the past, confidence is fully justified for those areas. Let's take the example of a 10-knot crosswind landing. If you have done it before and your proficiency, physical, and mental status are still reasonably good, you should feel supremely confident in your ability to make the landing.

Activities that fall outside of your demonstrated experience but are similar should be viewed slightly differently. A 20-knot crosswind landing should be approached with guarded confidence. It might be something you have not done recently, but you know the procedures and techniques involved. A little apprehension is natural, but you recognize the situation for what it is, and proceed with caution. You are careful to leave yourself an "out" (i.e., a go-around) if your skills falter.

When the new situation is out of your experience entirely, for example, a 25-knot crosswind with reported low-level windshear, confidence is not justified, unless you have been trained in the specifics of flying into windshear—and perhaps not even then, depending on the severity of the phenomenon and the capabilities of your aircraft. The cemeteries are full of overconfident flyers who thought they could hack it when their experience, skill, or proficiency did not justify the attempt. Self-awareness is the key and is discussed in detail in Chapter 4.

The hazards of visualization

Visualization is a performance-enhancing technique that has been in use for years in athletics, as well as other fields, such as sales and public speaking. The techniques for effective visualization are discussed in detail later in this chapter, but hazards are also associated with the technique. It uses the "mind's eye" to rehearse successful behaviors as a warm-up for the real thing. Psychologists and neurologists have a special explanation of how the appropriate neural pathways are activated and some other cool stuff, but let's just call it mental rehearsal, or "chair flying," for simplicity's sake. It is an extremely effective method for building habit patterns and preparing yourself to maximize available flight time. Mental rehearsal has been proven effective in everything from shooting free throws to public speaking, but there is an inherent danger with visualization as it relates to confidence. Although we can visualize ourselves doing everything perfectly, it's not real, and we must not let our subconscious convince us otherwise.

Aviators in general, and especially pilots, often assess their abilities based on their personal best day. As we fondly recall our aviation achievements, our minds naturally gravitate to what we have done well. This practice can lead to a presumptuous and dangerous overconfidence. We need to occasionally remind ourselves of our weaknesses as well as our personal bests.

The experience factor

It is no secret that experience builds airmanship skills, and further, that recent experience accounts for proficiency. In fact, in the eyes of many, it is the single greatest skill builder. The legendary Chuck Yeager reflected on the value of experience:

> I have flown in just about everything, with all kinds of pilots in all parts of the world—British, French, Pakistani, Iranian, Japanese, Chinese—and there wasn't a dime's worth of difference between any of them except for one unchanging, certain fact: the best, most skillful pilot had the most experience (Westenhoff 1990).

However, it is also common wisdom that certain types and levels of experience signal a warning flag to aviators and that some experience is more valuable than others. In a landmark study, *Flight Experience and the Likelihood of U.S. Navy Aircraft Mishaps* (1992), researchers found that "pilots with less than 500 flight hours *in model* (of aircraft) were at significantly greater risk for pilot error mishap factors" (Yacovone et al. 1992). Interestingly, the study found no correlation between *total* flight hours and accident rates, indicating that aviators transitioning to new aircraft are at increased risk as well as pilots who are checking out in their first aircraft.

The study summarized some obvious implications for training. "Less experienced pilots need special 'care and feeding.' However, equal attention must be paid to the more-experienced pilot who is transitioning to a new aircraft or returning to flying after a long absence from the cockpit" (Yacovone et al. 1992). This study has great importance and relevance to individuals who might be involved with transition training, either as students or instructors. It debunks the myth of experience or expertise in one aircraft as providing a shield against poor judgment or error in a new aircraft. This tendency is sometimes referred to as the "halo effect," in which expertise is viewed as universal and fully transferable between aircraft. When viewed through the lens of this study, it clearly is not. We must approach each new aircraft as if it were our first, and train each transition pilot as carefully as if they were new.

As the previously cited study points out, using total flying hours as a measure of a crewmember's airmanship capabilities might not be a valid approach. A more relevant number is "time in type," but even that number can be deceiving. Experience is a good teacher, but it doesn't always teach good habits, techniques, or procedures. Our experiential development depends largely on with whom we fly and what happens when we fly. Three critical events greatly shape our airmanship development.

The first role model

New flyers are a lot like kids. We are very impressionable and often take on the characteristics of those from whom we learn. Our parents used to tell us, "Choose your friends wisely." This advice holds a special value for aviators, where the flying environment is often defined by a relatively small circle of friends with whom we share our hobby or profession. Of all the influences on our development, perhaps none is as important as our first role model. This role model is not necessarily our first instructor, although it often is. Perhaps in no other endeavor is the importance of a good role model as important as in the flying game. The disastrous consequences of a negative example was seen in the case study in Chapter 2. I would like to share a personal story of a more positive example.

In my career, I have had a few role models, but I would like to tell you about the one who made the most impression on me, Lt. Col. Craig Wolfenbarger, who taught me how to be an instructor pilot. Although "Wolf" was an excellent pilot, I had flown with many others who were his equal in stick-and-rudder skills.

Two things made him special. First, he seemed to never make a procedural mistake, and second, he didn't take shortcuts. I flew with him on several occasions, and as far as I could tell, he never omitted a briefing item and never missed a checklist step. He put down his flaps at the same point on every pattern and never failed to adjust his patterns for drift. Although I have yet to fly the flawless sortie, he seemed to do it every time out. He was a machine. It was incredible. This level of competence set a standard that I might never reach but will forever strive to achieve. Although his instruction was also excellent, I can't say that I can remember anything he ever told me. However, I will never forget the way he flew. By the pure weight of his example, he changed the way I approached flying forever.

Wolf was just as thorough on the ground. Although many instructors reference the flight manual and regulations, he pulled out each and every one that applied to a critique item and showed me the words. If one of his students had a question, he either found it in writing, or he got on the phone to call the manufacturer to get the answer from the source. Craig Wolfenbarger was a professional flyer, and he took no shortcuts. He left his mark on many students who strive to carry on in his tradition of thoroughness and competence.

Some are not lucky enough or wise enough to get a role model like Wolf and are influenced by aviators who exhibit poor discipline, sloppy procedures, or questionable judgment. Some aviators even pride themselves on the number of shortcuts they can take or showboat their skills to younger aviators. They too leave their legacy—don't be a part of it. Select your heroes carefully, because in a very real sense, they become a part of you.

The first emergency

The second critical experience that leaves its mark on you forever is your first serious emergency. Until it occurs, you constantly wonder how you will react; afterwards, you never forget what you did. Your persona as an aviator is shaped by your actions in response to serious inflight demands or challenges. That is why it is so critical to be prepared to handle your *first* emergency. If you handle it correctly, you have taken a huge step towards self-confidence, as well as inspiring the confidence of others. You finally know that when it all hits the fan, you are prepared and can hack it. If, however, your first emergency is poorly managed, or you panic but survive through no fault of your own, you have sown seeds of doubt in yourself and others that might never completely disappear.

The first big change

The third vital experience comes the first time that you have to make a major change to your plans to accomplish your mission or flight objectives. It is the first time you really have to think on your feet in the aircraft. It might come as a result of unexpected weather, an aircraft malfunction, or some other event, but for whatever the reason, the script changes in flight, and you are left to your own devices to sort it out. You call upon the sum of all your knowledge and training, gather all your informational resources, assess the risks involved, and make a call. The control you exercise over changing situations is one of the major joys of flying. In flight, more than almost anywhere else in today's world, you are the master of your own destiny. This critical shaping event is also a crucial confidence builder, or destroyer, depending on how it is handled.

All three of these vital experiences have one thing in common: your decisions and actions control the outcomes. If you select a positive role mode, if you are prepared for the first big change or emergency, and you handle it well, you have set the course for good airmanship. Your future actions will be undergirded with well-deserved confidence.

Proficiency

Flying an airplane is not like riding a bike. If you haven't done it recently, you might not remember how. It is probably more like juggling bowling pins; if you're not proficient, you are likely to end up hurting yourself. In the previously cited Navy study on experience and mishaps, it was found that poor proficiency was as high a risk factor as low experience (Yacovone et al. 1992). In any 30-day period, pilots with less than 10 flight hours were found to be at far greater risk than those with between 10 and 30 hours. In an interesting side note, pilots with greater than 30 hours were also shown to be at high risk, indicating that proficiency may give way to fatigue or complacency in a high-workload environment.

All flying organizations, including the FAA, commercial airlines, the military, and even the local aircraft rental businesses have requirements to help pilots maintain their currency. While these requirements are designed to keep aircrew members from becoming nonproficient, they are not safety guarantees. Organizations, although well-intentioned, really cannot ascertain individual proficiency requirements with much precision; the variability between individuals is just too great. So once again, we come back to individual accountability and personal responsibility. It is our responsibility to track our proficiency and to ensure that we get what we need before we take an aircraft into the sky.

Personal currency

"Just because it's legal, doesn't mean it's safe." Although this statement is one of the more hackneyed clichés in aviation, it certainly applies to the relationship between "legal" currency and personal proficiency. Two conditions hint at the need for a personal approach to currency requirements. The first is a situation in which the organization does not have a requirement for the maneuver. For example, most all organizations have a currency requirement for landings, usually a certain number every 30 or 45 days. But have you ever heard of a currency for crosswind landings or low-visibility approaches? Probably not, but clearly there is personal proficiency concern here.

The second situation is when the organization's requirements don't work for you. For example, many pilots (the author included) need to fly more often at night than the regulations require to stay proficient. These individual requirements need to become part of your individual training plan or you risk loss of effectiveness and safety.

Because personal proficiency is such an individualized subject, it is difficult to generalize from either regulatory requirements or research findings in a way that is meaningful for everyone, but some overview of previous findings can be beneficial. Perhaps the most extensive study on individual proficiency was accomplished by Childs, Spears, and Prophet in 1983. *Private Pilot Flight Skill Retention 8, 16, and 24 Months Following Certification,* was a two-year study designed to look at skill retention levels, as well as to study how well pilots were able to predict their own proficiency. The results illustrate two important points. First, flight skill loss in important safety-of-flight items such as landings, unusual attitude recoveries, and crosswind takeoffs occurred quickly. Perhaps more importantly, pilots were not very accurate in assessing their proficiency at specific flight tasks. The report stated that "the lack of such a relationship (self-predictability of proficiency) is of ultimate concern from the standpoint of operational safety, since it suggests that the individual pilot is not able to diagnose specifically his own continuation training needs" (Childs et al. 1983). This fact is particularly scary.

The need for a conservative approach

We have maintained throughout this text that we as individuals are the best barometers of our own performance. Yet solid research findings indicate that we might not be very good at it either. What gives? Who can we trust when assessing our individual proficiency? Perhaps the best course of action here is to follow the advice given on the television show *The X-Files*: "Trust no one." Because the stakes of overestimating personal proficiency are so high, we must take a conservative approach if we have any doubts whatsoever. Don't let the fear of poor proficiency erode your confidence factor. Far better to get an extra ride with an instructor than to find yourself out there alone, sweating bullets and muttering the mantra of *The Little Train that Could*: "I think I can. I think I can . . ."

Patterns of physical skill loss

To help us make an educated guess at our proficiency, we need to understand the types of skills that we tend to lose quickest. In the previously mentioned study of general aviation private pilots, the greatest skill loss occurred in the least desirable, but perhaps most predictable places: landings and traffic pattern operations (Childs et al. 1983). Other areas of high skill loss across time included the accelerated stall, crosswind landing, crosswind takeoff, minimum controllable airspeed, and steep turns. A common theme runs through all of these maneuvers: finesse. It appears from these data that proficiency in physical hands-on flying disappears most rapidly in the fine motor skill events. Figure 3-2 illustrates common flight events that exhibit the greatest skill loss over a two-year period.

This pattern is also seen in the military environment with more experienced aviators. In a 1973 study at the Naval Postgraduate School, pilots performing simulated carrier landings were found to

Flight Tasks with Greatest Skill Loss Over 24 Month Period

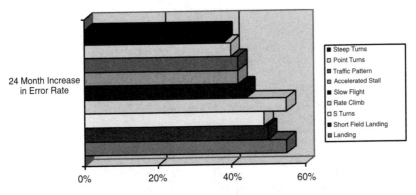

3-2 *Skill loss occurs rapidly in areas that require finesse. What skills deteriorate the fastest in your type of flying?*

be negatively affected by prolonged periods of nonflying (Wilson 1973). These trends come as no surprise to anyone who has flown much. Simply stated, flyers need lots of practice at regular intervals. This fact appears to be especially true for fine motor skill, or finesse, events.

But what about proficiency in the cognitive tasks like prediction, estimation, and decision making? Does our aviation mind lose its edge as fast as our hands?

Mental proficiency

Research shows that we lose our mental, or cognitive, piloting skills as fast or faster than our physical ones. In fact, since flying is a psychomotor process, it is by definition a blending of mind and body to achieve results. Pilots and other aviators must rely heavily on cue recognition and prelearned mental-response patterns, which are decidedly cognitive processes. If we fail to recognize a cue, for example a traffic pattern reference for flap lowering, then it can be the first in a long series of falling dominoes that will mess up our ability to maintain positive control of our flight parameters. Common mental proficiency errors include incorrect time estimations, poor or improper verbal communications, failure to monitor fuel requirements, radio frequency mistuning or misidentification, and improper control positioning based on failure to recognize environmental factors such as crosswinds (Childs 1983). Emergency procedure checklist items and regulatory guidance are also negatively affected by time away from the cockpit. Figure 3-3 illustrates examples of cognitive errors resulting from a lack of proficiency.

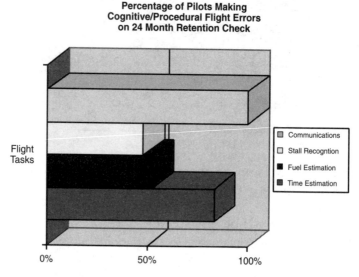

3-3 *Mental proficiency also suffers over time. Note the correlation between skills that require mental calculation and retention. Also note that these types of errors—fuel calculation, stall recognition, etc.—can hurt you.*

It's a given that frequent and disciplined flying that focuses on maintaining both physical and mental skills is the best prescription for avoiding poor proficiency. Unfortunately, that is not always possible. Many outside factors affect our ability to maintain a consistent flying schedule, but evidence shows that there are other ways to delay the loss of proficiency.

Chair flying: Why and how we do it

Chair flying is one of the easiest and most effective methods of self-improvement. It is a combination of visualization techniques, procedural practice, mental modeling, and cognitive conditioning. For our purposes, let's just call it chair flying. It can be done with a variety of props or none at all, but some kind of cockpit mockup might be helpful. Many military pilot training students fondly recall nights spent in a straight-back chair with a plunger stuck to the floor between their legs as a mock control stick, practicing their contact procedures for the next day. It was the ultimate low-tech simulator. But it is not kid stuff; considerable empirical evidence shows that chair flying works and works well for preserving, as well as acquiring, skills (Childs 1983).

Effective mental rehearsal requires very few tools—a checklist, a chair, and a willing and active imagination. The "student" needs only enough experience to follow the step-by-step actions of the flight mission. The briefings, checklist steps, communication, and physical movements should be mimicked exactly as if you were in the aircraft. For example, if you are simulating cutting off the throttle, your hands should make the appropriate movements. As you rehearse boldface emergency procedures, your fingers should reach to "touch" the appropriate switches. Making airplane noises is not necessary, but it can be fun.

You should also practice items that require visual cues, such as landings. Simply close your eyes and picture the landing aimpoint holding steady in your windscreen. Now imagine the aimpoint moving down in your windscreen a little bit, and "see" yourself drifting slightly high on glide path. Your internal dialogue might sound something like this. "Oops, I'm getting high, I'll pull back the power . . . and lower the nose a little . . . OK, back on glide path . . . nose coming up . . . power back in . . . that's better . . . here comes the runway . . . reduce power . . . trim . . . flare . . . flare . . . sweet touchdown." You can, and should, also practice "feel" procedures, such as recovering from a practice set of stalls. Imagine the buffeting, feel the recovery. You should also verbalize all radio calls and even the anticipated responses. Project yourself into the air, and rehearse every aspect of the flight that you can imagine. You can even critique your "mission," reviewing items that you might have overlooked. If you missed a checklist item while chair flying, you will likely miss it the next time you are in the aircraft.

A couple more key points about chair flying. Chair flying should be done in "real time." That is to say, if an approach takes 10 minutes to fly in the aircraft, it should take 10 minutes to fly in the chair. You need not "fly" the entire mission, but the events that you do fly should not be accelerated or slowed down. Secondly, as a rule you should not practice negative behaviors, unless you have a good reason for doing so. For example, I wouldn't chair fly bouncing in a landing, unless I needed to practice refused landings or go-arounds from the bounce. The subconscious is very powerful, and it pays to keep it filled with just the good stuff. Finally, it is probably important to find some privacy for your chair flying. It is difficult enough to overcome your own self-consciousness without having to worry about a significant other sneaking around the corner with a video camera.

Don't take this technique lightly. It is perhaps the single best thing, other than flying or getting in a simulator, that a pilot can do to prepare for the procedures, techniques, and skills required in flight.

Building a personal skill development plan

Individuals vary as to the natural attributes and learning curves that they bring into their flying careers. Therefore, it is understandable that there is great variance between aviators' skill levels when they reach standard measuring points in their development, such as check rides or evaluations. For example, there can be a large difference in skill level between two pilots who both just completed their private pilot checkout, graduated from undergraduate pilot training, or received their commercial or ATP ratings. Individuals also vary as to learning styles. Many are visual learners and learn best from demonstrations, while others benefit most from hands-on experience. Because of these individual differences in abilities, learning curves, and styles, continuous improvement plans for each of us will be considerably different. Chapter 15 contains an outline for a six-month improvement plan, which incorporates skill building with other areas of airmanship development.

Because of individual differences, training program developers typically orient their programs for the "average" student. Too advanced a program loses the students in the bottom of the class, while too basic an approach does not challenge or develop the top students. Individualized training programs do not suffer such constraints and can optimize personal development towards excellence. This applies to formal training as well, where students can optimize their training by helping instructors to understand their individual training needs and learning styles. But it is not possible to have an individual instructor forever, so if we wish continuous improvement, we must become our own instructors, evaluators, and curriculum developers. Unfortunately, few of us are trained, at least initially, to accomplish these tasks. We do have the advantage of knowing ourselves far better than any outside instructor ever will, and techniques for developing personal instruction can be learned.

A layman's course on personal program development

Instructional system development, also known as the ISD process, is a complex and mind-numbing maze of requirements and validation steps, caught in a seemingly endless feedback loop.[2] The complexity of the full blown ISD process should carry a warning label much like the stunt shows on TV: "These stunts are done by trained professionals — do not attempt them at home!" But the process can be simplified, and in its simplest form, training development can be broken down into six distinct activities, easily accomplished by normal folks. (For the purposes of this explanation, we will consider flyers as normal, although many might argue otherwise.) Additionally, we have the advantage of knowing our capabilities, limitations, and learning styles, allowing for an individual focus, which is critical for effective training. The condensed version of the ISD process is illustrated and described here.

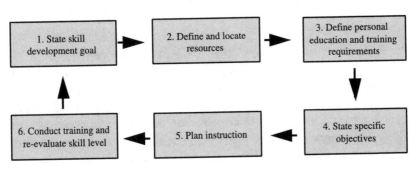

3-4 *Take charge of your own development. This six-step process closely mimics the model used by professional curriculum developers and training experts.*

Step one: Determine and state the skill development goal

For the purposes of this explanation, we will take the example of a pilot who wants to develop better instrument flying skills. We have identified instrument procedures as a potential weak area,

2. The author wrote his doctoral dissertation on an accelerated model of curriculum development for developing aviation human factors training and was left with the impression that some aspects of the ISD process, while valuable to other academic environments, can be greatly simplified and still be useful for aviation.

and our goal is simply to become the best instrument pilot that we can be. Having made this basic statement of purpose, we move to the next step.

Step two: Define and locate resources

Improving our instrument flying skills and proficiency will take some resources, which must be identified and planned for. For example, we might first need to brush up on our knowledge of instrument rules and techniques, so we need to find some good sources of information, such as the instrument flight rules contained in the FARs and any other associated regulations. These resources can be located in many places, including local training organizations or on the Internet, using the keywords "aviation," "FAA," or "FAR" for the search engine, and following the menus to find what we need. In addition, we should probably locate some advanced technique materials on specific types of maneuvers and approaches. Many "how to" books and pamphlets are available from the FAA, in libraries, or at local training organizations. We might also want to assess our current state of knowledge by taking an FAA practice exam, also available by writing to the FAA or downloading it off the Internet. Taking a practice exam allows us to see our weak areas and measure how far we have come at the end of our training.

Once we have located the materials we need to improve our knowledge, we need to ensure we have access to the flight or simulator resources to practice and refine the skills we wish to obtain. Do we have an IFR-equipped aircraft or simulator at our disposal? Where can we go to fly instrument approaches? Is there airspace to accomplish instrument proficiency airwork, such as "vertical S" maneuvers, and practice holding patterns? A hint here: Don't become upset because you do not have the ideal resources or situation at your doorstep. Use what is available. Chair fly if no flight or simulator resources can be obtained in the near term. The best place to begin is where you are. Remember, this is about improvement, not perfection. Don't use the lack of a perfect solution as excuse not to develop a personal program for improvement. This is the reason we have the "locate resources" step prior to defining the education and training requirements. Once we know what we have to work with, we can define realistic improvement requirements.

Step three: Defining personal education and training requirements

We have decided what it is we want to improve and what we have to work with. Now it's time to set about the task of defining specific training requirements for success. These requirements are nothing more than refined statements of the goal. Since our overall goal was to become the best instrument pilot we can, our education and training requirements might contain the following elements. (Please note, this is by no means a complete set of requirements, but it suffices for explanation of the process.) Under the heading of "education requirements" we have decided to begin with the basics, so we establish a requirement to memorize the four steps of basic instrument flying:

1. Set the picture.

2. Trim off the pressure.

3. Analyze the performance.

4. Make adjustments.

Beyond this, we can establish a requirement to know all holding and IFR approach rules and perhaps to better understand lost communications procedures. Care should be taken not to create too large a plan. Focus on small changes with immediate results, and the motivation of near-term achievement will help you to keep moving forward towards your goal.

Step four: State and prioritize specific objectives

This is the last level of refinement. It allows us to completely detail what we want to accomplish. For example, under the area of holding procedures, the objectives might be

1. Be able to draw random holding patterns as given by an approach controller.

2. Understand direction of holding pattern entry when approaching the holding fix from any direction.

3. Know the holding speeds and leg lengths above and below 14,000 feet msl, etc.

The second part of objective selection is prioritization. Some skills form a natural progression, and our training objectives might want to follow this logical flow. For example, we might want to build our instrument procedures in the same order in which they occur inflight, with standard instrument departures (SID) first, followed by en route procedures, holding, and finally approach procedures. Other methods might have to do with your current proficiency or simply your desires. Unless one skill builds directly on another, the order in which you accomplish your objectives is not critical.

Don't overdo this objectives step. If you only need to familiarize yourself with something, such as the legend on an en route chart, don't spend time or effort memorizing it. Conversely, if something does need to be committed to memory, such as lost communications missed approach procedures, the objective should require that you know them cold. When you've lost your radio in the weather on a final approach segment, it is no time to be looking for the procedures in the IFR Supplement. Once the objectives are accomplished and put in writing, we are ready for the next step, planning the training.

Step five: Planning the training

This step is merely the scheduling function of your personal improvement plan. First, you identified your goal and determined what resources were available. You then wrote specific training objectives to accomplish the goal, and now you need to set up an action plan to achieve them. If you are already in a training program or actively flying, it might simply mean adding a little prior planning to your sorties so that you get the opportunity to work on what's needed. If you need special instruction to accomplish the training, you might need to coordinate with an instructor who specializes in the area you intend to work on. The entire point of the planning phase, as well as the entire improvement plan, is that you take an orderly and systematic approach to improving your game.

Step six: Conduct training and reevaluate skill level

One of the unique attributes of taking a personal interest in your training is that you clearly understand the benefits you are seeking. Because you have taken time to organize and plan your individual improvement, you tend to take the training more seriously—you feel connected with the training. It is imperative that you continue to evaluate your performance and skill level, because this feedback determines not only your progress towards your current goals, but your future training plans as well.

This six-step process is fundamentally the same as that used by professional curriculum and training developers, without all of the "smoke and mirror" validation steps—steps that you don't require because you already have the best idea of where your weaknesses are and the areas you need to improve. The personal development plan validates your training at the personal level and will lead to rapid and predictable improvement in the areas *you* determine important. In a formal training program environment, you can provide meaningful input to your instructors. In day-to-day flying, it provides positive focus to your training, allowing you to make systematic and predictable improvements in areas you deem most important.

A final perspective on skill and proficiency

There is no substitute for being able to fly the aircraft or for being able to accomplish your inflight duties if you are not actually at the controls. None of the other airmanship pillars or capstones can overcome a lack of skill or proficiency. If you suspect that your individual capabilities are not what they should be, then you have a responsibility and moral obligation to get the problem fixed—or stay out of the sky. Nearly every pilot shop in the world sells a poster with the now-famous saying: "Aviation in and of itself is not inherently dangerous. But to an even greater extent than the sea, it is terribly unforgiving for any inattention, carelessness, or neglect." Don't neglect your skills or proficiency.

References

Childs, J. M., W. D. Spears, and W. W. Prophet. 1983. Private pilot flight skill retention 8, 16, and 24 months following certification. Atlantic City Technical Center: New Jersey Department of Transportation/FAA. p. 32.

Westenhoff, C. M. 1990. *Military Airpower: The CADRE Digest of airpower quotations and thoughts.* Maxwell Air Force Base, Alabama: Air University Press. p. 23.

Wilson, W. B. 1973. The effect of prolonged nonflying periods on pilot skill in performance of simulated carrier landing task. Masters thesis. Monterey, Calif.: Naval Postgraduate School.

Yacovone, D. W., M. S. Borosky, R. Bason, and R. A. Alcov. 1992. Flight experience and the likelihood of U.S. Navy mishaps. *Aviation, Space, and Environmental Medicine.* 63:72-74.

4

Know yourself

**by Richard Reinhart, M.D.,
and Jack Barker, Ph.D.**

Know thyself.
Socrates

A personal view from Urban "Ben" Drew

Aggressiveness is a natural—and some might say essential—part of most pilots' personalities. But it can set you up for a fall if you let it get the best of you. When you combine aggressiveness and expectations, you can end up fooling yourself and creating a dangerous situation.

We were on a deep escort mission in the Mustang (P-51), cruising just below a cloud deck. Our flight leader was getting ready to rotate back to the States in a few days and was aching for action. He wanted an aerial kill so bad he could taste it. Most of his squadronmates had air kills, and all he had was one confirmed on the ground. Well, this was his mindset when all of a sudden a flight of aircraft descended out of the clouds a few miles in front of us. The leader called out "drop tanks, bandits at 12 o'clock." Some of us weren't so sure they were really bandits, so we challenged his identification. "Are you sure they are bandits (enemy), or are they bogeys (unknown aircraft)?" "Bandits" he replied and rolled in and fired on one of the "Tail-End Charlies." The poor bastard never saw him coming and exploded in midair. When the rest of the "bandits" realized that they were under attack they wheeled to fight—and we could clearly see that they were British Spitfires.

When you combine aggressiveness, expectations, and a stressful situation, you are prearmed for trouble. If you recognize this in advance, you need to throttle back a bit and make certain the situation is really what it seems. This isn't just a problem encountered in combat. A pilot flying an instrument approach can expect to break out of the weather at a particular altitude, and when he doesn't, he often presses too far. Another example is the pilot that expects that he has the range to make a particular destination. To him, the fuel gauge might look a little fuller than it really is. Expectations and aggressiveness are a dangerous combination.

A World War II ace and recipient of an Air Force Cross, Ben Drew was the first pilot to shoot down two German Me-262 jet fighters. He is first adjutant general of the Michigan Air National Guard and a commercial and corporate pilot with more than 16,000 hours.

Understanding ourselves may be the most difficult system that we are ever required to learn. The human physiological and psychological systems are far more sophisticated than any machine ever invented. And then there is the problem of self-image. We all like to think of ourselves as great aviators. Greasing in those landings, right on the glideslope just about every time, quick to respond to airborne challenges and maintaining situational awareness. Yet who hasn't had a least one memorable episode or flight when you didn't have it all together. We all have bad days once in awhile. But why? And how can experienced pilots with years of practicing good airmanship end up as an NTSB accident file by flying a perfectly good airplane into the ground. "Controlled flight into terrain" (CFIT)—what a humiliating epitaph! But it happens. Basic airmanship is seldom even mentioned as a probable cause or contributing factor. But that doesn't mean it wasn't lacking.

This chapter looks at issues involved with understanding what is perhaps the most critical component of the airmanship model— yourself. It begins with a discussion of the physical self, a much overlooked part of our airmanship picture. The chapter concludes with a look into some psychological aspects of individual airworthiness.

Suggestions for improvement are made throughout, and examples and case studies illustrate the principles in action. To get at the important points, we must occasionally use technical terminology, but we try to keep it to a minimum and keep the discussion in the context of the flyer's operational environment.

The physical self: Medical airworthiness

Would you fly a plane if you found a mechanical problem during preflight or fly into known severe icing or a thunderstorm? Would you feel safe with your favorite auto mechanic working on your aircraft? These are dumb questions, because we all know better than to mess with a perfectly functional aircraft—an aircraft that is certified as airworthy.

But not so dumb when you ask the same kind of questions about yourself, when you're at the controls in less than ideal conditions, when you are hypoxic, or ill, or fatigued, or a multitude of other conditions. Are you airworthy—medically airworthy? Chances are you're more concerned about the performance of the airplane than you are about yourself and crewmates. There is a very human tendency, one that is very noticeable in the pilot community, that says we can rise above physiological challenges, no matter what they are. This "can-do" attitude, while admirable, conflicts with realistic self-knowledge and good airmanship.

On the other hand, complacency and apathy, born of boredom or overconfidence, has a similar effect—poor airmanship. However, there are other situations in which you are very conscientious about your personal physical performance, but you have little control over the fact that you are impaired or distracted. That situation occurs when you are subjected to the physiological hazards of flight, namely hypoxia, fatigue, self-medication, disorientation, etc. You are no longer medically airworthy and your performance is degraded, despite commendable airmanship. The insidious concern is that you and your crew probably are not aware that you are in such a condition. It is often referred to as "subtle incapacitation."

Medical airworthiness defined

So, what is this new term—medical airworthiness (or MAW)? Basically, it's the recognition that you could be the best pilot in the world, but if you aren't physically up to par, you are setting yourself up for problems. How we function, physiologically or psychologically, will have a direct impact (literally) on how well we perform in flight. If we are fatigued, suffering a hangover, coming down with the flu, under stress, or hypoglycemic (low blood sugar), we are not performing up to our expectations, and we won't be medically airworthy. Failure to perform as expected can be lethal in an aircraft, especially when coupled with unwarranted confidence.

Everyone—including aviators—makes mistakes, but why should we set ourselves up for a fall? Why establish conditions in which our effectiveness and safety are put at risk? The risk of a single error evolving into something more than surprise or embarrassment increases as personal airworthiness is eroded. Airmanship is compromised if you push the limits of your own ability to function in the physical sense.

When it comes to physical airworthiness, we are our own "mechanic." We should not look for someone else to determine if we are fit to fly. You are the one ultimately held accountable for being airworthy. It's an individual responsibility. Of course, if you have any questions on your health, you should see a flight surgeon or doctor, but it is your self-awareness that is the key to recognizing personal deficiencies.

Lack of medical airworthiness leads to those subtle and subjective conditions commonly referred to as impairment or subtle incapacitation. Even more insidious is an increased degree of complacency or apathy that typically accompanies poor physical function. Your personal inflight parameters become expanded or ill-defined, you begin to take more risks, and start to "tune out" from the events of the flight. Next you begin to lose situational awareness, your reaction time is prolonged, there is a change in the expected situation on final approach, and then . . .

When discussing airmanship, safety of flight, and fitness to fly, the pilot and crew must consider the medical airworthiness of everyone on the flight team. Practicing good cockpit resource management (CRM) skills will help identify losses of airmanship on the team and take measures to keep the situation "straight and level." But even good CRM skills can be impaired if you aren't fit to fly. We've all been there to some degree, so worn out or fatigued that we begin not to care about our performance standards. Yet as difficult as it is

to recognize this degraded performance in ourselves, it is doubly difficult to recognize in others—and next to impossible if you and your
crew are all in the same physiological boat.

If we could predictably prevent impairment, incapacitation, distraction, and disorientation, we wouldn't need this chapter, nor any
training in flight physiology. Or maybe it's because we have this
chapter, this book, and other training in physiology, that we have
some control in keeping accident rates low. But formal training is
only a small part of establishing medical airworthiness. To really
tackle this problem, we need to take it to a personal level.

So how can we reduce the risk of impairment and maintain our
medical airworthiness? Let's introduce each issue and see how it can
affect us. Keep in mind that virtually every one of the following situations can exist at the same time, compounding the cumulative effects. Each impairment, viewed separately, may be no big deal—but
combined? Then it's a different story, one we must be concerned
about or face the consequences of our oversight or overconfidence.

Topics in medical airworthiness

In this section, we introduce some of the more common conditions
that can impair an aviator or flight crew and make them less than airworthy. This text is not meant to be a final or definitive source on
these topics. That is left to you. Good publications describing these
problems at length are noted in the reference section.

Hazards at altitude

This is a true story. An airline captain, very professional and experienced, had the added responsibility of checking out newly purchased aircraft before acceptance by the company. Part of that
checkout was to evaluate "engine-out" performance, with one scenario being a loss of cabin pressurization. The day after one of these
checkout flights, he called the flight surgeon, very concerned that he
had had a stroke, because he had experienced dizziness, blurred vision, weakness, and generalized discomfort. He explained to the
flight surgeon that it occurred during the engine-out sequence.
Asked what the cabin altitude was at the time, he stated "only about
15,000 or slightly more—no big deal. Should I come in?"

He didn't have to. He was symptom-free when he called, and he was
hypoxic in the aircraft. The scary part was that he didn't recognize

the symptoms. This pilot didn't have a military background, which requires aviators to get altitude chamber "flights" at regular intervals, designed to help flyers recognize their personal symptoms of hypoxia. Even though some training in high-altitude physiology is required for commercial flyers, it's a topic that is taken for granted in the civilian world. No matter what type of aircraft you fly, you need to be aware of the hazards of altitude.

The fact remains that hypoxia is experienced by *every* flyer to varying degrees with a variety of symptoms. The problem with hypoxia is that few aviators respect how it can impair mental and physical performance, without even knowing there is impairment. Fatigue, visual disturbance, poor reaction time, a feeling of diminished motivation, or an attitude of being willing to accept a less-than-safe situation are common. By the time hypoxia is recognized, it probably is too late because the pilot is unable or unwilling to take corrective action. The pilot's time of useful consciousness (TUC) or effective performance time (EPT)—both meaning the same thing—is a matter of minutes. It is less if it occurs in a rapid decompression. This is critical because, by definition, TUC or EPT is the amount of time that the pilot can determine that there is a problem and get out of harm's way. The pilot's thinking after this point is neither useful nor effective.

Practically all airplanes have pressurized cabins "for our comfort." True. But consider what the cruising cabin altitude is for the larger airliners—about 6000 to 8000 feet! You are slightly hypoxic even at that altitude, especially if you have been at cruise for several hours. Since the brain and vision are especially susceptible to hypoxia, you will be further impaired at night after a long trip, such as just before you begin a descent for a complicated approach and landing. Fortunately, by the time you are well into your descent, you are at an altitude that will not affect you, and, of course, any strange feelings during cruise usually will be dismissed. If your hypoxia didn't interfere with your decision making, you won't realize how close to your limit you really were. If a bad decision was made or you had trouble reading the instruments and charts, it's doubtful that hypoxia would be considered the cause. We just don't think about it at normal cabin pressures—but it's there.

A wise man once said that "time heals all wounds." Not so with hypoxia. As you get older, you are less tolerant to the effects of hypoxia or all the other physiological situations about to be discussed.

Near vision is greatly compromised, because many pilots won't wear reading glasses unless, in their minds, they really need them. More on that later.

Another altitude-related problem that can evolve into incapacitation is an ear or sinus block. Congestion from a cold, hay fever, or flu puts you at risk of developing a block. But how much of a cold, flu, or allergies, affect your medical airworthiness to an unsafe degree? There is no straight answer, but you, as the individual flyer, must know how these disorders affect you. Often occurring during descents, an ear block can be minimized by beginning corrective measures as soon as you notice any blockage or "fullness feeling" of the ears. That means doing a valsalva maneuver (holding your nose and blowing hard against that resistance), moving your jaw around, etc. Don't wait until it's a full-blown block, because incapacitation might be right around the corner. Take the aircraft back up in altitude to reverse the block, then descend at a slower rate.

Another altitude-related concern is decompression sickness (DCS), commonly referred to as "the bends." It is not a common occurrence, but it happens, especially when flying after scuba diving, a sport that is becoming increasingly popular. Even if you don't develop the bends, it's possible that your passengers might, and you don't need medical inflight emergencies. For more information on DCS, see your local flight surgeon or physiological training officer.

Orientation

Mention the word *orientation* to a pilot and the word *disorientation* immediately comes to mind. Furthermore, to a pilot, disorientation is normally equated with spatial disorientation (SD), when, in fact, SD can be subdivided into different types. The danger of orientation problems is greatest when we are negatively affected by other physical problems, and it can be a killer.

Orientation needs no definition—it's a relative term, describing where we are compared to something else. Obviously, disorientation is the opposite—not knowing where we are. Disorientation can create severe airmanship challenges. At least five different kinds of disorientation are inherent in aviation, and spatial disorientation often has a double meaning. Here's a brief description of each:

1. *Postural disorientation:* Postural orientation describes our position relative to gravity as a result of our *proprioceptive*

sense of relative motion of our muscles and joints. Are we up or down or sideways at any given time during flight, or a combination of all? How we sense motion from that position is determined also by our *otolith* system within the inner ear (a part of the *vestibular* system). This group of sensory inputs gives us the ability to fly "by the seat of our pants" without visual references. When these inputs give conflicting and erroneous signals to our brain, we experience postural disorientation.

2. *Positional disorientation*: Positional disorientation simply means that we are lost, if even for a moment. It is often equated with lost situational awareness. Humans are creatures of habit. We bathe ourselves the same way every time. Change that routine and you can get lost on your own body! If you don't believe me, try changing the sequence of your actions in your next early-morning shower. You will likely have to stop and think about what to do next! What might be amusing in the shower is not nearly as funny in flight. Many of us fly the same route, enter the same published approach routing, and talk to the same controllers as a matter of routine. It becomes ingrained in our mental processing. Change that routine and for a moment we are confused, not sure where we are and what to do next. We have disrupted our expected procedures. We're temporarily disoriented. We often lose our positional orientation.

3. *Temporal (time) disorientation*: As a result of increased activity, our perception of time is expanded. We think we have more time to accomplish a task, but we don't, especially when multiple tasks are coming quickly. The best example is a situation in the military community—the decision to eject from an aircraft. Because so much is happening in a compressed time window, the pilot often takes too much time to "punch out" due to expanded perception of time. The mind works so fast during that crisis, the pilot thinks he or she has time to think things out and try to recover. Discussions with many survivors of ejections point this out. They often take 20 minutes to describe an ejection sequence that took only seconds. Captain Mike (Goose) Gossett, a B-1B ejection survivor, once described in incredible and surrealistic detail the receding glow from the instrument panel as his ejection seat went up the rails on a

bad night in Rapid City, South Dakota. It is an extreme case, but it illustrates how our mind can switch to overdrive and confuse our temporal orientation. Conversely, a boring and uneventful flight without mental activity makes the passage of time drag on, which is another form of temporal disorientation.

Before going on, consider how postural, positional, and temporal distortion could conceivably occur at the same time, resulting in an incapacitating loss of SA. To add more confusion to this complicated phenomena, let's look at spatial and vestibular disorientation, two types that are often confused with each other but occur in distinctly separate situations.

4. *Spatial disorientation*: Most pilots would now think we are talking about how our semicircular canals are disrupted, but that's not quite what we mean here. Spatial disorientation, or the "vection illusion," is the motion or position of the body in a space relative to nearby objects as seen in our peripheral vision. It's very noticeable in a car wash or sitting in your car at an intersection waiting for the light to change to green when the car next to you is creeping forward. You swear your car is moving forward, and you slam on the brakes of your motionless car. Your peripheral vision is a powerful sense, and even though you are not moving, if you see something in space next to you move, you have the sensation of moving. Flying through snow or rain gives you the same spatial disorientation of relative motion. It can confuse some aviators, who have always been taught a more generalized definition of spatial disorientation. So if SD is not SD as we know it, then, what is the other SD?

5. *Vestibular disorientation*: We have our own set of gyros in our head—our vestibular system and its three semicircular canals. Each represents an axis of position. Their role is to determine motion in three directions, often in combination. If these canals are not sending signals that are consistent with our other senses (visual and proprioceptive), as in an uncoordinated turn, we will have vestibular disorientation. Our gyros are "tumbled," and we have the sensation of being in a turn when we are not turning—a very powerful sensation, often difficult to overcome. This sensation will make any flyer airsick if the inner ear is "tumbled" enough.

Pilots can get physically ill in a simulator if the motion felt by the body is just a fraction of a second off from the peripheral visual reference of position.

This variation in definitions is not meant to confuse you, but because these last two are two of the most powerful forms of disorientation, we must distinguish between them. And we haven't mentioned visual illusions! Furthermore, your tolerance to any of these forms of disorientation will be seriously impaired if you are not—you guessed it—medically airworthy. For example, if you are tired or hypoxic, your ability to overcome conflicting orientation cues will be compromised, if not disabled, and might cause you to turn the controls over to your crewmate or an autopilot if you have one. Even subtle disorientation compromises your level of airmanship.

These are examples of how something in flight—hypoxia and disorientation—can be taken for granted and subtly cause impairment. The remaining topics are less well-defined, but part of airmanship is being aware of these impairments, recognizing when you are at risk, and managing your flight accordingly. The following case study points out that disorientation can affect anyone in the cockpit, and the consequences, if survived, can be frightening.

Case study: The terrified passenger (NTSB 1994)

While en route to his destination, the pilot of a small multiengine aircraft and his passenger watched as the weather deteriorated from clear skies to "ceiling obscured, 100 foot overcast, RVR 1200." While these were definitely not the conditions forecast during the briefing he had received three hours earlier, he was a competent and confident instrument-rated pilot and he continued to the destination, prepared to shoot the instrument approach. Upon arrival at the destination, the pilot decided to shoot the approach, but he took a few moments to brief his passenger on some procedures. He showed the passenger, who, although he was not a licensed pilot, did have some flight instruction and exposure to previous IFR trips, what the approach lights would look like and assigned him the task of looking for them. The pilot also explained the elements of the approach and what would determine whether they landed or missed the approach. In a nutshell, the pilot was doing everything he could to maximize the team by giving the passenger the tools to contribute.

The approach went smoothly aside from an inadvertent descent in VMC below glideslope intercept altitude by about 200 feet, about which approach control provided a gentle reminder. During the descent through the clouds (tops 800), the pilot concentrated on his instruments, calling out altitudes at 100-foot intervals, and ended the descent at the decision height of 220 feet msl. With no runway in sight, he reconfigured for the missed approach and climbout. Suddenly, the passenger yelled something about "going down," grabbed the yoke and pulled—putting the aircraft into a steep nose-up pitch, setting off the stall horn, and then into a dive. While fighting for the yoke, the pilot finally got the passenger to let go, recovered to a normal attitude and resumed the climbout, all at only a few hundred feet from the ground in essentially zero-zero conditions.

The pilot stated that he was unaware of any problems that existed before the passenger took the yoke. He further stated that if he hadn't had the few extra feet he gained at the onset of the missed approach, recovery would have been almost impossible given the extreme nature of the maneuvers that were induced. Neither the tower or approach control commented on any altitude excursions, perhaps because the elapsed time was very short and sampling by radar might not have shown anything particularly unusual. The pilot presumed that the passenger was a victim of spatial disorientation and stress. The reporting pilot stated that "the stress of the situation might have been reduced if I had continued to verbalize during missed approach segment. In the future I will communicate more during the entire procedure. I have never been through a more terrifying experience."

Sleep

Most likely, no one needs to tell you that sleep is essential or that sleep debt is an impairment. No matter what duty and rest-period regulations or policies are in effect, the pilot must respect the need for sleep. It is a straightforward requirement, but studies show that aviators are notoriously bad at managing their rest periods. When we should be resting, we are often doing other things. So a discussion and understanding of sleep is important.

When we need food, the brain tells us we are hungry. If we need water, we are thirsty (but often well after we are significantly dehydrated). So, if we need sleep we are _____. Fill in the blank. The problem is that there are two kinds of sleepiness, and we often don't respond to this cue as well as we respond to hunger or thirst. We just press on, thinking we can tough it out.

One kind of sleepiness is physiological. The body must have sleep, the amounts varying with people. If sleep debt gets big enough, the body WILL sleep because of its physiological need. It might be in the form of microsleep, lasting anywhere from a few seconds to a few minutes. But you are asleep and probably not even aware of it.

The other kind is subjective sleepiness. It is how we feel. It's a signal that is easily overcome by many factors, such as stimulating activities or conversation, caffeine or other stimulants, and the intense motivation to get the job done. It gives pilots a false sense of security, which often leads us to accept challenges we might be unable to complete. We act like we are alert, so we must be, right? We may be fooling ourselves, but we can't fool Mother Nature, and the physiological need for sleep will soon take over, often in the middle of a demanding part of the mission that we took on when we felt OK.

NASA's sleep studies in its Fatigue Countermeasures Program indicate that most of us suffer from sleep debt on a relatively frequent basis. The only fix is—you guessed it—sleep! Just as thirst or hunger must be solved with fluid or nutrition, there is only one cure for sleep deprivation. Getting enough sleep becomes the challenge. How to achieve an alertness level that is safe will continue to be a controversial and difficult task when it comes to organizational policy-making and scheduling. That's why we as individuals must accept the final role in ensuring a rested state no matter what the policies are.

How sleep occurs, and techniques to manage sleep and alertness are simply topics that are too expansive to be discussed in this text, but the cure has been identified. Suffice to say, only you can understand and manage your sleep needs. No one can do it for you.

Jet lag

Closely related to sleep, yet distinctly different, is the effect on a body passing though several time zones. It is commonly called jet lag, but the scientific term is "desynchronosus," if you want to impress your flying buddies. They mean the same thing; the body's many circadian rhythms have been disrupted. One of those rhythms is the sleep cycle. Others are the many biological rhythms that keep our bodies running effectively. Mess them up and you feel messed up. The more time zones you pass, the more messed up you will be.

Global flying is here to stay, and the aviation community will always struggle with the best ways to cope with jet lag. Coping techniques abound and are very individualized. One key point to remember is that exposure to sunlight remains the one most effective element in adjusting to local times. The other is to schedule trips so that the body doesn't have time to become out of synch. But now we are talking about situations that are often out of our individual control, such as scheduling.

We can, however, control other factors that exacerbate jet lag. Jet lag's symptoms are many, the most significant being fatigue and sleep debt. We must be disciplined enough to eliminate other self-induced causes of sleep debt during the critical readjustment period. Those factors that can be controlled must be a high priority.

Fatigue

When discussing the physiology of flight and safety, nothing stirs a flyer's emotions as much as the subject of fatigue. No other topic generates such a variety of responses, mostly negative. If you were to ask the one thing about flying that concerns aviators the most from a physiological point of view, it would be fatigue. When it comes to airmanship, fatigue is the most consistent physiological cause of impairment.

Since everyone has been fatigued, definitions of its symptoms are not necessary. There are two major problems with fatigue, beyond its symptoms. The first is that fatigue is very subjective, difficult to quantify, and therefore difficult to regulate. If you can't quantify or regulate fatigue, how do you determine how much fatigue is safe? When should you call in "sick" or ground yourself because of excessive fatigue?

The second problem is that fatigue is not caused solely by lack of sleep or overwork. It has multiple causes, many not clearly respected or recognized, but all having a cumulative effect on performance. For example, we have already mentioned three: sleep debt, hypoxia, and jet lag. Just about everything we talk about in being medically airworthy has a resulting fatigue. Temperature extremes, noise, stress, hypoglycemia, caffeine withdrawal, illness, and after-effects of alcohol can all cause fatigue. All of these factors can be occurring at the same time, and when the level gets too high, we are no longer in control. The trick to effective fatigue management is to determine what causes the combination of our personal fatigue. In other words, fatigue is a function of multiple situations, many that are controllable. That doesn't mean that a long day or night of flying can't be fatiguing in its own right, but rather that we must consider the other controllable causes when coping with fatigue and managing rest periods to maximize performance.

Managing rest periods is not one of the pilot community's strongest attributes. Despite adequate time to get rest, aviators often use that time for other activities, ranging from pleasure flying or flying with military reserve units, to commuting long distances both in the air and by car, to prolonged sightseeing in unique locations, to just about anything else you can imagine. It is one area that is easy to improve—with a little willpower and common sense.

Fatigue has many symptoms that are similar to hypoxia, including channelized thinking, fixation, impaired reaction time, irritability, and that dangerous attitude of willing to settle for less in performance. One technique to maximize fatigue recognition is to mentally make a note of how you feel when you are fatigued for any reason (hopefully at home and not behind the wheel or in the cockpit). Notice how your attitude, thinking processes, and general feeling of being "down" are surprisingly beyond your power to overcome. Now mentally put yourself in a cockpit and try to imagine whether or not you would consider yourself prepared to handle an airborne emergency. Good airmanship is admitting your potentially unsafe level of fatigue to yourself and your teammates, so all can be wary of unsuspected impairments of performance. Fatigue is a killer, and it demands our respect.

Self-medication

Aviators like to play doctor, perhaps as part of the personality profile that wants to remain in control. And once a self-diagnosis has been made, literally hundreds of over-the-counter (OTC) medications are available to be used in "treatment." The aviator now has two concerns: the symptoms of the medical disorder and the side effects of the medication, both being unpredictable in flight.

Although flight surgeons around the world will cringe when they read this, it is unrealistic to expect flyers to ground themselves for all minor ailments. Part of airmanship is understanding whether or not you are fit to fly. Knowing you are not at your physical peak as a result of a minor ailment is part of the answer. You can compensate for your decreased performance level by flying more conservatively and taking extra precautions. Letting your crewmates know you aren't up to par is essential, but occasionally pride gets in the way of this critical step.

Many medications have side effects that are known only to those who read the small print on the package. It's common that most people seek medical attention because they have symptoms that interfere with their work or play. Most labels appeal to that need, often using the effects of the same medication (or chemical) in two different forms. For example, the side effect of most OTC antihistamines used for colds, allergies, and flu is sedation. So, what do you think is used for sleep medications? Right—an antihistamine. Which means you treat a cold with a sleeping pill! Next time you are in the drug store or grocery store, read the label of the ingredients of sleep medications, allergy medications, and cold and flu remedies. Diphenhydramine (an antihistamine) is a common ingredient in all three. Think about that before you self-medicate with one of these.

Also consider the fact that many OTC medications contain alcohol in substantial amounts, often up to 50 proof. There is also the dangerous rationalization that if one pill or swallow helps, two or three has got to be better. How about using a couple of medications in combination just to be sure to really knock out the symptom? Well, the symptom might not be the only thing that gets knocked out.

Airmanship means that you know what you are doing to your mind and body. It's up to you to be familiar and knowledgeable about self-medication, the illness, and its effects on your performance. The bottom line is this: It's your call on flying with minor ailments, but unless you are a physician, don't self-medicate and fly. Let's take a closer look at some of these minor ailments.

Common illnesses

As just discussed, how we feel, how we consider the significance of any illness, and how motivated we are to treat ourselves has a great impact on our fitness to fly. Often we minimize the impairment from a simple cold because we feel we can overcome those symptoms ourselves. That's not to say that we shouldn't fly every time we have the sniffles. But we must be aware of how a common illness can affect our performance and then realistically balance the risk to safety and airmanship.

For example, a common cold is a viral infection of the respiratory system. While we might be able to tolerate and cope with symptoms such as the cough, sniffles, and congestion, the other side of the coin is that our body is working double- time fighting off the infection. As a result, there is a general sense of fatigue and malaise along with subtle impairment. The pilot must weigh real-world priorities, noting that some colds or flu are worse than others. The best way to deal with this situation is to check with your flight surgeon or aviation medical examiner.

Vision

Vision is, no doubt, the most important sense we use in flight. Yet there is often an internal conflict between seeing effectively and the nuisance and image problems associated with wearing glasses. This subject of vision is very broad and for our purposes of relating it to airmanship skills, it should be apparent that every aviator must clearly visualize what's going on around the aircraft and within the cockpit. Many pilots take short cuts and underestimate how their vision is compromised by wearing designer-type sunglasses, becoming hypoxic or fatigued, getting older, and other conditions that affect vision.

Probably the most important dilemma with vision is the reluctance of older pilots to wear reading glasses. Until these self-conscious pilots find themselves unable to read the approach plates or have to write in key data with a felt tip pen, their glasses remain in the pocket. After all, the FAA does not require that reading glasses be worn—just possessed. Yet, very often, the pilot who is over 45 has impaired near vision but is able to pass the FAA medical exam. Put that same pilot in a cockpit at 7000 feet cabin altitude, at night, after a five-hour leg, and the near vision has deteriorated to a dangerous degree. Age and experience do not always equal good airmanship.

Miscellaneous impairments

Several other situations can interfere with our performance. While most flyers have a healthy respect for the effects of alcohol on performance, there is often a lack of awareness of how impairing a simple hangover can be. For hours, and perhaps days, after your blood alcohol is zero, your performance and mental processing can be impaired, perhaps significantly. You might be legal with the 8- or 12-hour bottle to throttle rule, but you might not be safe. Indeed, your airmanship will be sorely tasked. Your body takes a beating from alcohol, and it needs time to repair itself.

Poor nutrition generates more than just appearance changes or decreased stamina. It makes you more susceptible to illness and the physiological effects just discussed. Hypoglycemia, for example, is not a disease. It's a controllable condition of feeling fatigued and shaky as a result of missing a meal or substituting a Coke and Twinkie from the vending machine for lunch. Like the computer programmers say, garbage in, garbage out. Eat right and let your body take care of you.

Unresolved stress is a common yet powerful cause of distraction and impairment. Airmanship takes a back seat to sloppy performance and adherence to procedures when we move past our peak of performance and into the high-stress, low-performance regime. Stress is also physically fatiguing. Some stress is important to the aviator. After all, the complete absence of all stress is a condition known as death. But too much stress can interfere and becomes distress, and then we must find ways to cope. Part of the solution is to get regular exercise, address problems in a forthright manner, and talk out unresolved problems with someone who understands but will be candid. Hauling a bunch of unresolved stress into the air is not smart or safe.

We could go on. Books are written about each of these issues and you should read them as you grow in your appreciation of this pillar of airmanship. Suffice to say, airmanship is eroded with even the simplest of physiological impairments. Socrates said to "know thyself" is the key to wisdom. It's easy to say but so easy to minimize. Pride, unrealistic expectations of performance, and control of our personal medical airworthiness are often in conflict. How do we keep a balance?

It's said that you learn nothing the second time you're kicked by a mule. Learn from personal mistakes and lessons, tune in to your body, and take care of it. It is sound advice for anyone but especially for aviators. Self-awareness is essential to good airmanship.

Psychological airworthiness

You have determined that you are "medically" airworthy in the physical sense and can now take to the skies safely knowing that you are the epitome of health. Right? Wrong! Physical health is only half the picture. How well do you understand and appreciate your psychological health? Many factors not directly related to our physical health can still impact safety of flight. We do not intend to create aviation psychologists out of every aviator, but a general awareness of factors that have an impact on our psychological airworthiness are worthy of discussion. The following example illustrates how a combination of inadequate medical and psychological airworthiness can affect aviation safety.

A midair collision occurred between two F-15 fighters piloted by highly experienced Air Force pilots. The accident board determined that one of the pilot's judgment was impaired due to excessive personal commitment, fatigue, and illness. The mishap pilot was a flight commander, standardization/evaluation flight examiner, squadron phase briefing monitor, and project officer for an upcoming headquarters inspection visit. When not at work, this pilot kept busy by working on a masters degree and instructing a Bible study class. To top it all off, he had flu symptoms the day before the mishap. Clearly this pilot was not physically or psychologically fit to fly. Yet he chose to accept the mission as scheduled. This was a top-notch pilot and officer who clearly was exercising poor airmanship, because he did not understand the limitations he needed to place on himself. Like most high achievers, he thought he could hack it—and didn't want to admit to himself or anyone else that he might be overtasked. I suspect that this unhealthy trend is more prevalent then we care to admit in all circles of aviation. The lesson is: Even a great person, working his or her hardest to excell, can be a second-rate airman due to poor self-awareness.

What causes an otherwise logical human being to make an illogical decision about his or her own capabilities? Sometimes it is the harddriving type A personality like the one just described. But far more frequently, it is a hazardous attitude that can creep up on any of us if we are not aware of the types and tendencies of these demons of the aviation mind.

Hazardous attitudes

Airmen run the gauntlet of personalities from the meek computer geek flying his Mooney Ovation to computer shows across the country to the confident and cocky fighter pilot flying her F-16 to the bomb range. Researchers have yet to determine the perfect pilot personality profile, and it's doubtful that they ever will. But no matter what your personality type, you can succumb to any of the various hazardous attitudes that affect both performance and flight safety.

The goal of this section is to introduce you to some of the behaviors that are indicative of the hazardous attitudes and discuss how to recognize and avoid them. These descriptions are likely not complete or conclusive, and you might know of many others, but the important point is to recognize *any* attitude that might have an impact on airmanship and have a plan to counteract it. The following are descriptions of some of the more common hazardous attitudes.

Get-home-itis (also known as *get-there-itis* or simply *pressing*): How many times have you pushed yourself because you had to get to your next destination on schedule? The tragic crash of seven-year-old Jessica Dubroff on her attempt to be the youngest aviator to pilot a cross-country flight is a possible example of this attitude. Despite bad weather, which included freezing rain, drizzle, and high winds, the instructor pilot elected to take off in their overloaded Cessna 177. It is noteworthy that a United Express commercial aircraft delayed its departure due to poor weather at the same time. Why would an experienced pilot not wait for better weather? Was it possible that the pressure from the crowd in Wyoming waiting to see the departure or the desire to arrive in Indiana to awaiting crowds and news coverage influenced the aviator's decision? Possibly, and there are many other examples of accidents in which the flyers were in a hurry to return home or get to their destination.

How do you avoid this trap? First, be honest with yourself when you really want or need to get to a destination. If you find yourself taking risks that you normally would never consider, ask yourself why. When you find yourself saying that you *must* get somewhere, ask yourself what price you are willing to pay for your arrival. Don't let your desires write checks that your aircraft or skills can't cash. Delay your flight if under normal circumstances you would not go. It is always better to get home a day late than to never see home again.

Anti-authority: Some aviators see the relatively unsupervised environment of flight as their chance to strike back against authority. As aviators, we are required to abide by the "rules of the skies." It is tempting at times to think that a little deviation from the rules won't hurt anyone, but it can, it has, and it will do so again until such time as we all have the discipline to comply. Noncompliance in aviation is not the act of a rebel, but rather the act of a child—someone who is immature, unknowledgeable, or both. The regulations are not for someone else, they are for all of us. If you decide to aviate without following the rules, you are endangering the lives of others and casting real airmen in a bad light. Avoid this behavior by simply following the rules unless an emergency dictates otherwise. Although they often err to the side of conservatism, the rules are usually right.

Machismo: Many pilots believe they can do anything in the air, and they take too many chances in a vain attemt to prove something to someone. While justified confidence is crucial to safe and effective operations, the macho pilot is often all boast, which can place him or her in a position of having to attempt a maneuver merely to prove prowess. You might be the best pilot in the world, but if the last time you landed your aircraft in a 30-knot crosswind was a decade ago, it's time to swallow your pride and divert. If you take chances, you must be prepared to accept the consequences. A sure-fire way to avoid this behavior is consider what you would say to the safety board following the results of your decision if all does not go well. If you are comfortable with your answer, then go for it. Otherwise, consider an alternate course of action. You have nothing to prove but your good judgment and airmanship. Be smart and remember the two most dangerous words in aviation: "Watch this!"

Invulnerability: Some aviators—usually pilots—feel bulletproof. They think disaster or bad luck always happens to the other person. These wishful thinkers are the prime candidates to run out of fuel, fly into thunderstorms, or ice up and crash. This attitude often develops over time, perhaps after we have had several close encounters with bad weather, mechanical problems, etc., and still survived. The more experience we have in the air, the more susceptible we are likely to be to this hazardous attitude. After all, if you survive 10,000 hours flying without a scratch, you must be leading a charmed life. Wrong! A thunderstorm does not care if you have 100 hours or 10,000 hours of flight experience; it can and will kill you despite your experience. Many of us have been lucky and survived a situation that we should have avoided. A smart aviator learns from these mistakes and avoids the same situation in the future. But others develop the attitude that says, "I made it once, I'll succeed again." You might make it, but then again you might not. Why chance it? Many pilots have lost their lives thinking that it couldn't happern to them. Their luck ran out, and yours might too.

Impulsiveness: There are many examples in which aircrews have acted in haste, later regretting their actions. An Airbus departing from London had engine problems and the crew took prompt action to shut down the engine. The only problem was they were a bit too quick in analyzing the situation and shut down the *wrong* engine, making themselves the world's largest multiengine glider. The result was devastating and could have been avoided if the crew had recognized their impulsiveness. To be certain, there are rare situations in some aircraft that require immediate action, but the majority of us have time to think before acting. A C-141B military flight instructor insisted that his students wind the manual clock before taking any action concerning an aircraft malfunction. It might be a bit extreme, but the point is well taken. Think before you act, and give your mind a chance to overcome the excitement and adrenaline of the moment.

Resignation: Sometimes we put ourselves in a situation in the air in which we feel there is no way out, and we give up. This attitude is often referred to as "what's the use" syndrome, or resignation. The best way to avoid this behavior is to never get yourself into a situation without an "out." Flying in the mountains we learn to have an out and avoid things such as box canyons. We train to recover from stalls and other flight conditions. The more training you have and the more you work on polishing your skills, the more "outs" you have. The more you know about your aircraft systems and operating procedures, the better your chances that you can handle any situation. Even if you have never spun an aircraft, you can easily recover from a spin if you are aware of the recovery procedures.

Regardless of your experience and training, you might find yourself in a novel situation; then you have two options. You can give up and not work at a solution and let fate take its course. Or you can realize that you are never helpless and that you can make a difference. A classic example of a crew that avoided this hazardous attitude was United Flight 232 (see Chapter 11), a DC-10 that experienced a complete loss of hydraulics and flight controls. One pilot on this remarkable flight commented that as he sought solutions to this previously unseen problem, he recalled the words of his flight instructor, who told him "Never give up—try something, anything, but don't ever give up." The crew of Flight 232 never said "what's the use?" Instead, they realized that they could—and eventually did—make a difference.

Complacency (been there, done that): Flying can sometimes become routine, and we tend to let our guard down after we find ourselves on the same route of flight for the hundredth time with nothing new on the horizon. This complacency can result in inattention. Brigadier General Chuck Yeager has often stated that complacency is the number-one enemy of experienced pilots. New instructor pilots are a very vulnerable group to complacency. Once you become an instructor in an aircraft, aviation becomes more routine than when you were struggling to master the machine. This attitude probably was influential in many of the gear-up accidents that continue to plague both the military and general aviation. In 1983 two separate C-5 Galaxy aircraft landed gear-up (all 28 wheels) due to pilot error. Both crews had at least one instructor pilot on board. Avoid complacency by continuing to challenge yourself even on routine flights. During lulls in the flight, play the "what if" game (what if my engine quit now?) to avoid boredom and inattention that can result in complacency.

Air show syndrome: This hazardous attitude occurs when a pilot decides that "it's time to make a name for myself and impress all those earthbound people." Tragically, aviators often succumb to the temptation to "show their stuff" to friends and families. Far too often, the "show" the family and friends see is not the one intended. A Navy F-14 crash in Nashville, Tennessee, was caused in part because the pilot was performing a maximum-performance takeoff into the weather while his family was watching. The results of this unnecessary maneuver were the deaths of two Navy crewmen, two civilians, and the destruction of an expensive fighter.

On another occasion, two Air Force pilots on a cross-country flight in their T-37 did an unapproved air show for the family of one of the pilots, who had the tragic opportunity to videotape their son's death. Avoid this behavior by developing, briefing, and flying a safe plane for all flights. The crowd won't know the difference, unless you end the show by auguring it in.

Emotional jet lag: This hazardous attitude is particularly dangerous to aviators who pride themselves on being perfectionists. When an error does occur, these perfectionists can't seem to put it out of their minds, and their brain stays at the point of the mistake, dwelling on the cause or fretting about what others are thinking. The obvious problem with this is that the aircraft keeps moving while their thinking does not. As much as many of us hate mistakes, they are a part of life—and flying. We all make mistakes, but while you're in the air is not the time to dwell on it. You cannot allow your mind to get behind the aircraft. If you do, much bigger—and perhaps deadlier—mistakes are likely to follow. The result of emotional jet lag will probably be a loss of situational awareness and possible compromise of safety. Stay caught up and worry about mistakes when you are debriefing on the ground.

The hazardous attitudes discussed here can affect aircrew members flying alone or with others. Some attitudes specifically have an impact on flight safety when pilots—especially experienced or high-ranking pilots—are flying as a passenger or as an additional crewmember.

Excessive professional deference: If we are flying as a copilot or passenger with a pilot who has more experience than us, or in the case of military aviation, a higher rank, we are sometimes hesitant to call attention to their deficient performance. When we do call out deficient performance, it tends to be vague. Instead of the first officer on probation telling the captain that she is 15 knots slow on final approach, he often cloaks the criticism with a comment like "you're a little slow." This vague coment is not particularly helpful and can, if fact, be dangerous because it might mask the severity of the situation. No matter whom you are flying with, if you notice something wrong, speak up, be specific, and worry about the consequences later. Most pilots appreciate the help in the cockpit, and it's better to say something before it's too late and you can no longer speak.

Some pilots do not appreciate input and become angry with assertive crewmembers. Consider the plight of the first officer on this small multiengine commercial aircraft in an ASRS report from his captain:

Case study: To be or not to be—assertive (NTSB 1994)

We were on the first flight of the day after a subfreezing night. After we were cleared for takeoff, I stood the throttles up to let them stabilize. As we scanned the engine instruments, we noticed that the number two engine showed idle EPR, but all other instruments duplicated the number one engine readings. I determined that the problem was with the #2 EPR gauge and called for autothrottles. The First Officer refused to put autothrottles on saying that he thought we had engine ice problems. Sky conditions were clear and the weather did not favor icing conditions. I told the First Officer that it was a gauge problem. He was talking about another accident that had occurred from icing on takeoff. At this point we were accelerating through 35-40 knots. Instead of trying to argue with the First Officer during takeoff, I elected to reject the takeoff. After exiting the runway, running the appropriate checklists, and contacting company mainte-nance, I discussed with the First Officer why I should have pressed on with the takeoff. All engine parameters were nor-mal except for the EPR gauge. Lack of training or under-standing of the conditions by the First Officer led to the rejected takeoff, complicated by Cockpit Resource Manage-ment training, which added to his assertiveness in refusing to do his required duties during takeoff roll [emphasis added].

This captain just doesn't get it. Perhaps he was right in his estimation that it was in fact a gauge malfunction, and perhaps the first officer should have had a better grasp of the situation, but to attack the CRM training and assertiveness of his first officer is ludicrous. By ba-sically telling his first officer to "shut up and color," he has likely eliminated any backup that this first officer will provide in the future. Consider the trade-off here. The captain would seem to prefer to avoid one rejected takeoff to an assertive and proactive partner in the cockpit—a very bad trade indeed. One has to wonder how much this first officer will be willing to say next time. The lesson here is that not everyone welcomes our input, but that it is *their* air-manship problem if they react like this captain did. Continue to avoid the hazardous attitude of excessive professional deference, even if you do run into a dinosaur like this captain once in a while.

Passenger/copilot syndrome: Closely related to excessive profes-sional deference is the hazardous attitude that manifests itself in a behavior that believes that no matter what we do, big brother will bail us out. Sometimes copilots (or passengers) flying the aircraft as-sume that the pilot in command or instructor will catch and correct any mistakes that are made. This attitude puts a lot of pressure on our partner in the cockpit. Even as a student, your instructor expects certain competencies from you, and if you are just along for the ride, your unexpected inadequacies might be more than the instructor is prepared to handle. That's bad news for both of you. If you are qual-ified, think and act like it. Don't fly expecting the other crewmember to catch your mistakes—fly as if you are the pilot in command or in-structor; after all, someday you will be.

The hazards of psychological stress

In addition to knowing and preventing hazardous attitudes, you must also be aware of psychological stress levels. Although physical stress was discussed in a previous section, the burden of severe psychological stress alone can exceed a flyer's capacity to cope. Life is inherently stressful, but gauging when too much stress will affect your ability to aviate is a key to knowing yourself. Consider the following mishap.

An accident investigation suggested that personal stress might have been a causal factor in the crash of a Navy fighter. That could be the understatement of the decade. The pilot had recently gone through a divorce and was experiencing complicated difficulties with his ex-wife. Such a situation is certainly a serious stressor, and one for which many aviators have taken a few days off. But this pilot had a few more difficulties. His new girlfriend was pregnant, and his son was in trouble with the law. To top it all off, he was experiencing financial difficulties and his house and car were at risk for being confiscated for bad debts. While many aviators are able to effectively "compartmentalize" outside stresses by locking them away while airborne, one has to wonder how many closets of the mind this fellow would have needed to hold all of these stressors at bay. Perhaps one more than he had available. Hopefully most who are reading this won't experience anything like this unfortunate young man, but we all need to look at ourselves and determine if we are experiencing stress that might affect our ability to concentrate and aviate. If we are, the better part of valor might be to take yourself off the schedule, and fly another day when things have calmed down.

Although it is often overlooked, it is important to point out that "good" stress—like getting married, buying a new house, or finding religious enlightenment—can be as dangerous as bad stress, such as a divorce, death in the family, or excessive work pressures. Think of stress like a timeline that goes out in both directions. Any deviation that goes too far in *either* direction can upset your psychological equilibrium—and your airmanship. Just as deciding not to fly is a good decision after a personal tragedy, you might consider not flying after becoming a new father or getting that long sought-after promotion at work. The bottom line is that every aviator must take stock of his or her own psychological stress and make the go/no go decision.

A final word on personal airworthiness

Airmanship is a personal responsibility, one that rests on the physical and psychological boundaries of ourselves. If we fail to appreciate this fact, we undercut any effort towards individual improvement. Although it is critically important to realize when we are impaired, one solution is much easier. Stay physically and psychologically healthy. It not only increases your performance potential, it makes it much easier to recognize impairments when and if they occur.

Maturity is one of the marks of an airman. The ability to admit to yourself and others that you are not fit to fly is an essential key to good airmanship. But proactive measures can help prevent these necessities. Work out, eat right, sleep right, avoid overindulgence, relieve stress, don't self-medicate, and stay healthy. Take care of your mind and body, so when you call upon them in flight, they can take care of you.

References

NTSB ASRS. 1994 Aeroknowledge [CD-ROM]. Assession number 133773.
NTSB ASRS. 1994 Aeroknowledge [CD-ROM]. Assession number 180477.

5

Know your aircraft

Knowledge is of two kinds. We know a subject ourselves,
or we know where we can find information upon it.
Samuel Johnson

A personal view from Bernie Hollenbeck, Colonel, USAF (Retired)

The part of knowing your aircraft that the manuals don't often address is the human design problems that are inherent in just about every cockpit. Certain switches and instruments almost seem as if they were designed to see if they could get a pilot to screw up. Let me cite a couple of examples to illustrate what I'm talking about.

In one fighter aircraft, the generator switch sits right beside the electronic countermeasures (ECM) switch. The two switches are the same shape and size and easy to confuse. Now add in to this equation a checklist step that calls for the ECM switch to be turned off while taxiing into parking, and you've set the stage for inadvertent electrical failure. This wouldn't be so bad if you didn't need electrical power for nosewheel steering on the taxiway. Another example is a three-way toggle switch for a brake system, where the bottom position is antiskid ON, the middle position is antiskid OFF, and the top position is the PARKING BRAKE! We have flattened many a tire from pilots failing to stop at the middle position during a high-speed brake malfunction, and I suspect we will continue to do so. My point is that most all aircraft have hidden traps that seem to lure pilots into error—and while the "old heads" are generally aware of them, the new guys are not. That's where good mentoring becomes essential.

Mentoring is a process where experienced flyers take new arrivals under their wings and show them the "ropes." A critical part of good mentoring is ensuring that inexperienced flyers are acquainted with inherent cockpit design problems and how to avoid them before they have to experience them firsthand. Although the flight manual is a great place to start learning about your aircraft, there is more to know—and sometimes it takes another flyer to learn it from.

Bernie Hollenbeck was an F-4 instructor pilot who has flown 25 aircraft types.

An aviator's ability to develop a personal relationship with his or her aircraft is one of the key indicators of excellence in airmanship. This ability is not merely being able to handle or operate a given type or model of aircraft, but rather a genuine desire to make the machine an extension of yourself—a real attempt to bond human and machine into a single functional unit. Like other social arrangements, this relationship must be based on knowledge, understanding, and trust.

It is indeed a special kind of trust when one becomes willing to put his or her life in the hands of another. We do it every day in aviation. Would you trust your life to an acquaintance whom you just met and barely knew? Probably not, and yet every day hundreds of aircrew members take to the sky in aircraft that they are not completely familiar with or knowledgeable about. This casual attitude often comes back to haunt unprepared airmen in a multitude of ways. Fortunately, the structure of aircraft knowledge is well defined, and the study of systems, procedures, and techniques have been established by generations of aviators. But more sophisticated levels of aircraft knowledge go beyond these minimums, like ergonomic considerations and maintenance histories. But let's start at the beginning, when we first shake hands with a new friend. Consider the tragic death of former New York Yankee Thurman Munson, who had just checked out in his new Cessna Citation and decided to take some friends for a ride.

Case study: He didn't see it coming (NTSB 1980)

Thurman Munson had always been a quick study, what most of us would call "a natural." While enjoying a hugely successful baseball career, he had found a new interest—flying. He began his training, like many new pilots, in a Cessna 150. Between February and April of 1978, he trained in single-engine Cessnas, but he quickly sought out a greater challenge and moved up to twin-engine aircraft. He completed his private pilot check ride on June 11, 1978, less than four months after he began training. Four days later, he received his multiengine rating in a Beech BE-60 "Duke" aircraft. It appeared that flying came to Thurman Munson as naturally as baseball had, but he was still a rookie in this game, and his logbook showed less than 100 hours when he got checked out in the Duke.

A year later, he decided to make the leap to flying jet aircraft and purchased a Cessna Citation on July 6, 1979. He received his type rating in the Citation 11 days and 10 instructional flights later, having logged a total of 23.2 hours and 32 landings in the twin-engine corporate jet. During his Citation training, his instructors described him as an above-average student.

Following his checkout, Munson flew his new jet back to Ohio to show his friends, and on August 2, he invited two friends to see his new jet at the Akron-Canton Airport near Canton, Ohio. From the beginning, the approach to this flight showed evidence of poor discipline and a lack of knowledge of the applicable regulations. After boarding the aircraft, the passengers were not briefed on the location of the shoulder harnesses, the operation of the emergency exits, or procedures to follow in an emergency. Although both passengers were certified pilots, neither had flown a turbojet-type aircraft, yet one occupied the right seat. The passengers said that they were not even aware of the pilot's intention to fly until he requested and received clearance for local traffic pattern operations.

The initial takeoff was normal, and the Citation entered a left downwind for a touch-and-go to runway 23. The pilot was already getting behind the aircraft and allowed his airspeed to creep up to nearly 200 knots—well above the 174 KIAS gear-down limit. After accomplishing a normal touch-and-go, the pilot retracted the gear and flaps and then pulled the right throttle back to demonstrate the jet's single-engine climb capability. Altitude deviations of nearly 500 feet were noted by the passengers on this pattern. After leveling off in the pattern, the pilot shoved the throttles full forward to demonstrate the acceleration capability of the aircraft. Clearly, he enjoyed showing off his prized possession.

After another normal touch-and-go, the pilot offered the aircraft to the unqualified right-seat passenger and proceeded to talk him into flying a zero-flap touch-and-go. The approach airspeed had been set in the airspeed "bug"—a moveable airspeed reference indicator—at 93 knots and was not reset for the faster no-flap approach speed, but the pilot worked the throttles for the right seater and kept the aircraft well above the referenced speed. The passenger flying the aircraft was only handling the yoke, rudders, and trim. Touchdown was accomplished approximately halfway down the runway, and the aircraft "ballooned" about 5 to 10 feet in the air shortly after touchdown. At this point, Munson took control of the aircraft and applied takeoff thrust to complete the touch-and-go. At no time during any of these pattern operations did either of the passengers see the pilot reference a checklist.

During the fourth takeoff, the tower advised the Citation to enter a right pattern for runway 19 because of other traffic. Once again the pilot overshot normal traffic pattern altitude and found himself high and hot. He pulled the throttles to idle to dissipate airspeed and altitude. The passengers recalled that the throttles were reduced to a point at which the landing-gear warning horn sounded and that the pilot silenced the horn. All of these actions are indicative of a pilot who was behind the aircraft.

After a few more instructions from the tower to sequence traffic, the pilot turned base leg, still in a clean configuration. The right-seat passenger noticed that the aircraft was approaching glidepath and said, "I don't think you want to land this aircraft with the gear still up." Munson lowered the gear. About this time, perhaps because of the increased drag, the aircraft began to slip below the VASI (vertical altitude slope indicators) glidepath and drifted slightly left of runway centerline. The passenger cautioned the pilot about getting low and reminded him of possible downdrafts near the approach end of runway 19. The pilot had his hands full, and he was beginning to sense something was wrong. One passenger recalled, "I could see in his face that he felt there was something wrong . . . I sensed the airplane sinking and I could sense through the expression in Thurman's face that the aircraft was out of control." What was wrong, of course, was that the pilot had neglected to lower the flaps but was flying at flaps-down airspeed. The right seater observed that the airspeed was "nailed right on the bug" at 93 knots, approximately 20 knots below the appropriate no-flap speed.

The sink rate continued. The driver of an automobile traveling southeast on Interstate 77 observed the Citation flying "extremely low in a gradual descent, barely clearing the trees. He said "the aircraft was going very slowly, resembling the landing speed of a light small aircraft." He further observed the wing "wobbling" as it passed in front of him, finally disappearing from view in a cloud of red flames and black smoke. Thurman Munson had flown his last approach.

The aircraft touched down about 100 yards short of the runway on slightly rising terrain. The aircraft rolled across rough terrain and eventually came to rest against a stump after plowing through a clump of small trees. A fire erupted immediately. Both passengers were able to free themselves from their seats and attempted to pull the pilot from the burning aircraft, but the New York Yankee was helplessly pinned between the seat and the instrument panel. After several heroic attempts to extricate their friend, they were forced to evacuate the aircraft to save their own lives. The official cause of death was listed as "asphyxiation due to acute laryngeal edema and due to inhalation of superheated air and toxic substances." Thurman Munson's fundamental lack of understanding of his aircraft had resulted in a hideous death — certainly an unfitting end for an American sports hero.

What happened here, like in most accidents, was an amalgam of several factors. Poor discipline, distraction, and unfamiliar circumstances all played a role in the tragedy. But at a more fundamental level, the pilot was simply not yet equipped to recognize what was happening to him. He was late to recognize the sink rate, and we will never know if he ever realized the nature of his error. Simply put, his lack of knowledge and "feel" for his new jet put him in a high-risk condition. He exacerbated this condition by his cavalier attitude towards preflight briefing requirements, lack of precision in the airport traffic pattern, failure to use a checklist, and his ad hoc attempt at flight instruction for his passengers. He had created conditions ripe for disaster. Using an appropriate metaphor, he attempted a veteran play with only a rookie's experience and knowledge.

What to know and how to learn

Learning complex aircraft systems, flight characteristics, procedures, and techniques can be a daunting undertaking, even for the most professional and motivated aviators. Fortunately, there is a systematic approach to learning and understanding aircraft systems, which we briefly review in the following section. For this approach to be effective, however, we must overcome a few human tendencies. Most flyers have little desire to delve deeply into aircraft systems, preferring to skip right to the operating procedures, to figure out "how to fly this baby." Patience is a virtue, and the first step should be to obtain a general understanding of the aircraft as a whole.

Educational psychologists speak of different levels of learning, and one method for evaluating comprehension depends on the learner's ability to *use* the acquired knowledge when it is needed. "Inert knowledge," or knowledge that cannot be recalled under stressful conditions and put to use, is the nightmare of every aviator. Critical action procedures that can be recited at groundspeed zero but disappear in flight are not helpful and might even create a false sense of security. So how can we be certain that the material we are studying will be subject to recall when needed? The answer to this question involves a multitude of factors, both physiological and psychological, but the way in which we go about learning new material is a significant piece of the puzzle. Although we never can be completely certain of instantaneous recall under severe stress levels, part of the answer lies in our ability to reinforce certain critical knowledge items while simultaneously keeping the ability to integrate aircraft knowledge into our whole airmanship picture.

Where to start

Begin systematic study with a thorough reading of the "general description" portion of the aircraft's flight manual. Don't try to memorize everything or expect too much from this initial step, but keep in mind that this material prepares you for a more thorough study later on. Seek merely to understand the general operation of the aircraft and the interrelationships of the various systems. Much of this material is dry and occasionally boring, but it is essential groundwork for what is to follow.

After the initial reading of the aircraft description, it is time to move into the essential material that we wish to imbed into the deepest crevasses of our brains—the critical action emergency procedures and aircraft operating limits. This is life-saving knowledge, so we study it hard and we study it often. There are three distinct ways to study operating limits and emergency procedures, and I recommend that you use them all. First, read the material and commit the critical action steps and operating limits to memory. Second, use the aircraft as a training tool. Get in the cockpit and repeat the emergency procedures while you physically go through the actions. To reinforce the various system limitations, touch the instruments and recite the limits. Finally, as you begin to fly the simulator or aircraft in training, practice the various procedures—simulated if required, under the watchful eye of an instructor. These steps reinforce the "book learning" and make it far less likely that the hard-earned knowledge will fail you when you need it most. When you are certain that you know this information *cold*, you can begin to divert more study time into other areas, such as flight characteristics, communication and navigation equipment, normal procedures, and techniques.

The key to expertise: Continuous and systematic study

The key to becoming an expert in a given aircraft is to have a continuing plan for study *after* your initial checkout and training program is complete. There are many ways to accomplish this plan, but a combination of two techniques, systematic review and experienced-based inquiry, has worked well for many aviators. Systematic review means following a specific plan for reviewing the technical and operational data on your aircraft on a regular and continual basis. An annual plan is usually the norm, with specific chapters and topics projected against a given month. Many flyers use copies of initial qualification tests or other means to assess their comprehension and retention of the material. Merely reading through the review material is helpful, but if you can construct a few study questions prior to the study, it often helps focus your time and attention. For example, "What would I lose if I lost my ac power?" or "How many ways can I communicate with the equipment I have on this aircraft?" A final note here is that it is often helpful to create and keep a written record of your study, if for no other reason than to remind you when you have been out of the books for a while.

The second study method, experienced-based inquiry, uses actual flight occurrences to trigger further study. It is a very useful technique because it attaches automatic relevance to the material. An example of how this works would be a pilot who has an encounter with turbulence or windshear, who would then study everything he or she could find on those two subjects. This method often leads to seeking information that lies outside of the standard technical or operational flight manuals. Serious pilots are not deterred by an immediate lack of information and will call or write the manufacturer, FAA, or other flying training organizations for more information.

A combination of a structured review process and experientially focused study provides aviators with a solid plan for continued growth in understanding their aircraft. The key to effective study remains in the hands of the individual, because only you know the knowledge areas that frighten you most, and only you can address them systematically. Keep in mind that many aviators get "bit" by knowledge areas that they are comfortable with, which breeds complacency and inhibits a structured approach to reviewing known information.

Unwritten knowledge and ergonomic "gotchas"

A great deal of important information is not contained in the flight or operational manuals. Some flight-handling characteristics and ergonomic considerations are best learned by talking to other flyers who are more experienced in the aircraft type. Still other sources of information are to be found in mishap databases or from the manufacturer. The point is simply to understand that the sum of all aircraft knowledge exists in a variety of locations, and the serious pilot seeks them out. Furthermore, realize that the knowledge base on an aircraft type is constantly growing and dynamic. With aircraft knowledge, to stand still is to fall behind, and the only way to stay current is to continually seek new information with an inquisitive and questioning attitude.

Understanding your aircraft as an individual

Even if you know everything there is to know about your aircraft type, there is still more to a complete approach to understanding your aircraft. Aircraft are different from one another in many small but often important ways. One important difference between aircraft are the individual maintenance histories. An aviator must be familiar with what work has and has not been done on an aircraft. A review of the maintenance record should detail whether the aircraft has had all required inspections or is on a "waiver" of some type. Secondly, check for recent major overhauls, such as military "phase inspections" or other major tear-downs. The chances for problems occurring after these are significantly greater than for normal flights. Third, look carefully at the forms for parts that might have been replaced since the last flight. You are, in effect, a test pilot for these replacement parts. Fourth, view with suspicion any "could not duplicate" writeups in the aircraft forms, meaning that the maintenance personnel on the ground could not recreate the conditions that caused a malfunction on the last flight. Expect the malfunction to reoccur and have a plan to handle it, collecting further data if possible.

Finally, ask the maintenance personnel and other aviators who are familiar with the aircraft if there is anything special about the aircraft that you are about to fly. This simple question might reveal a wealth of valuable information, some of which might be vitally important, and some that might simply lead to greater peace of mind. An example of this type of individualized information could be something as important as a scratched windshield that makes night landings more difficult or something as mundane as the cause of an annoying rattle that would have worried you for an entire flight if you hadn't been advised of its origin. When the breadth and depth of your aircraft knowledge is sufficient, you are then prepared to function as an asset, rather than a liability, in a team environment.

Using your knowledge as part of a team

Each of us flies as a part of a team. This fact is true whether we fly a single-seat general aviation aircraft, a solo reconnaissance mission, an air superiority fighter, or as part of a traditional crewed multiengine jet. Other parts of this book go into the rationale for this statement, but suffice to say that in nearly all circumstances, the sum of our personal knowledge is still only a part of a larger equation for decision making and airborne problem solving. As such, we can become either a valuable asset to the team or a dangerous liability, depending on the usefulness of information and action we can provide from our personal skill and knowledge base. Consider the differences in the following two case studies as you reflect on your own level of personal knowledge.

Case study: The value of knowing your aircraft (Hughes 1995)

The mishap aircraft was a large crewed transport on a transoceanic mission, cruising at flight level (FL) 220, with the autopilot engaged. Flight conditions were clear, with no turbulence, and cloud tops were at 6000 feet. Suddenly, the aircrew heard an explosive sound and a white mist rushed into the cockpit. The aircraft pitched down and began a slight right turn. Both pilots felt pressed up into their shoulder harnesses while several unrestrained passengers reported

5-1 *Maintenance personnel can be a wealth of knowledge. Here, early flying cadets learn about servicing their aircraft under the watchful eye of maintenance instructors.* USAF Air Education and Training Command Archives

being lifted up in their seats. At this point the autopilot was still engaged. Yoke movement was erratic, and the aircraft was experiencing moderate to severe buffeting.

Once the autopilot was disconnected, the aircraft buffeting stopped, but a moderate level of airframe vibration continued. Aircraft control was sluggish. Primary aircrew members and both crew chiefs donned oxygen equipment. The pilot, still unsure as to the nature of the problem, initiated a 500- to 1000-foot-per-minute rate of descent at 200 knots indicated airspeed (KIAS). The loadmaster provided the first concrete evidence of the nature of the problem when he informed the pilot of a large hole forward of the wing on top of the aircraft fuselage. The aircraft had experienced an explosive decompression that had blown a gaping hole in the top of their aircraft. But this crew was knowledgeable and well trained and not about to panic even though they were over open ocean with a catastrophic emergency on their hands.

The flight engineer (FE) scanned the aircraft engine instruments and fuel gauges. He reported number three and four main tank gauges indicated zero fuel. All other instrument indications were observed as normal. The pilot took in this information but could not accept it at face value. He knew that if this were in fact true, he should have been able to feel some lateral imbalance in the flight controls, which he did not feel. The pilot reported this observation back to the flight engineer. Meanwhile, the loadmaster scanned the right wing and reported no visible leaks. The engineer then realized that the number three and four fuel gauges were off-scale low and likely inoperative. He advised the crew and pulled the fuel gauge circuit breakers to prevent any chance of electrical shorts in the fuel system, as the tech order recommended. An additional flight engineer was riding on this flight and he quickly came forward and relieved the primary FE while the primary FE went into the cargo compartment to assess further damages.

Prior to the decompression, the navigator had provided the aircraft commander (AC) with information on the closest emergency airfield relative to their flight track position. This was his normal mode of operations, because he wanted the pilot to be able to take an immediate vector towards safety in the event of a severe emergency; this time it paid off. The AC quickly confirmed this information with the navigator and initiated a 5-degree bank turn towards the airfield, which was 250 miles away. The copilot (CP), after initially backing up the aircraft commander and ensuring the aircraft was under control, attempted to declare an emergency using the high frequency (HF) radios but was unable to directly contact any air traffic control (ATC) agencies. He then used other radios to contact a nearby U.S. military aircraft to relay information and passed emergency information and diversion intentions through them. This critical step activated search and rescue as well as ground-emergency action plans that might have been required upon arrival at the emergency base, if they made it that far.

By this time the primary FE had had an opportunity to survey the damage to the aircraft and informed the AC of associated damage to components in close proximity to the hole. The damaged area included mission right outboard and center fairings, damage to electrical wiring bundles and cannon plugs, bent and torn air-conditioning ducting, and, worst of all, possible damage to the flight control cables. The air-conditioning ducting had been blown upwards into the flight control cables, and the potential for flight control loss loomed as a very real possibility. The FE attempted to remove pieces that were entangled in the throttle/condition lever control cables with marginal success. He then returned to his duty station.

The crew used the aircraft sextant, a small navigation telescope used to "shoot" or view celestial bodies for over-water navigation, to look at the outside and further assess damage to the top of the aircraft.

This unconventional approach produced valuable information, as the crew noted that the number one high frequency (HF) antenna was missing and number two was disconnected from the aft mount and most likely nonfunctional. As a result of this observation, they realized the nature of their high frequency radio problems and wasted no more effort there.

Now that the aircraft was under control and the situation had been stabilized, the crew began to prepare for their approach and landing. The crew initiated a controllability check passing through 9000 feet msl while in a continuous descent. The aircraft commander elected to fly the approach at 150 KIAS using 50 percent flaps, with a planned touchdown at 140 KIAS. During the controllability check, the AC slowed the aircraft to 140 KIAS in 5-knot increments and lowered flaps to 50 percent in 5-percent increments. For every 5-knot airspeed change or 5-percent flap change, the pilot made an S-turn using 5 degrees of bank and made a slight pitch up and down to determine controllability. No control difficulties were noted other than slow control response. This crew knew its emergency procedures well and wanted to leave no stone unturned.

As a result of the copilot's initiative to contact a relay aircraft to alert the emergency base of their plight and intent to divert, a rescue aircraft and helicopter were launched from the emergency airfield to escort the mishap aircraft into the field. When the aircraft was 30 miles from the airfield, the landing gear was lowered, and the AC intercepted a precision approach final with ground radar monitoring. The landing was uneventful. Reverse thrust was not used due to the possibility of foreign object damage (FOD) due to the loose HF antenna. After landing and clearing the runway, a crewmember deplaned to determine if further taxiing was safe. From this assessment, the AC elected to taxi the aircraft to parking. Engine shutdown was uneventful. Flight time from decompression until landing was 1 hour and 12 minutes.

A summary of this outstanding performance notes several key points about the depth of preparation and knowledge present on this crew. The reactions of the entire crew were systematic and methodical, yet not restrained by a lack of imagination. The use of the relay aircraft and the navigational sextant are clear indications of creative problem solving based on a thorough knowledge of the aircraft. The pilot's assessment that the aircraft did not feel "wing heavy" led to the elimination of a red herring that could have distracted the crew from other critical action procedures. After the explosive decompression, the aircraft commander maintained aircraft control while the rest of the crew, including the additional crewmembers and the crew chiefs riding as passengers, worked in union to assist other passengers and analyze the situation in detail. The damaged systems included air conditioning, hydraulics, flight control cables, bleed air, oxygen, and throttle/condition level cables. A thorough aircraft systems knowledge and analysis was required to take the appropriate corrective action. During individual interviews, both pilots and both engineers stated that mission-oriented simulator training (MOST), a crew resource management tool, enhanced crew coordination during this mishap. Although an explosive decompression and subsequent structural/systems damage occurred, the synergistic knowledge of the crew proved more than adequate to handle the challenge.

In an ideal world, all aircrews would respond with this level of competence and confidence, but the following case study tragically illustrates that this level of performance has not yet been achieved by all aircrews.

Case study: The price of poor aircraft knowledge (Rouse 1991)[1]

On February 2, 1991, the crew of Hulk 46, a B-52G flying in Operation Desert Storm, took off from a location in the Indian Ocean at 1216 local time. It was part of a three-ship "cell" formation of B-52s that carried tons of explosives destined to rain down upon the heads of Saddam Hussein's fielded forces. It was the crew's fourth combat mission, although it was their first daytime mission over hostile territory. The tactical situation required that the earlier missions be flown at night. The aircraft commander (AC) provided a brief, if understated, overview of what the mission entailed. "The mission itself was pretty much what we were flying every day, take off, two air refuelings (A/R), fly in country, release weapons, fly out, get a third air refueling, (and) come home . . . that was pretty much the same type of mission we had been flying. We were just flying at a different location in country [this time]." The planned duration for the mission was 15.5 hours. Tragically, it was the last combat mission for this aircraft and crew. As the case study progresses, we see a chain of events that began months before the February mission, which led up to the fiery crash of Hulk 46 just 15 miles short of its destination.

1. All information in this case study was obtained through this report or the B-52 technical orders. No information came from AFR 127-4 "Aircraft Mishap Investigation," which contains privileged information for mishap prevention use only.

Although many factors were involved with the mishap, in the final analysis the accident revolved around the B-52's electrical system. Some background information is provided to assist those not familiar with this aircraft to better appreciate the progressive events in the case study. The B-52, like most aircraft, operates off of a direct current (dc) and alternating current (ac) electrical power supply. Dc power is provided by the batteries and is typically limited to some lighting and emergency equipment. The primary power source is ac power, provided by four generators—each run by the odd-numbered engine (1, 3, 5, 7) in the B-52's four engine pods. Although some redundant capability is built into the system, a single generator is not capable of carrying the electrical load for many operations. Although it is not an extremely complicated system, the crew must possess or be able to recall or locate specific items of knowledge about this system to operate safely, especially under emergency conditions.

The preflight and takeoff had gone smoothly, with only a short maintenance delay for an aft battery light, which was quickly fixed by maintainers. What the crew did not know, however, was that this aircraft, tail number 59-2593, had two recurring maintenance problems, a critical lack of information that would come back to haunt them later in the mission. The first was a number one fuel quantity gauge failure, which had been written up five times previously, beginning in November, more than two months earlier. Maintenance personnel had attempted to fix the problem in a variety of ways, including draining water from the main tank and cleaning fuel gauge connectors and probes. The gauge did not fail on every mission, but it had failed on *every other* mission since November, and no discrepancy was noted on the last mission, which was flown on January 31. Because of the maintenance rules in effect, the fuel gauge writeup was not reflected as being a "recurring" discrepancy; therefore it was not in the forms reviewed by the crew on this day.

The second bit of individual aircraft knowledge that was unavailable to the crew during their preflight forms review was the recent troubles with the number five engine. A month earlier, on January 2, the number five engine had rolled back to below 40 percent power and the oil pressure was reported as low. Maintenance could not duplicate the malfunction (CND) and the corrective action read, "Ran engine. No defect noted! Can not duplicate malfunction." Another pilot testified that he had had a similar problem with the aircraft during engine start on a later date, but that maintainers had adjusted burner pressure/fuel control and that seemed to correct the problem. Neither the malfunction nor the correction was documented in the aircraft forms.

These two existing, but unknown (to the crew), aircraft malfunctions proved instrumental in the gradual unwinding of the situation. They are both significant in that they affect engines with generators. The first, a simple fuel gauge malfunction, was misdiagnosed by the crew. The second malfunction caused an underexcitation relay to cut the number five generator off line at the worst possible moment, ample proof that Murphy was alive and well. As the B-52 and the crew of Hulk 46 lumbered into the morning sky on February 2, 1991, the sequence of events resulting in their demise had already been set into motion.

It didn't take long for the first malfunction to manifest itself. Between the first and second air refuelings, the pilot noticed a fuel gauge fail indication on the center-of-gravity fuel level advisory system (CGFLAS) and the number one main tank fuel quantity indicator was reading zero. The copilot recalled, "I pulled and reset the number one main tank fuel circuit breaker for that gauge and that corrected the problem." The B-52 tech order specifically warns against this procedure, stating, "A malfunction of any fuel quantity gauge may indicate a failure that would, with proper sequence of events, allow the introduction of high voltage electrical power into the associated fuel tank. Therefore, if any fuel quantity gauge occurs, pull the applicable fuel quantity indicator circuit breaker." It seems to indicate that the initial action of the crew was correct, but to *reset* the circuit breaker is a different matter entirely, as indicated by the warning that follows the previous text.

WARNING

If any fuel quantity gauge malfunction occurs, pull the applicable fuel quantity indicator circuit breaker. The gauge will not be removed or changed and the circuit breaker will not be reset until proper inspections and repairs have been made (USAF 1991).

Perhaps the crew felt that the need for the fuel quantity information on the combat mission justified the risk associated with violation of the warning, or perhaps it indicates a lack of knowledge or respect for the existing procedures, a trend we see repeated on this ill-fated flight.

The next bit of bad news occurred somewhere between the first air refueling and the post-target turn. (The aircraft commander and gunner recalled the malfunction occurring between the first two air refuelings, but the copilot recalled the generator malfunction as occurring after the post-target southbound turn. The timing of the event does not significantly alter the course of action that the crew took.) The number three generator dropped off line, meaning that it was no longer supplying ac power to the aircraft systems. The copilot attempted to reset the generator twice, but it failed to come back on line. The aircraft commander did not consider this malfunction to be especially worrisome but was aware that it could lead to further problems, as indicated by the following testimony: "The loss of a generator at Eaker (the crew's home base) on our Air Force bombers was a typical occurrence and we pretty much take it as it is . . . If we have another problem we'll discuss it at length but with this one—it's just one generator, we'll press on."

The crew completed its combat delivery and turned back towards its forward operating location in the Indian Ocean. By this time, day was wearing into evening. The crew likely ate their inflight meals and began to settle in for the long ride home, which was punctuated by one more air refueling. The aircraft commander used this time to "clear off to the tenth man position" to get a little rest prior to the final refueling, standard procedure on long missions. The aircraft commander returned to his seat just prior to the A/R and completed the second air refueling as planned with two KC-10 Extenders providing the gas. During this A/R, the number one fuel gauge began to act up again. The crew pulled and reset the circuit breaker once again. When the number one main fuel tank quantity dropped 7000 pounds below its number four counterpart, the pilots began to discuss the possibility of a fuel leak but took no checklist action at this time. Although the tech order points out that "the most positive indication of a fuel leak may be an abnormal change in the pitch or lateral trim requirements," the pilots did not take advantage of this cue because they failed to disconnect the autopilot to check for a difference in feel. Following this discussion, the copilot checked off for his turn to rest.

Shortly thereafter, the aircraft commander noticed that the number five engine was displaying "abnormal indications." The engine readings were somewhat lower than the other seven, and it caused the aircraft commander enough concern that he had the copilot awakened to come back into position, so that they could work the problem as a team. The crew discussed the indications: low readings on the number five engine instruments across the board, the engine responded only to decreasing throttle movements and not throttle advances, the copilot testified that the number five generator was still on line. They determined that they would not elect to shut the engine down, and with that, the copilot went back into the bunk to rest again.

The aircraft commander made a determination that it might be beneficial if Hulk 46 landed last in the cell, because of the possibility that the multiple malfunctions might cause the B-52 to have to close the single runway on landing at their island destination. The cell leader agreed and a position change was accomplished to place Hulk 46 in trail for the nighttime landing.

As the aircraft entered a holding pattern and began its 20-degree bank turns, the number one and two engines suddenly flamed out. The copilot returned to his position and the aircraft commander informed him that engines numbers one and two had flamed out because they "had run out of fuel." (Although it was impossible to definitively tell what caused the flameout from this report's testimony, it is likely that the cause was at least partially self-induced. The author bases this determination on the facts that the crew themselves were unsure of fuel remaining and the multiple actions taken with the fuel panel during the flight.) After a short disagreement about the situation with the fuel system, the copilot immediately began to work the problem without referencing the tech order. This might have proven to be a critical error, because the amplified checklist contained in the tech order directs a step specifically designed for the case of flameout due to fuel starvation, which was highly likely given the fact that the two engines flamed out simultaneously. This checklist step directs some specific throttle actions designed to "purge the fuel control unit of air and the engines of fuel," which would significantly increase the chances of a successful airstart (USAF 1991). As luck would have it on this day, the number two engine restarted, but number one, the engine with a generator on it, would not.

The situation now was growing rapidly worse. The crew was preparing for a six-engine approach, considering engine number five, which had failed to subidle conditions, as unreliable. Ironically, the most crucial product being produced by the number five engine was not thrust but the electrical power flowing from its generator. Although the number five engine hovered precariously close to the generator cut-out speed, the crew seemed not to appreciate the possibility of its loss. A good deal of time was spent preparing to land with the loss of two engines, but none on the possibility of having to rely on a single generator. The copilot did advise the crew to reduce the electrical loads after seeing high amperage readings on the two remaining generators, which were running off of the number five and seven engines, but apparently did not recognize the impending loss of number five due to undervoltage.

Unknown to the crew, the aircraft began to operate on a single generator shortly after they began their descent out of the holding pattern. The number five engine fell below generator operation speed and tripped the underexcitation relay. Interestingly, the master caution light did not illuminate to indicate the new malfunction. (It appears from testimony that the master caution light may not have functioned during the entire flight, which should have been evident during preflight operations.) Had the crew been aware of the single generator operation, they might not have elected to lower the flaps, which greatly exceeded the capacity of the lone remaining generator, causing complete ac power failure.

At 2136:10Z (Z stands for Zulu time, or Universal Coordinated Time), Hulk 46 called to advisory control, "Sir, Hulk 46 is declaring an absolute emergency at this time, we have no power to the aircraft and I need minimum vectors to get me home." After moving a preceding tanker out of the way, the controller gave the stricken B-52 an initial heading for the approach. "Hulk 46 heavy, fly heading 170." When Hulk 46 responded, the crew made the final mistake in the long sequence of events. The pilot of Hulk 46 replied, "Sir, I'm gonna need no-gyro vectors."

A "no-gyro" approach is required when the aircraft has lost the capability to determine its heading due to a failure of the aircraft heading system(s). In this case, the pilot still had some heading information available, including a C-2A heading indicator and a magnetic standby compass. The no-gyro approach is completed by beginning and stopping turns based on a radar controller's instructions and is frequently a very abrupt approach, as the pilot attempts to precisely "turn left/right" and "stop turn" in accordance with the controller's directions.

When the B-52 lost all ac power, it also lost the capability of the electrical fuel boost pumps to provide positive fuel pressure to the engines. Although the engine-driven pumps should have continued to provide pressure, they are susceptible to cavitation, which can result in flameout. To help prevent this eventuality, a critical, three-step "Fuel Management Without Boost Pumps" checklist is included in the emergency procedures section of the B-52 tech order. It was neither referenced nor remembered. The third step in the procedure reads as follows: "3. Maintain aircraft in as nearly level attitude as possible and avoid abrupt changes in speed or direction." To emphasize the point, a **WARNING** was included immediately below the checklist step.

WARNING

Changes in flight attitude or acceleration forces may cause main tank boost pumps to become uncovered allowing air to be drawn into the system, thus causing engine flameout (USAF 1991).

Over the next several minutes, the radar controller responded to Hulk 46's request for a no-gyro approach by providing two turns and stop turns to align the aircraft with the runway.

As the aircraft continued in towards the runway, the crew was also talking with a duty instructor pilot who had come on frequency to provide any technical support that he could from the ground. As he was advising the crew to ensure that they had done the boldprint emergency procedures, the aircrew suddenly cut in. "Sir, standby, standby . . . " The duty instructor, sensing the urgency in the aircrew's transmission responded, "We're not in a rush, guys, we're not in a rush." The voice from the cockpit replied, "Yes sir, we are. We don't have any engine power at all." Hulk 46 had just experienced total engine flameout at low altitude over the Indian Ocean, less than five minutes from a would-be landing.

It would seem that the lack of information and aircraft knowledge had done just about all it could to ruin the day for this B-52 aircrew. The accident investigation board determined that, based on testimony taken from both pilots, there were several deviations from tech order directives, including the decision to reset the number one fuel tank quantity indicator circuit breaker (twice), an incomplete and incorrect accomplishment of the fuel leak detection procedure, improper positioning of fuel panel controls, failure to follow the prescribed engine air starting procedure, and finally, the failure to accomplish the fuel management without boost pumps procedure or to abide by the associated warning. The effects of poor procedural knowledge and analysis did not end there.

Transcripts of radio transmissions indicate that the call to the duty instructor pilot indicating the flameout of all engines came at 2141:32. We can assume that this radio call roughly approximated the actual time of the engine failures. The copilot recalled that the aircraft was near 1800 feet when the flameout occurred. The aircraft commander's testimony indicated that he was focused on maintaining airspeed and did not initially notice the flameout. When the copilot said, "Pilot, we need to get out of the aircraft," the pilot's response was "Why?" When the copilot shined his flashlight on the engine instruments, the aircraft commander suddenly realized why the aircraft had gotten so quiet.

The radar navigator, whose ejection seat is propelled out the *bottom* of the aircraft, was much more insistent about his desire to leave. He stated emphatically "Let's just get the hell out." But it was more than 30 more seconds, at 2142:09, when the crew made a radio call stating their intention to abandon the aircraft. "Hulk 46 is abandoning the aircraft at this time, no airspeed, we are losing everything." Based on surviving crewmember testimony, the ejection occurred between 600 and 800 feet, descending rapidly, at about 170 knots indicated airspeed.

Stop reading right here, turn the book over, look at the second hand on your watch, and imagine yourself descending in dark silence for 37 seconds. Take a moment; do it now. This delay most likely prevented the navigator and radar navigator from a successful ejection. We will never know if they attempted to eject or not, since neither the aircraft nor bodies were ever recovered.

The B-52 tech order is clear on low-altitude ejections.

WARNING

Do not delay ejection to below 2000 feet above the terrain in futile attempts to start engines or for any other reason that may commit you to an unsafe ejection or a dangerous landing. Accident statistics show a progressive decrease in successful ejections as altitude decreases below 2000 feet above the terrain (USAF 1991).

This final error was the most tragic and is perhaps the best illustration of the price one can be forced to pay for failing to adequately understand the nature of the aircraft. Throughout this case study, there were bits of knowledge that these crewmembers certainly knew at one time to qualify to fly the aircraft in the first place. But the level and depth of the knowledge was not sufficient for them to be able to use it when it was needed.

A final perspective on aircraft knowledge

No one can ever be sure that they know enough to handle every situation that may come their way in flight. Even the most knowledgeable aviators experience inner doubts about their systems and procedural knowledge. Perhaps that is the mark of airmanship relative to aircraft knowledge—the fear that we might not know enough, which in turn fuels the quest for greater understanding. The ironic implication of this factor is that those of us who consider ourselves knowledgeable enough are likely at the greatest risk.

True airmanship appears to be revealed by a systematic and disciplined approach to continuous learning. Take time to establish a system that works for you. Stick with it until it becomes a habit, and then share it with others.

References

Hughes Training Inc. 1995. Aircrew coordination workbook. Abilene, Tx.: Hughes Training Inc., CS-61.

National Transportation Safety Board (NTSB). 1980. Aircraft accident report NTSB-AAR-80-2. Washington, D.C.

Rouse, D.M. 1991. Report of aircraft accident investigation B-52G serial number 59-2593. February 3.

United States Air Force (USAF). 1991. Accident Investigation Technical Order 1B-52G-1-11.

6

Know your team: Teamwork and crew resource management (CRM)

by Pete Connelly
NASA/University of Texas
Crew Research Project
and Tony Kern

Four brave men who do not know each other will not dare to attack a lion. Four less brave men, but knowing each other well, sure of their reliability and consequently mutual aid, will attack resolutely.

Ardant D'Picq

A personal view from Brig. Gen. Steve Ritchie, USAFR

I feel very fortunate to have received so much of the credit that really belongs to so many others. There were literally thousands of military and civilians involved with my success; the crew chiefs, the maintainers, the weapons loaders, and the quality control in the manufacturing process of the aircraft and the weapons systems. Many fighter pilots could have done what I did, but we had a unique opportunity in the Vietnam air-combat arena, and there were some reasons for our success at that time. You already know what they are: preparation, teamwork, discipline, dedication, education, training, communication, enthusiasm, attitude, will, determination, integrity, and spiritual development—those were the ingredients that enabled us to be successful.

In the final analysis, it's people in a wide array of support functions who are dedicated to excellence and high quality that make it possible for us to win rather than to lose, to succeed rather than to fail, and, sometimes, to live rather than to die. A few years ago Admiral Hyman Rickover commented that "survival for America requires the revival of excellence." He said, "Internal mediocrity can destroy us just as surely as anything external." I wonder if he knew how right he was. General Patton said "We fight with machinery, but we win with people." There is no way that we could have done what we did in the Gulf War, without trained, ready, and motivated people. Because people make the difference, learning to operate within a team may well be the single most important thing we can do to contribute to success.

On July 8, 1972, we downed two MiG 21s in 1 minute and 29 seconds, but it would never have happened if the team hadn't functioned perfectly. Every morning before takeoff, I called the Air Force and Navy controllers on the secure phone to give them our names, call signs, and a brief description of our mission. We even flew up to Korat just to meet with the controllers and pilots. I thought that it was important to be on a first-name basis with those who were providing critical information from more than 100 miles away. On July 8, 1972, it paid off.

About 30 miles southwest of Hanoi, we began getting calls from ground controllers that there were two "Blue Bandits" (MiG 21s) in the area. At approximately 5000 feet on an easterly heading, we received a "heads up" call. "Heads up" meant the MiGs had us in sight and were cleared to fire. When we got that information, it was 40 to 60 seconds old, and I still had no visual on the MiGs! At this point, the controller who was 150 miles away looking at the developing combat situation on his radar scope dispensed with the normal radio procedure and announced, "Steve, they are two miles north of you." I made a hard left turn and picked up a "tally ho." The battle was over in less than two minutes, and we were fortunate enough to engage and destroy two MiG 21s.

This mission was a classic example of teamwork! All of the elements for success came together. The radar and computers worked perfectly. The call from the controller more than 100 miles away watching the battle develop came at the precise moment. Three perfect Sparrow missiles worked beyond design specifications. There was split-second crew coordination. On this mission, and on others to varying degrees, everything that I ever learned or experienced during my then-30 years, came together in just a few seconds. It required drawing on every life experience during that 89 seconds of time. Years of preparation, teamwork, and discipline made the difference for us on that memorable day in 1972.

Brigadier General Steve Ritchie is a Vietnam fighter ace, the only American pilot to shoot down five MiG 21s. He has won the Air Force Cross, four Silver Stars, 10 Distinguished Flying Crosses, and 25 Air Medals, and is the 1972 Winner of Colonel James Jabara Award for Airmanship.

Teamwork and CRM

Good teams are made up of good individuals. Ironically, teamwork is an individual responsibility, like much of the rest of airmanship. The concept of teamwork has evolved throughout the history of flight from a simple relationship between a flyer and his mechanic to something far more complex. As aircraft grew in size and importance, other crewmembers were added, either in the form of wingmen for military fighter operations or as additional crewmembers in larger commercial and military aircraft. Teamwork has evolved from these simple relationships into one of the most extensively researched and sought-after aspects of modern aviation, and for good reasons.

Teamwork now involves myriad relationships between a flyer and multiple sources of information, equipment, and people. The skills associated with aviation team development are defined in many current cockpit/crew resource management (CRM) courses throughout commercial, corporate, and military aviation. This chapter does not attempt to recreate the instruction contained in these excellent courses, although some descriptions of military and commercial CRM programs are presented. Instead, this chapter seeks to highlight key aspects of team building, to include traits of successful teams, leadership, and followership. In addition, it addresses two categories that are sometimes viewed as CRM-exempt: single-seat fighter aircraft and general aviation.

The value of teamwork

Teamwork seeks to improve performance in three primary ways. First, it adds to the pool of knowledge and expertise available to confront various situations. Second, it allows for a synthesis of ideas and skills to make new options available to the group. Finally, expert teams are capable of synergy, in which the complementary capabilities of individuals create a level of performance for the group that exceeds the sum of the individuals. Teamwork is an individual aviator's insurance policy.

How does an individual prepare to be a good team member? How can you best apply your individual skills to the team effort? Like everything else in airmanship, it begins with personal competence and expertise. By mastering the knowledge and skills associated with your own individual job, you have taken the first essential step towards good teamwork. Only then are you ready for the next step—refining team skills by understanding and practicing the techniques inherent in aviation team building.

Advantages and hazards of the team

The French general Ardant Du Picq highlighted the two most important advantages of teamwork with his introductory quote to this chapter—increased reliability and mutual support. These elements are critical to successful flight operations in all environments, where rapidly changing conditions can leave an individual's situational awareness in the dust. Buoyed by the knowledge of competent, and perhaps even expert, teammates, an individual in a team environment can act with more confidence and initiative. Each person can afford to be less cautious—not unsafe, but acting with an undergirding of assured mutual support that allows a more aggressive approach to mission accomplishment. Everybody benefits from good teamwork, but few fully understand its nature and intricacies as it applies to the complex, and often hostile, environment of flight.

Hazards are also involved with team operations, and they should not go unmentioned. Interpersonal relationships can adversely affect operations and decision making on the flight deck or in the cockpit, resulting in a team performance that is actually *poorer* than individual performance. This ironic decrease in performance can be due to conflict between team members or, at the other extreme, due to a phenomenon known as "groupthink," in which conflict avoidance takes precedence over the task at hand. Because of these inherent pitfalls, teamwork development should be approached systematically, looking for the inherent pitfalls associated with team operations.

Personal responsibilities: When charisma cloaks incompetence

Good social skills are no substitute for deficiencies in individual knowledge or skill. Because of the inherent trust involved in a mutual support operation, each member of a team must be fully prepared to do his or her job and be open and honest about any current lack of proficiency. This is a matter of personal integrity and, if not done, can create an extremely dangerous condition in which others count on you for skills and capabilities that you do not possess, setting the stage for a big surprise that can end up in personal embarrassment on a lucky day and disaster if the winds of fortune are not favorable. When others count on you in the inherently dangerous environment of flight, you have a moral obligation to respond with competence. An unskilled or undisciplined aviator adds very little to the team effort. In fact, matters can be made much worse by an incompetent team member who masquerades as a competent one, in hopes that the team will cover for personal inadequacies. This type of individual is a parasitic liability to the safety and effectiveness of the whole group.

Characteristics of effective teams and team members

Successful aviation teams have clearly identifiable traits, and extensive research has outlined dozens of attributes towards which we can direct our improvement efforts. Perhaps the most complete list comes from the NASA-Ames/University of Texas Crew Research Project under the direction of Dr. Robert Helmreich. Based on years of aircrew research, this project has created a team assessment tool for evaluation of inflight or simulator performance, called the NASA/UT LINE/LOS Checklist. The following characteristics of effective team performance are paraphrased from the LINE/LOS checklist. (The LINE/LOS Checklist is reprinted in Appendix A and is also referenced extensively in Chapter 13.)

Team management and crew communications

Good teams establish an environment for open communications. These skills are outlined in Chapter 3. They are exemplified by team members who listen with patience, don't interrupt, and make appropriate eye contact and nonverbal communication. Good teams also conduct thorough, systematic, and interesting operational briefings. The leader ensures that the content of the briefings are relevant and not merely dogmatic. Expectations are set for how deviations will be handled. When you listen to an effective team conducting a preflight or postflight briefing, there is appropriate social conversation and often humor involved. This is where the seeds of success are sown.

Other communication traits of successful flight teams include the ability of all team members to ask probing questions and voice their concerns or information with determined persistence and assertiveness. Events, changes, and decisions are verbalized to all crewmembers in a timely manner.

Situational awareness and decision making

Successful teams distribute tasks and workload in a fair and equitable manner. Leaders must take great care to allow adequate time for task completion and communicate clearly when there are deviations from standard operating procedures. Nothing can undermine effective teamwork more than an unequal distribution of work or allowing a floundering crewmember to go it alone. Teamwork means pitching in—even if it means helping to do someone else's job. Small bits of assistance during periods of high workload seldom go unnoticed by the person being helped and can greatly facilitate the team-bonding process.

Conversely, during periods of low workload, good teams keep talking and sharing information and maintain alertness and appropriate vigilance. Some aviators refer to this as a "watchful posture." Good teams use "slack time" to prepare for periods of increased workloads, such as approach and landing or other high workload taskings. Time is an asset, and effective crews manage it carefully. In addition to interacting with each other, effective teams are also in tune with their machines.

Automation management

Automation advances were designed to reduce aircrew workload and improve performance, but unfortunately they often lead to new challenges and levels of complacency. Dozens of automation-related accidents and incidents are in the files of the NTSB and military safety agencies. The techniques and practices of effective crews are important to understanding the role of the machine in the team. It is important to remember that automation does not necessarily mean high technology. A simple VHF radio can also be a friend or foe, depending on the skills of the operator and the demands of the situation.

Effective teams are very careful about their procedures with automation. They establish specific guidelines for dealing with automated equipment, to include when such systems will be used and, even more importantly, disabled. Verbal statements and acknowledgments are used to ensure that the entire team is kept "in the loop" as to the current status of the systems in question. Machines don't always require human inputs to change. Good teams understand the need for monitoring and cross-checking systems at regular intervals, and they establish timelines for reviewing the status of automated equipment.

Perhaps most importantly, effective and safe airmen understand that automation, like the human team member, is not infallible, and good crews never rely exclusively on a computerized system to maintain the safety of flight. When updates or programming demands could reduce situational awareness or create a work overload, the level of automation is reduced or the system is disengaged.

Other attributes of effective teams

Several other traits typical of successful teams do not fall into any of the previous categories. Feedback, conflict management, and proactive measures to fight poor performance have all been found in flight crews who consistently perform above average.

Feedback

Feedback should be timely and accurate, and it should be made a positive learning experience for the entire crew. Comments can come from any crewmember, and they are typically specific, objective, and constructive. Feedback is usually most effective if it is given as soon after an observed behavior as the workload permits. Because it is a normal part of a good team's interaction, feedback should be received and accepted nondefensively.

Conflict

Good aircrews experience conflict, but they handle it appropriately. Conflict is to be expected and even encouraged in the cockpit and ensures that all opinions and inputs are being surfaced in response to a problem or inflight challenge. Avoiding conflict is the first dangerous step towards "groupthink," a phenomenon that seeks group harmony over all else. This tendency can result in accepting the first reasonable idea that is proposed to deal with a particular situation.

Good crews understand that conflict is merely the by-product of effective assertiveness and communication—a normal part of the flight routine. It allows individuals and teams to remain focused on the challenge at hand by not taking criticism or conflict personally. The difference between constructive conflict and destructive argument is a fine line, and the team leadership must be prepared to step in as required to avoid this escalation. Open argument between team members can cause unnecessary distraction and result in some team members withdrawing from further crew interaction.

The most-conservative response rule

Occasionally conflict resolution is ineffective, and airmen need an effective measure to deal with these situations. The most-conservative response rule provides a ready solution by simply stating that in the absence of agreement between team members, the default solution will be for the team to "fail-safe" to the more conservative of the options being discussed. The decision to rely on the most-conservative response rule must occur before the conflict arises, or it merely becomes another part of the argument. Many effective teams brief this rule as a standard operating procedure at the formation stage of the team during preflight briefings.

Employing the most-conservative response rule does not mean that flyers will always be taking conservative approaches to all situations—that approach is too limiting and inappropriate for the flexibility required in many airborne operations. It simply means that in the absence of agreement in flight, the team will automatically fail-safe to the least hazardous of the options under debate. It is often a difficult step for pilots in command to take, because they perceive it as somehow infringing on their divine right to command. Nothing could be further from the truth. In fact, the most-conservative response rule allows the airborne commander more security and freedom to discuss and propose possible courses of actions, with the insurance of an assertive crew and a fail-safe tool to ensure reasonableness. Uninhibited by fears of pushing his or her crew into a questionable course of action, the pilot in command is able to use his or her experience, creativity, and leadership to actively address the challenges of the mission.

Proactive measures

Good teams don't wait for problems to happen before taking action. They learn to watch each other and take preemptive actions against such hazards as fatigue, overload, complacency, and stress. A gentle reminder to a complacent crewmember, a bit of humor, and social discussion, can all be effective methods for dealing with these problems. But an underlying assumption is at work here. Good crews watch each other, as well as the specific automated systems that fall into one's individual area of responsibility, which means that individual team members must be able to detect subtle shifts in performance in teammates. These performance shifts might be noticed because of action or inaction, a change in voice inflection, or nonverbal communication like posture or visible symptoms of fatigue.

Proactive steps can avoid a serious crew problem known as "subtle incapacitation." Subtle incapacitation is described in detail in Chapter 4 as a situation in which a part of the team is suffering diminished performance, but the rest of the team is unaware of it. It can result from myriad physiological problems, such as fatigue, hypoglycemia, or hangover, but it is often undetectable by the rest of the crew or even by the person who is experiencing the phenomenon. Individuals experiencing subtle incapacitation might still feel that they are performing their duties, running their checklists, etc., but their capacity is reduced. Often only the careful observation of a tuned-in teammate can help them snap out of it.

The two-challenge rule

Although most good teams can identify and address problems of diminished performance or subtle incapacitation before the problem becomes dangerous, occasionally it occurs when workloads are high for all members of the team, and it goes unnoticed. An even more dangerous situation occurs when it happens to a pilot who is actively controlling the aircraft. Incapacitation can also occur suddenly, like the pilot of a commuter aircraft who had a heart attack on final approach to Martha's Vineyard. The NTSB was shocked to find that the copilot never took control of the aircraft and allowed the aircraft to crash without intervening to correct the situation. This finding indicates that the copilot was likely unaware of the problem with the pilot in command. Subsequent recreations of the incident in simulator studies showed that a large percentage of copilots were unable to recognize the problem and intervene in time to save the aircraft.

One method of dealing with occurrences of incapacitation is the use of the two-challenge rule. In a prebriefed agreement between crewmembers, if one crewmember does not respond to two verbal corrections or "challenges," another crewmember will automatically assume control of his or her duties. An example of the two-challenge rule goes something like this: A copilot or first officer notices the pilot in command drifting slightly below glidepath on an approach to landing. He or she states: "Pilot, we are 50 feet low." Observing no correction or response, he or she quickly repeats the challenge with greater assertiveness. "Pilot, we're 100 feet low, we need to get back on glidepath!" If the copilot still observes no response, he or she takes the prebriefed action. "Pilot, I have the aircraft." The two-challenge rule takes longer to describe than to accomplish, and this entire exchange would likely take less than 10 to 15 seconds.

The importance of prebriefing this response is obvious. The challenger cannot afford a lengthy debate on whether the boss will get mad at him or her for taking the aircraft, and the criticality of assuming control of an aircraft in a potentially hazardous situation demands a coordinated assumption and release of controls. In the previously mentioned study, crews that utilized the two-challenge rule did not suffer any losses due to an incapacitated pilot at the controls. Establishing precautions like the two-challenge rule is good airmanship and bears the mark of effective team leadership.

Laying the groundwork: Leadership

Good leaders establish the team concept early—as in the first few *minutes* that a new team is together (Ginnett 1987). Dr. Robert Ginnett, the director of research for the Colorado campus of the Center for Creative Leadership, calls team formation "a critical leverage point" (Weiner et al. 1995) and points out that effective leaders typically address three areas during this initial formation process: boundaries, norms, and authority. He found it interesting that "tasks" were not often discussed by effective leaders, perhaps an indication of the trust and competence that also helped establish an effective team early.

Ginnett found that effective leaders spent time early expanding or breaking down typical "boundaries" or dividing lines between team members and even between the team itself and other parts of the larger organization. Instead of a "you do this and I'll do that" approach, effective leaders speak in terms of "we" and often refer to the implications of the team's actions on others. For example, an aircrew might take certain fuel transfer actions to make it easier on the maintenance personnel who will recover the aircraft. These simple steps set the tone for effective interaction and synergy.

Establishing norms for the group is another critical first step. The three norms established most often by effective crew leaders were an emphasis on safety, communication, and cooperation. Some very simple techniques and statements can establish these norms in early team briefings. The leader can simply state, "We need to all keep an eye on each other. Don't expect me to know everything, so speak up if you see or hear something that is important to us. We need and expect everybody's input to get the job done safely and efficiently." With this simple preface, a leader has set the tone for a safe and cooperative team environment.

A good team leader is not afraid to take charge, but he or she operates with an *appropriate* amount of authority based on the qualifications and capabilities of the team. Although it is impossible to know all of your team members as well as you know yourself, team leaders (and all team members) should seek the same types of information on their teammates as they do about themselves. What are their qualifications, strengths, and weaknesses? How much experience do they have? Do they seem distracted? Are there any hidden agendas? Does any member of my team appear fatigued, angry, or frustrated?

Armed with these impressions, a leader can move up and down the authority scale between fully "in command" and autocratic to a more democratic and participative leadership style. Ginnett suggests good leaders are highly unlikely to be totally passive or laissez-faire and are able to adjust their leadership style according to their team and the individual circumstances of the situation. Leaders establish their authority by demonstrating competence, disavowing perfection, and by engaging the rest of the team in the briefings and actions of the mission at every opportunity. But even with all of these qualifications, you still need good followers to be a good leader.

Followership: A forgotten art

Followership is often thought of as a temporary necessity on the way to a leadership position. Nothing could be further from the truth. In many ways, being a good follower or teammate is much more difficult than being a good leader, because you must often fit your own personal style of action to the leader's, as well as to the situation. Doing so can be difficult in any social setting, but it is increasingly so in the time-compressed confines of the cockpit or flight deck. Followership connotes flexibility and adaptability to both situations and people and takes considerable skill and practice to optimize. It may be a forgotten art.

Followers come in different types, but certain identifiable attributes are typical of good followership and lead to team cohesion and synergy. A good aircrew team member should possess competence, communication skills, assertiveness, and predictability. Perhaps the most elegant and appropriate description of follower types comes from Kelly's model of followership illustrated in Fig. 6-1 (Kelly 1988).

Although it has been much neglected, it is important that we have some way to talk about followership in the detail that we talk about leadership. Kelly outlines five types of followers.

Sheep

Sheep-style followers are neither active or critical thinkers. They lend very little to the airborne decision-making process. Sitting quietly at their crew positions, they tend to complete their checklist duties in silence, taking up space. Sheep are incapable of and uninterested in providing much input to the crew; they are merely human automatons.

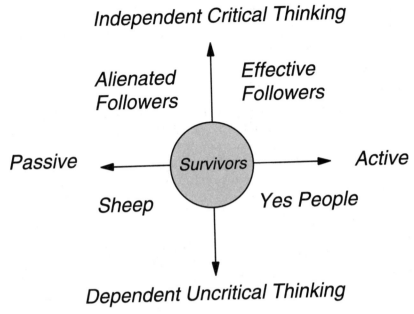

Independent Critical Thinking

Alienated
Followers

Effective
Followers

Passive   Survivors   Active

Sheep   Yes People

Dependent Uncritical Thinking

6-1 *The neglected art of followership is critical to airborne operations. This model shows patterns of followership that dramatically affect teamwork. Where do you fit in?* Kelly 1988

Yes people

Perhaps more dangerous to effective teamwork are the "yes people." They are very active but merely political creatures intent on backing up whatever position is taken by the boss. They can usually be recognized by their incessant agreement with every statement made by the pilot in command—their heads forever nodding in agreement like the little stuffed dogs sitting in the back of a car. Yes people are dangerous for two reasons. First, they inhibit crew decision making by their inability to think for themselves, and, more importantly, they can add to the illusion of omnipotence of a pilot in command who might be a poor decision maker.

Alienated followers

The opposite of yes people are alienated followers, who are typically bright and critical thinkers, but are uninterested in contributing due to some animosity with other crew members or the organization at large. Alienated followers are often described as "festering wounds" and might even secretly wish the team to fail because it "would serve them right." Because this type of follower can completely destroy effective teamwork and endanger the mission or crew, they must be identified and removed from the team as quickly as possible if their attitude and behaviors are not corrected.

Survivors

Survivors represent Kelly's fourth style and are often referred to as "company men." They are typically mediocre performers who adjust their style and shift into all quadrants—not to create a more effective team, but to avoid making waves. While survivors are not as dangerous or disruptive as yes people or alienated followers, they are not as effective as they could be, and they are often unpredictable.

Effective followers

Effective followers are essential to safe and efficient flight operations. They are active without being yes people and are not afraid to challenge the prevailing mode of thought. In the following example, an effective follower demonstrates the value of a good follower.

During the early 1980s, the United States undertook a mission, code-named El Dorado Canyon, to bomb Mohammar Quaddaffi's Libyan headquarters as a measured response to continued Libyan terrorist attacks against U.S. military personnel. During the long cruise from the United Kingdom to the target, a KC-135 navigator who was not on the lead aircraft, noticed that the strike force package was falling well behind the mission timing schedule. Reluctant to speak up at first and break radio silence, he double-checked his data. As he understood the mission, the current airspeed would result in the Air Force strike missions arriving *after* other critical components in the plan—a fact that could result in the loss of surprise and jeopardize the safety and success of the entire operation. Luckily, this young officer was an effective follower, and he broke radio silence to advise the strike team leader of his observations and calculations. After a few careful cross-checks, the flight lead ordered a substantial speed increase and the potential timing error was avoided. Although history seldom records the actions of assertive and active followers, they can often have dramatic influences on the success or failure of the team.

This example highlights three critical attributes of effective followers; good listening skills, assertiveness, and communications abilities. To be a good follower, one must by necessity be a good listener. Active listening means focusing your attention on the sender and being able to gather meaning from voice inflection, body language, and even things left unsaid. Although there are entire college communication courses on the art of listening, the essence of the art can be stated in two simple words: "listen hard." Practice applying yourself to every word said to you. Listen to each message as if it were the combination to the vault at Fort Knox.

Assertiveness is a mandatory prerequisite for good followership. Although assertiveness can be fostered by an effective team leader and enhanced by an open organizational climate, it remains an individual responsibility for safe and effective flight operations. There are many obstacles to assertiveness, including peer pressure, fear of superiors, and fear of being wrong. But a chain is only as strong as its weakest link, and a nonassertive team member can be the unraveling of an otherwise effective team. Often a crewmember with a vital piece of information won't make himself or herself heard if the idea is initially rejected by the pilot in command or flight lead. One can only wonder how many debriefings take place in the great aircrew lounge in the sky, where the specters of a human-error crash whisper, "I had the answer, but you wouldn't listen." Make your presence felt. Ask questions, offer suggestions, and get in the game.

Communication: The key to successful teams

Communication is the glue that holds a team together. Because of the nature of aviation, communication is inextricably tied to the type of flying being accomplished, and cultural norms often dictate inter-cockpit communication styles. Aviation researchers Foushee and Kanki broke down aviation communication into 18 types and drew several conclusions regarding effective aircrew communications (Foushee and Kanki 1989).

By analyzing the various types of communications taking place in the cockpit, the researchers were able to make certain inferences about the relationship of communication to crew performance. Ineffective crews communicated less but had a greater incidence of response uncertainty, which indicates lack of information upon which to act. Other common communication characteristics of poor crews were frustration or anger and embarrassment, along with a lower incidence of agreement. On the other hand, crews who talked openly about flight conditions or problems, either real or potential, seemed to be more effective. Aircrews that utilized more commands, inquiries, and observations had fewer errors than did those who did not. Clearly, effective teamwork relies on the quality as well as the quantity of communications.

These findings correlate well with findings on cockpit leadership, followership, and assertiveness. Although the discussion on traits of effective teams can be long and complex, it can also be summed in a single sentence used in many modern CRM courses. "Good teamwork is authority with participation and assertiveness with respect."

Cockpit resource management (CRM): Formal training for teamwork

CRM means maximizing mission effectiveness and safety through effective utilization of all available resources. That's it. What makes CRM unique as a training program is the environment and target audience for which the training is designed. CRM is designed to train team members how to achieve maximum mission effectiveness in a time-constrained environment while under stress. As popular as CRM has become, many people still do not understand its basic principles.

The fog that often surrounds CRM deals directly with the definitions of a few key terms we have used above. "Maximizing mission effectiveness" can be broken down into three areas for a better understanding. Mission effectiveness can be maximized by

1. Achieving the objective. It could include reaching your destination, putting bombs on target, or delivering cargo on time.

2. Preserving resources. It means not crashing airplanes or killing or injuring crew members, saving fuel, and preventing crew-induced aircraft damage, like over-Gs or structural damage.

3. Training efficiency. Improved training efficiency means better student-instructor rapport and interaction and more effective communication during flight and ground instruction. It also means using available training time and fuel efficiently.

By "all available resources," we mean hardware, software, printed materials, people, the environment (wind, sun, terrain), time, fuel, etc. Research has demonstrated that many crew members cannot even identify all of the resources at their command, let alone access them in a time-stressed emergency situation.

Finally, "team members operating in a time-constrained environment under stress," represents all aviators. In a nutshell, CRM training is designed to produce team members who consistently use sound judgment, make quality decisions, and access all required resources under stressful conditions in a time-constrained environment. To understand the importance of CRM to airmanship and team building, it is important to understand the reasons it was developed and the results this training has achieved.

A brief history of CRM

Aviators have been making poor judgments since the day Icarus decided to check out the maximum service ceiling of his new wings. In an Air Force Inspector General report titled "Poor Teamwork as a Cause of Air Craft Accidents," published in 1951, data from 7518 major accidents between 1948 and 1951 (now that's a database!) determined "poor organization, personnel errors, and poor teamwork" resulted in the majority of aircraft accidents. It further stated that "the human element . . . and effective teamwork is essential to reducing the accident rate." The IG report even went as far as recommending a "teamwork training program" but, unfortunately, neglected to add a suspense date for completing the task. It was three more decades until we got around to it.

The aviation community refocused on the need for some type of human-factors training following the much-publicized crash of a United Airlines DC-8 in Portland, Oregon, in December 1978. Attempting to ascertain the nature of a possible landing gear problem, the aircrew allowed the aircraft to completely run out of fuel while circling near the landing field on a clear night in good weather. This accident resulted in the amendment of Part 121 of the FARs and allowed airlines to train what is now called CRM.

Following CRM implementation, air carriers began to notice dramatic decreases in their accident rates. Military application of these principles lagged behind the civilian sector, but in the mid-1980s, the Naval Safety Center and the Air Force's old Military Airlift Command (MAC) began to implement airline-style programs with good results. The popularity of these programs grew throughout the 1980s and early 1990s to the point at which nearly everyone has "a program." In fact, CRM has been deemed so important by both sectors that it is now mandated for nearly all flight crew members in commercial and military aviation, and separate levels of CRM are conducted for initial awareness, instructor upgrade, and evaluators. Some military programs even require aircraft-specific CRM programs, so that flyers can efficiently integrate previously learned human-factors skills into new weapon systems. A very brief mention should be made of the content within the excellent CRM programs that exist within most commercial and military organizations today. Although all CRM programs are uniquely designed for the needs of the organization, most address some or all of the following elements:[1]

1. Situational awareness. A desired end state of CRM training is a high state of situational awareness. Tools for preventing lost situational awareness, cues for recognizing lost situational awareness, and techniques for recovering from lost situational awareness are covered under this area.

2. Group dynamics. Group dynamics include command authority, leadership, responsibility, assertiveness, conflict resolution, hazardous attitudes, behavioral styles, legitimate avenues of dissent, team building, and desired traits.

3. Effective communications. Effective communications include common communication errors, participation of all crew members, cultural influences, and communication barriers such as rank, age, and position. It also stresses coordination with other participants, interface concerns, listening, feedback, precision, and efficiency of communication.

4. Risk management and decision making. This element includes risk assessment and risk-management styles, process, tools, breakdowns in judgment and discipline, problem solving, evaluation of hazards, and management of regulatory deviation during emergencies.

5. Workload management. This area covers overload, underload, complacency, management of automation, available resources, checklist discipline, and standard operating procedures.

---

1. These curriculum items are taken from the U.S. Air Force Instruction (AFI 3622-43) on CRM, but they are representative of many other programs.

6. Stress awareness and management. This area includes sources of stress, benefits, hazardous effects, and coping techniques.

7. Mission planning, review, and critique strategies. This area covers premission analysis and planning, briefing, ongoing or midmission review, and postmission critique.

8. Physiology and human performance. This element includes cognitive processing, anomalies of attention, stress and stress management, behavioral styles, mental attitudes, and fatigue effects.

As you can see, airmanship and CRM go hand in hand. In fact, it can be said that airmanship is the individual structure on which CRM builds. Although CRM programs are now mandated for nearly all commercial and military pilots, many pilots still feel that it does not apply to them and have been reluctant to buy in to the cultural shift towards greater training for teamwork. One example of this can be seen by military fighter pilots, some of whom feel that CRM simply doesn't make sense for them.

Your wingman is your copilot: Teamwork and CRM for the fighter jock (Wagner and Diehl 1994)

What does teamwork and CRM have to do with flying a single-seat fighter? The answer to that question can unfortunately be found in many military mishap reports. In one mishap, the lead fighter had his three-ship in a wedge formation, with weather closing in, presenting the mishap pilot with a "box-canyon" effect that required a rapid decision. To make matters worse, the overcast ceiling limited the altitude available, forcing the flight below the minimum altitude of 1000 agl established for the first-look at this area during the exercise.

The Number 3 aircraft requested a hook turn to avoid the weather and did a route abort away from an area of high terrain. As the mishap pilot (MP) called for and then executed his turn to abort, he chose the wrong direction, encountered high terrain, and impacted a ridge. Although the Number 3 pilot was not officially to blame, he could have been explicit about going the other direction as he saw the conflict developing and perhaps saved his teammate's life. Perhaps he felt that he was being a good wingman by letting the lead direct his own element. Or perhaps he lacked the assertiveness to speak up.

The flight lead had quite a load of personal "baggage." Most of the squadron knew that the MP had two full-time civilian jobs and that his personal life was abound with conflict. He was summoned to appear in court three days after the exercise started. The state was threatening to garnish his wages for child support. He was also significantly in debt, with several major credit cards maxed out. The day of the mishap, the signs of stress were showing through. His preflight briefing omitted several important items, and he forgot his ejection seat harness on the way to the jet. The bottom line was that the other pilots in the squadron were in the best position to call a "knock-it-off" before this guy became a statistic. Perhaps they thought that they were being good wingmen by allowing the lead to direct his own flight. Or perhaps they lacked the assertiveness and listening skills of good followers.

Poor communication contributed to another mishap. During the preflight briefing for a routine cross-country, the flight lead sent Number 2 off to get the latest weather. Number 2 learned the weather was going down, and by the estimated time of arrival, it would definitely be instrument conditions. Number 2 dutifully wrote this information on the DD Form 175-1 weather briefing sheet, but when he got back to the briefing, lead was winding up by saying that he wanted to see a "sharp" overhead pattern on arrival, a maneuver that would require visual conditions.

Without comment, Number 2 handed the lead the form with the note about the weather going down. Lead ignored this note and told everybody to "step." Lead first learned about the poor weather on arrival at the forward base. He then announced they would be doing a five-ship aircraft single launch and recovery. Nobody challenged this decision even though they knew that wing standards said a four-ship was the maximum for these circumstances. During the approach, the weather continued to deteriorate with rain and turbulence. Number 5, who was on his first deployment, got distracted and forgot a critical step-down altitude. He hit a tree-covered ridge line but managed to eject successfully. The ending to this ugly scenario is almost humorous. The jet ended up in a nearby farm house (which the taxpayers bought), making it perhaps the only time that "buying the farm" was literally true, but not fatal.

The investigation revealed another intercockpit coordination problem. Number 4 had unintentionally flown through the TACAN course and decided to come back with a hard turn—too hard, evidently, for the inexperienced guy in trail. Number 5 didn't anticipate such a vigorous correction, overreacted, got into buffet, and became extremely disoriented. When Number 4 was asked why he did not warn his team member with a radio call, he said he didn't want to look bad to the "guys listening on the ground." Another avoidable statistic.

Remember, fighter pilots almost always fly and fight as team members. Furthermore, the fighter "team" includes not just the people in the flight, but all the folks with whom you communicate and coordinate, including AWACS, ATC, Command Post, maintenance, FACs, tankers, etc. Of course, these guys are not usually seated next to us, and, furthermore, time is typically very critical in the fighter game. Ergo, fighter jocks need this CRM stuff as much (and maybe more) than folks flying the heavies.

The Naval Safety Center was the first organization to reach the conclusion that fighter pilots could benefit from a CRM-type program. Aircrew coordination training was introduced into their A-6 training wings in 1989. Although they faced the same cultural barriers (fighter guys don't need CRM), their mishap rate dropped dramatically in the years following, as illustrated by Fig. 6-2. The 1991 mishap rate increase shown occurred during Operation Desert Storm, but the rate quickly dropped back to normal levels after combat operations ceased.

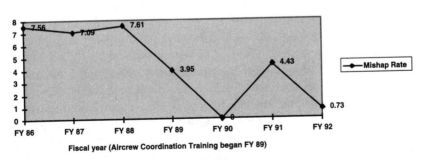

**A-6 Mishap Rate Before and After
Aircrew Coordination Training Implementation**

Fiscal year (Aircrew Coordination Training began FY 89)

6-2 *Cockpit resource management (CRM) or aircrew coordination training, as it's called in the U.S. Navy, has had dramatic results in both the civilian and military arenas. Teamwork effectiveness is dramatically improved when CRM principles are seriously implemented.* U.S. Navy Safety Center, 1993

Although charts and statistics are often cold and impersonal, ask anyone who has lost a buddy to a human-error accident what these statistics mean. People who would have perished are still flying today because some Navy fighter pilots were mature enough to learn something new, and the Navy had the vision to provide the training. Airmanship is about personal responsibility. Remember your responsibility to the teammates on your wing.

The last great vacuum into which CRM must flow is the field of general aviation. The following section demonstrates how CRM principles and concepts are more applicable to the private pilot than you might imagine.

CRM in general aviation

The CRM principles used by senior commercial and military pilots are also extremely useful for the general aviation flyer. Consider the following case study, and see how good CRM principles might have avoided the disaster.

Case study: In over his head

It was a beautiful day for flying in southern California—clear and 15 miles visibility. The pilot of a single-engine Piper Archer and his two passengers lifted off the runway at Torrance Airport on a flight to Big Bear, a resort area in the mountains west of Los Angeles. It was 1140 a.m. The Piper—call sign Nine One Foxtrot (91F)—turned east and began a cruise climb to 9500 feet.

While he had only 231 total hours of flying, the pilot of 91F was careful. People who had flown with him described him as being "old maidish" with his preflight checklist and sometimes "too careful" with the rules. Flight instructors who had flown with him described him as a diligent, attentive student, but a "VFR pilot who liked to look out," more inclined to navigate by visual reference to the ground than radio navaids. Before he left for Big Bear, he discussed the route with a fellow pilot who was familiar with the route. He was advised on how to avoid the Los Angeles Terminal Control Area (LAX TCA) by using freeways as geographical boundaries. However, having recently moved to LA, the pilot of 91F was not as familiar with these highways from the air. On this morning, the pilot of 91F was squawking 1200, had the chart out, was making a couple of gentle turns in the climb over Long Beach, and was preparing to enter some of the most heavily congested airspace in the world without air traffic control (ATC) clearance.

Seven minutes later, Aero Mexico 498, a DC9 inbound for LAX, called in level at 7000 feet. Flight 498 had been airborne only 27 minutes since its departure from Tijuana. The last leg of the day for 498's crew, the 40-minute flight from Tijuana was a busy one. The approach controller was busy too, vectoring airliners and calling out traffic. Just as Flight 498 completed its "in range" call, the controller called out unknown traffic to them, "traffic at eleven o'clock, one mile, altitude unknown." Flight 498 was then cleared to slow to 190 knots and descend to 6000 feet.

Eleven minutes after 91F left Torrance, a single-engine Grumman Tiger checked in with the same controller handling Flight 498, requesting flight following. As the controller was giving the Grumman a transponder code, the cockpit voice recorder on Aero Mexico 498 recorded these words, the last from the crew of Aero Mexico 498: "Oh (expletive deleted), this can't be!" As the Grumman read back the squawk code, the leading edge of Aero Mexico Flight 498's horizontal stabilizer hit 91F's cabin area broadside. The occupants of the smaller plane were decapitated. Both aircraft fell 6500 feet onto the Los Angeles suburb of Cerritos.

Many warning devices now found in modern cockpits have the names of accidents written all over the instrument; for example "take-off configuration warning," "ground proximity warning," and after Cerritos, "terminal collision avoidance system" (TCAS). But these warnings are largely found only in airliners, as well as in some military aircraft. Only one warning system is common to airliners, military aircraft, and the Piper—and that is the human one. The motto of FlightSafety International, an aviation training corporation, says it all. "The best safety device in any aircraft is a well-trained pilot."

Crew resource management (CRM) is a term that is familiar to many military and airline crewmembers, but it is not well understood by many general aviation (GA) pilots. Some GA pilots might have heard the term and felt it applied only to airplanes flown by "crews" and was not relevant to single-pilot type operations. As mentioned earlier, the military has heard this argument for years from its fighter pilots. Furthermore, some general aviation operations that are flown by more than a one crew person, such as charter, corporate, air ambulance, or even flight instructor/student sorties, may regard CRM as something that only works for the airlines or the military. While CRM came about as a result of human factor-based accident records and pressure from the FAA and NTSB, no such thing has happened in general aviation, despite a substantial record of human factor-based accidents and incidents. Human error occurs as often—or more often—in the general aviation environment as it does in the more structured flight environments of commercial and military aviation. We all share the same sky, and sometimes—as in the Cerritos midair collision—aviators share the same point in space and time.

Airline and military crews are trained in CRM, but general aviation pilots are currently not. Are the CRM skills taught to airline and military crews transferable to the general aviation pilot? Are some CRM skills universally useful? Perhaps CRM training should be a part of every pilot's training, starting with the private pilot's license and continuing through ATP.

Resource management

Crew resource management, originally called cockpit resource management, was defined about 15 years ago by NTSB member John Lauber as the use of all available resources—software, hardware, and liveware—to accomplish a safe and efficient flight. In this context, the single-engine Cessna pilot shares several common resources with the turbojet airline pilot. In terms of hardware, both have access to radios, flight controls, emergency equipment, and some even have autopilots. In terms of software, both have access to charts, airplane manuals, checklists, and flight logs. In terms of liveware, both have access to ATC, ground assistance personnel such as dispatchers or flight service, and finally, other crewmembers or passengers. From the files of Aviation Safety Reporting System (ASRS) comes an example of a general aviation pilot who made excellent use of liveware resources during an emergency, demonstrating that CRM principles can be a lifesaver for any type of flyer.

I was flying on an IFR flight plan in VMC conditions. The one passenger on board was my wife, who occupied the right front seat. The aircraft was in a cruise climb configuration approaching 6000 feet on a clearance to 10,000 feet. We were over the ocean about 10 minutes into our flight after departing the airport. Passing through 6000 feet, I noticed the vacuum gauge needle was flickering just above the green arc. The engine then made a grinding noise for about three seconds and the propeller came to an abrupt stop.

I immediately decreased airspeed 35 knots to 85 knots IAS, turned toward my departure airport, closed the cowl flaps, and alerted Departure [control] of the engine failure and my course deviation. I then passed to my wife the emergency procedures laminated cards to read aloud to me while I turned the fuel valve and mixture control to off, maintained airspeed, and alerted ATC that I was shutting down the transponder.

Dividing my time between keeping the airport in sight and monitoring the airspeed, I instructed my wife to ensure her seat belt and shoulder harness were tight. Air traffic control then handed me off to the airport tower. I informed the tower I was with them on an emergency approach. I was cleared to land "any runway." The active was runway 24. My altitude, airspeed, and progress to the airport were good, and I elected to fly an abbreviated downwind and base to 24. I lowered the gear on final to help lose altitude. I maintained 85 knots until short final, when I lowered the flaps to enable me to slow the airplane and land in a normal landing attitude.

The gear held. The aircraft rolled to a stop after a smooth, normal, uneventful, landing. Twelve minutes had passed from the time the engine shuddered to a grinding halt and we rolled to a stop on Runway 24. I believe the successful outcome of this flight can be attributed to the fact that I remained focused on the task at hand, flying the airplane, and used all of the resources available to me, including my wife, who read the emergency procedures. I also believe that I was very lucky to have been presented a situation which was equal to my level of skill (ASRS 1993b).

This pilot had clearly thought about his approach to an emergency situation long before this incident occurred. His use of all available resources and timely decision making are clear examples of CRM principles in action. In this case, the concept of a husband-and-wife team takes on a new dimension. CRM principles can also help identify the potential for problems before they occur.

Red flags and error chains

Several major U.S. carriers employ a CRM/human factors safety theme as the subject for the recurrent training of their pilots. While these themes are focused on the crew concept, they also contain concepts that would be useful to the general aviation pilot or two-person corporate/charter crew. United Airlines (UAL) has used the "red flag" concept and combined it with error-chain recognition. Pilots universally recognize the red warning flag when it appears in a cockpit instrument. A red GS (glideslope) flag on an ILS approach inside the outer marker is a significant warning to the pilot and carries with it specific implications. Just as pilots are trained to recognize technical red flags and know what procedures to follow, UAL trained its pilots to recognize human factor red flags as well.

UAL used FlightSafety International's paper on error chains, "Reducing the Human Error Contribution to Mishaps Through Identification of Sequential Error Chains" (Schwartz 1990) to help its pilots understand the concept of error chains and how to stop a potential incident or accident by identifying a "red flag" in an error chain. Once the "red flag" is identified and corrective action is taken, the link is broken, and the incident/accident is prevented. The FlightSafety paper further stated, "There is seldom one overpowering cause, but rather a number of contributing factors or errors, hence the term 'error chains.' The links of these so-called error chains are identifiable by means of eleven clues. These clues are: ambiguity, fixations or preoccupation, empty feeling or confusion, violating minimums, undocumented procedure, no one flying the airplane, no one looking out the window, failure to meet targets, unresolved discrepancies, departure from standard operating procedure (SOP), and poor communications." The following case study illustrates an error-chain accident that is commonly used in CRM training courses.

Case study: Tiger 66

One of the classic CRM training films is the tragic story of the crash of Tiger 66, a Boeing 747 in Kuala Lumpur, Malaysia. Several "red flags" formed the error chain that killed the crew of Tiger 66, any one of which—if identified—would likely have prevented the crash. It was a clear case of miscommunication—a setup that could have trapped a single Cessna pilot, a DC3 charter crew, a military F-111, or this B747 cargo crew.

As Tiger 66 descended into Kuala Lumpur, confusion between the approach controller and the crew led to a misinterpreted clearance. As the crew began its en route descent, the first ATC clearance in the setup was "Tiger 66, descend five five zero zero." The crew read back " . . . fifty-five hundred." The next clearance was "Tiger 66, descend two seven zero zero." The read back was " . . . two thousand seven hundred." The 747 leveled at 2700 feet mean sea level (msl), and tracked inbound towards the airport. Then came the killer clearance, the second-to-last red flag. "Tiger 66, descend *two four zero zero*, cleared NDB33 approach."

The pilots were intially angry at receiving the nondirectional beacon (NDB) approach, which might have distracted them just enough to miss the intent of the controller. "NDB—that SOB!" snapped the first officer. The captain replied to the controller's instruction, "OK, *four zero zero*." The captain had misinterpreted the controller's "two" as "to," and perhaps due to the distraction of the surprise NDB clearance, neither pilot realized that they were descending into the terrain indicated on the approach chart. But their fate was not yet sealed. The last of the red flags were 11 "pull-up!" warnings from the ground proximity warning system (GPWS) that went unheeded by the crew. The initial approach altitude for the NDB33 was 2400 feet. Tiger 66 flew into a hill, level at 400 feet msl.

Error chain analysis can also be an effective tool for general aviation pilots, and accident reports can be obtained for this purpose from the NTSB, or in many cases from your local safety office.

Workload management and distraction avoidance

Workload management is another tool trained in the military and airlines that can be used by general aviation pilots. In general, workload management is a way to prioritize tasks so that a loss of situational awareness does not occur through overload or distraction in the cockpit, which can lead to a self-induced incident/accident. The skills associated with workload management reflect how well an aviator or a crew manages to distribute tasks and avoid overloading individuals. It also considers the ability of the crew to avoid being distracted from essential activities and how work is prioritized. Secondary operational tasks are prioritized to allow sufficient resources for dealing effectively with primary flight duties. The principles of task prioritization and distraction avoidance are all CRM/human factors tools that are extremely useful to the GA pilot.

Breaking the error chain

Back in the 1970s, Eastern Airlines lost an L1011 TriStar in the Florida Everglades. It's a classic accident involving a crew that got distracted with a minor landing gear indicator light while the airplane slowly descended into the ground and crashed. From the files of ASRS comes a report from a GA flight instructor whose recollection of that incident raised a red flag and broke an error chain that prevented another controlled flight into terrain (CFIT) accident:

Today I had a problem with my landing gear on takeoff. It failed to retract; in fact it didn't even leave the down-and-

*locked position. We tried cycling the handle a few times but
no response. We checked circuit breakers, electrical busses,
etc., all to no avail. After we determined that the gear was
down and locked, we called the tower and returned for [an
uneventful] landing.*

*What I found so interesting about this situation was how
closely the events mirrored the air carrier crash in the Florida
Everglades. My student was flying and I began working on
the problem and then so did he. After a while I realized that
no one was really flying the plane, and I told my student to
fly the plane and I would work on the problem. Thank good-
ness we broke the chain or the situation could have turned
into something worse — such as unexpected contact with terrain (ASRS 1993a).*

General aviation pilots can also fall prey to the same hazardous attitudes that confront military and commercial aircrews. Although
these attitudes are discussed in detail in Chapter 4, the following example clearly points out the need for general aviators to heed the
message as well.

"Time Pressure as a Casual Factor in Aviation Safety Incidents: The
Hurry-Up Syndrome," by McElhatton and Drew, is a 1993 NASA
study on the phenomenon better known to most pilots as "get there-
itis." The study defined the hurry-up syndrome as "any situation
where a pilot's human performance is degraded by a perceived or
actual need to 'hurry' or 'rush' tasks or duties for any reason" (McElhatton and Drew 1993). While this report dealt with data obtained
from air carrier NASA reports, many of their findings have significant
implications for the GA pilot as well.

The top two hurry-up incidents listed in the NASA report were deviation from ATC clearance or FARS, at 48.0 percent, and deviation
from company policy or procedure (SOP and checklists), at 20.8 percent. The majority of incidents had their origins in the preflight
phase (63 percent), the next was in the taxi-out phase (27 percent).
What was even more significant was the finding that human errors
that occurred in the preflight phase manifested themselves later in
the takeoff phase. For example, a distraction during preflight causes
the pilot to omit an important checklist item, which has deadly consequences in the climb-out. A general aviation accident in Cockeysville, Maryland, in April 1984, is an excellent example that had its
seeds in preflight errors with a pilot who was in a hurry to go.

On the morning of the accident, the pilot of the twin-engine AeroStar picked up the airplane, which had been upgraded with new Machen 656 Superstar conversion engines. He received a 30-minute dual familiarization flight, a short review of performance charts and flight manuals, and two partial weather briefings. He topped off the fuel tanks, filed a flight plan, and took off with his passenger for Florida.

Nine minutes after takeoff, the flight, 79R, was cleared by Washington Center to climb to FL180. Eight minutes later, Washington Center received 79R's last transmission: "OK, Mayday, lost engines, lost engines, dropping fast." Radar data from Baltimore Approach Control showed that 79R descended from 16,900 feet to 2300 feet in about 90 seconds, an average descent rate of more than 9700 feet per minute. Witnesses saw the aircraft banking from left to right, roll inverted, and impact the ground nose down.

Witnesses at the airport described the pilot as nervous prior to takeoff and said his "hands were shaking." The instructor who had given the pilot of 79R the familiarization ride testified to the NTSB that "the pilot said he felt uncomfortable and nervous." The NTSB examination of the wreckage found the electric fuel boost pump switches in the off position and mixture control on rich. The airplane flight manual required the boost pumps to be on during climb above 10,000 feet. So did the takeoff checklist. When the boost pumps are off during climb above 10,000 feet, fuel starvation will result, and the engines cannot be restarted with the mixture rich.

In its summary, the NTSB concluded that the pilot never turned on the boost pumps as required by the checklist and flamed out both engines in the climb (NTSB 1985). The pilot was nervous and uncomfortable about flying to Florida. Self-induced psychological stress over his minimal experience contributed to his unease. Once the dual flame-out occurred, the situation exceeded the pilot's capabilities and caused him to lose aircraft control. At a descent rate of more than 9700 feet per minute, he was probably not capable of completing the 28-step checklist, "Engine Failure During Flight," or "Restarting a Feathered Engine."

Before this flight ever left the ground, there were psychological "red flags" present that contributed to the error chain, including fixation and preoccupation, confusion, undocumented procedures, and omission of critical checklist items. The pilot's eagerness to hurry up the preflight phase most likely was the cause of the fatal dual engine flame-out and subsequent loss of control.

Another major finding in this NASA report was that high workload was cited in 80 percent of all incidents (McElhatton and Drew 1993). It is very significant because procedural errors are less likely to be detected in high-workload situations. When the pilot is in a rush to complete a multitask situation, it is more likely that something important will be forgotten. Checklists are likely to be delayed or ignored. An emergency medical service (EMS) pilot submitted this incident to ASRS.

> *I failed to remove the rudder lock during preflight. I recognized the situation after liftoff from runway 32. I requested to return and accomplished normal landing without incident. Cause: Medevac flight that rushed to get off the ground. Preflight inspection was incomplete. Corrective action: Our Chief Pilot discussed with me the importance of thoroughness when performing inspections and checklist items, regardless of the pressure to hurry (ASRS 1994a).*

The NASA study made several specific recommendations to pilots to avoid hurry-up error:

1. Maintain awareness of the potential for the hurry-up syndrome in preflight and taxi-out phases of flight and be particularly cautious if distractions are encountered in these phases.
2. When pressured to hurry up, particularly during preflight, take time to evaluate tasks and their priority.
3. If a procedure or checklist is interrupted, return to the beginning of the task and start again.
4. Strict adherence to checklist discipline is a key element of preflight and pretakeoff (taxi-out) phases.
5. Defer paperwork and other nonessential tasks to low-workload operational phases.
6. Positive CRM techniques will eliminate many errors.

It ain't over 'til the chocks are in

"General Aviation Landing Incident and Accidents: A Review of ASRS and AOPA Research Findings" (Morrison et al. 1993) is another NASA study, the findings of which have significant implications for all pilots in terms of error chains, workload management, and the hurry-up syndrome in the landing phase of flight.

The study found that 63 percent of reported landing incidents resulted in aircraft damage, with prop strikes the most frequent, followed by damage to gear doors and fuselage. Thirty-three percent of ASRS landing-incident citations involved some loss of aircraft control, with gear-up landings being the second most frequent, coming in with 31 percent of citations.

The study classified results of landing incidents both in terms of errors of commission (pilot actions) and errors of omission (pilot inactions). In order of frequency, errors of commission were improper control usage, destabilized and unstabilized approaches, misselection of runways or taxiways, and delayed initiation of go-arounds. Errors of omission included failure to extend the landing gear, failure to monitor instruments prior to landing, failure to execute prelanding checklist, failure to maintain attention outside the cockpit during landing, and failure to perform adequate preflight planning and preparation. The authors make the point that "the association between gear-up landings and the first three errors of omission is obvious."

In their analysis of contributing factors cited by reporters, the authors found that ASRS reporters' assessment of their own underlying human and environmental incident causes listed distraction in 45 percent of their reports. Distraction by other aircraft or ATC communications accounted for 57 percent of distraction citations. Next in line is improper operating technique, i.e., failure to use a checklist, misselection of gear or flap switches, and failure to check gear indications on short final. In terms of error chains, the study found that a single ASRS report "often involved more than one kind of error." In terms of pilot inexperience, the study found that "time in aircraft type appeared to be a much more important variable in landing incident occurrences than total flight hours. . . . Specifically, it is the pilot with less than 200 hours in type who appears most at risk" (Morrison et al. 1993).

The study paints a clear picture of the classic landing phase accident or incident profile: 91 percent of the pilots involved were acting as the flying pilot, and of those, 80 percent were flying solo when the incident occurred. In 29 percent of the incidents, minor equipment problems were distractions. A third of those equipment problems were landing-gear system problems. And regardless of time of day, 91 percent of these incidents happened in VMC conditions. If you add inexperience to these profiles, the risk factor is multiplied.

The study makes the following conclusions and recommendations: Aircraft time in type and loss of directional control are major risk factors. The majority of landing incidents involve aircraft damage, with gear system problems and gear-up landings going hand-in-hand. And finally, the study recommends that general aviation pilots need CRM. "Distractions and related attention failures are the key factors in many landing incidents and accidents. CRM is needed in general aviation, especially in single pilot operations. General aviation pilots should receive formal instruction in how to cope with multiple attention demands, prioritize tasks, maintain cockpit discipline (use checklists and adhere to standard operating procedures)."

From the files of ASRS comes a report that is a real-life example of the landing incident profile drawn by this study. While this pilot was apparently flying solo, "Murphy" was definitely an unseen passenger:

> *It was near the end of a very long day. I was completing a three-hour flight on an aircraft with a total of 6 hours since major overhaul and 3 hours since annual inspection. The last 1.5 hours was at night and over rough terrain. There was no moon out, and therefore, [it was] very dark. Although the weather was VMC, I had to keep a close eye on the gauges due to lack of horizon. Needless to say, the last leg of this trip was very stressful.*

> *When I finally saw my destination airport, I noticed that my descent from cruise was going to leave me too high and fast for a straight-in approach. Therefore, I slowed the aircraft down to flap extension speed and lowered the gear and flaps nearly simultaneously. Knowing I needed to lose altitude quickly, I immediately side-slipped the aircraft until short final. Once there, I initiated my flare for landing. The next thing I heard was the ticking of the prop and the scratching of the airplane fuselage on concrete. My initial thought was that I did not put my gear down. However, I remembered doing so because I needed the drag. I checked the gear selector. It was in the down position. Then I remembered that I had never verified that the gear had actually come down. How could this have happened? I realized that three systems must have failed for this incident to have occurred.*

> *First, the actual gear system must fail. This mechanical system is not foolproof. Indeed, on this night, the electric motor*

[that] drives the hydraulic pump did fail. Therefore the gear was only partially extended. Second, the pilot must fail. It was a long day, I was tired, stressed, and hungry and I was trying to salvage a poorly planned approach. And finally, the gear warning system must fail. This is another mechanical system which is prone to failure. This final system failed along with the previous system, on the same approach. As a pilot, and not a mechanic, I can only improve on the second system. I've determined the most important element [that] could have avoided the human error was to have flown a complete landing pattern and stick to my habit pattern. To fly a pattern appropriately and successfully, I would have lost altitude before descending into the landing pattern. Before this incident, I flew the pattern only if other traffic was in the area. However, I now realize that the pattern serves a purpose other than keeping aircraft sequenced—it helps distribute and organize tasks required for landing. Each element has its proper place (ASRS 1994b).

This pilot's experience highlights the importance of habit patterns and situational awareness, as well as the hazards associated with stress and distraction. Each of these areas are covered in detail in good CRM programs.

The flight instructor as team builder

The role of the general aviation flight instructor is crucial to the development of CRM concepts and techniques in each and every student he or she trains. From a student's first flight lesson through the airline ATP check ride, the flight instructor has many opportunities to reinforce good technical skills, teamwork, and good human factors skills. For the student pilot, the technical skills are spelled out, but in terms of CRM skills, there is nothing directed for the flight instructor to teach beyond his or her own personal style. The airmanship model and existing CRM programs gives the general aviation instructor a solid foundation to use as a guide until greater standardization occurs.

The manual *FAA Flight Instructor for Airplane Single-Engine Practical Test Standards Guide* (1991) has only two references to CRM or human factors. The cockpit management task requires the applicant to exhibit knowledge in proper arranging and securing of essential materials and equipment in the cockpit, orderly maintenance of records on flight progress, use of safety belts, shoulder harnesses, seats, and briefing occupants on emergency procedures and use of safety belts. The human factors task requires the flight instructor applicant to exhibit "instructional knowledge of the elements related to human factors by describing control of human behavior, development of student potential, relationship of human needs to behavior and learning, relationship of defense mechanisms to student learning, relationship of defense mechanisms to pilot decision making and general rules which a flight instructor should follow during student training to ensure good human relations" (USDOT/FAA 1991). For the civilian student pilot, rated pilot, or instructor, these are the only mentions in the FAA training manual for flight instructors of CRM or human factors. Furthermore, neither task, cockpit management, or human factors, specifically addresses CRM issues, such as workload management.

CRM has become commonplace in commercial and military aviation. Much has been learned, a lot of which could be used by the GA community. TCAS might have prevented the Aero Mexico 498 collision with the Piper Archer over Cerritos, but, unfortunately, it was not developed in time. But while high-tech warning systems such as TCAS are generally exclusive to airliners, the human warning system is still common to all flyers, and abundant knowledge, procedures, and techniques are available to share. But the motivation to improve must come from the instructors and students in the general aviation community.

A final perspective on teamwork

Recent management fads have centered on the ultimate power and reliability of the team, and although aviation teamwork is essential, it does not remove final authority or responsibility for safe operations from the pilot in command. Nor should we wish to do so. A cockpit is not a democracy, but neither should it be, as Webster defines it "a place of especially bloody, violent, or long continuous conflict" (Webster's 1990). Somewhere between these two extremes lies optimal performance, and the goal of this chapter was to provide several key areas for self-improvement.

Good teamwork begins with effective leadership, usually demonstrated early in team formation. Good leaders establish norms of safety, compliance, communication, and cooperation. A good leader does not pretend to be a know-all, do-all, omnipotent leader but rather establishes an environment conducive to good team operations. This environment fosters assertiveness, honesty, and trust. But effective teams also need good followers.

Followership might be a forgotten art, but it is not a lost one. The keys to effective followership are individual competence and expertise, good listening and communicative skills, critical thinking abilities, and assertiveness. Research findings also indicate that teams communicate differently and that these differences affect performance. Several suggestions for improving team communication are made based on these significant findings.

Finally, although formal teamwork training takes the form of CRM in the military and commercial sectors, these skills permeate throughout airmanship. Aviators need not wait for their annual training or upgrade to learn and practice CRM skills. General aviation enthusiasts have also seen how they can employ CRM and airmanship principles to their flight operations. In fact, CRM has been shown to be effective—and perhaps essential—for all types of aircraft. Team operations are complex, but if we arm ourselves with knowledge and teamwork skills, the group environment can be optimized to outperform any individual effort.

References

Aviation Safety Reporting System (ASRS). 1993a. Heads up, somebody! *NASA CALLBACK.* Publication 169. Moffett Field, Calif.: NASA-Ames Research Center.

————. 1993b. A well-planned response. *NASA CALLBACK.* Publication 173. Moffett Field, Calif.: NASA-Ames Research Center.

————. 1994a. If misfortune knocks . . . don't answer. *NASA CALLBACK.* Publication 178. Moffett Field, Calif.: NASA-Ames Research Center.

————. 1994b. The pattern serves a purpose. *NASA CALLBACK.* Publication 187. Moffett Field, Calif.: NASA-Ames Research Center. December.

Foushee, H. C., and B. Kanki. 1989. Communications as a group process mediator of aircrew performance. *Aviation, Space and Environmental Medicine.* 60: 56–60.

Ginnett, R.C. 1987. First encounters of the close kind: The formation process of airline flight crews. Unpublished doctoral dissertation, Yale University, New Haven, Conn.

————. 1995. Groups and leadership. *Cockpit Resource Management.* Eds.: E. L. Weiner, B. G. Kanki, and R. L. Helmreich. San Diego: Academic Press.

Kelly, R. E. 1988. A two-dimensional model of follower behavior. *Leadership: Enhancing the Lessons of Experience.* Eds.: Richard Hughes, Robert Ginnett, and Gordon Curphy. Homewood, Ill.: Irwin Press. p. 229.

McElhatton, J., and C. Drew. 1993. Time pressure as a causal factor in aviation safety incidents: The hurry-up syndrome. Proceedings of the Seventh International Symposium on Aviation Psychology, Ohio State University, Columbus, Ohio.

Morrison, R., K. Etem, and B. Hicks. 1993. General aviation landing incidents and accidents: A review of ASRS and AOPA research findings. Proceedings of the Seventh International Symposium on Aviation Psychology, pp. 975–980.

National Transportation Safety Board. 1985. Accident/incident summary report: Cockeysville, Md.: April 28, 1984. NTSB-AAR-85-01-SUM.

Schwartz, Doug. 1990. Reducing the human error contribution to mishaps through identification of sequential error chains. *Safetyliner.* pp. 13–19.

U.S. Department of Transportation/FAA. 1991. Flight Instructor for Airplane, Single Engine. Practical Test Standards. Washington, D.C: U.S. Department of Transportation and Federal Aviation Administration.

Wagner, B., and Alan Diehl. 1994. Your wingman is your copilot. *Flying Safety Magazine.* Albuquerque: USAF Safety Agency. *Webster's Ninth New Collegiate Dictionary.* 1990. Springfield, Mass.: Merriam-Webster, Inc.

7

Know your environment

Nature's laws affirm instead of prohibit. If you violate her laws you are your own prosecuting attorney, judge, jury, and hangman.
Luther Burbank

A personal view from Robert A. Alkov, Ph.D.

Good airmanship should encompass the art of listening. Aviators, whatever their experience level, should be willing to listen to and heed the advice of others. Critical to flight safety is maintaining situational awareness. A pilot must continually gather information to create and maintain a mental model of the flight environment. To do this, he or she must get feedback from various sources, including the aircraft's instruments, other aircrew members, other members of the formation, ATC, etc. This data must be perceived and organized into meaningful information on which to base decisions. A flight instructor of mine once said, "If you're not making decisions every second in flight, you're in trouble." Of course the old adage "Garbage in, garbage out" holds for our personal data processor as well as for computers. A good flight manager must consider all inputs, allow for the environment, resolve conflicting information, and make reasoned decisions.

The ego is a stumbling block for many aviators. Pilots are selected, in part, for their self-confidence. A good pilot must be in control of the situation or at least give the appearance of control to reassure passengers and crew. However, self-confidence can turn into self-delusion about one's own abilities.

Therefore, pilots exceed their own performance capabilities, those of their aircraft, or both. This is often due to overfamiliarity with and feeling comfortable in an aircraft and flight environment, leading to complacency. A good aviator knows his or her own physical and mental limitations and never tries to exceed them. This implies a certain degree of introspection and self-awareness. Being comfortable with oneself means never trying to "show off." The cockpit is not an arena for self-aggrandizement. A true professional maintains a conservative attitude toward flying. In assessing risk, mission urgency, timing, and safety must all be considered, but safety must be paramount in peacetime. "It's better to arrive late than not at all."

The decision-making process for many senior aviators often involves a rush to judgment based on past experience. This rush can lead to a mistaken assumption based on partial information, resulting in the formation of a false hypothesis. Thus the knowledge that a particular engine has had a history of problems can result in shutting a good engine down when sudden loss of power is encountered. Once an expert has made a bad decision, it is often difficult for him or her to reverse that decision because of ego. The person rejects any information that would contradict a bad decision and accepts data supporting the judgment, no matter how statistically improbable. Good airmanship means a willingness to review decisions and change them when needed. It implies suppressing the ego and keeping an open mind. It also means accepting input from any useful source.

A retired Naval Reserve aviator, Robert Alkov is a human factors instructor at the Southern California Safety Institute.

At first glance, knowledge of the flight environment appears pretty simple. Big sky, little airplane—thrust, lift, drag . . . what's the big deal? However, the scope expands considerably as we look at the various ways in which our operational environment can affect safety, effectiveness, and efficiency—in short, our airmanship. Consider the following events, each of which identify a different area of the airmanship environment.

- A Southern Airways DC-9 crashed in New Hope, Georgia, after losing both engines following an inadvertent thunderstorm penetration, killing 70. Analysis of the crash revealed that a major contributing factor to the crash was the captain's overreliance upon his airborne radar, which had identified limitations for such operations (NTSB 1978).

- A Cessna TU-206G and a General Dynamics F-111D collided in midair about 11 miles northeast of Cannon AFB in Clovis, New Mexico, killing all aboard both aircraft. Perhaps because the weather was clear and the visibility was 30+ miles, neither aircraft requested the radar advisory service that was available from nearby air traffic control facilities (NTSB 1982).

- A 24-year-old military F-15 pilot with only 513 hours in the aircraft accomplished a high-speed flyover of the airfield immediately after takeoff. At the departure end of the field, he executed an 8.5-G pull-up, severely overstressing the aircraft in its external tank configuration and literally ripping the wings off the Eagle. The investigation revealed that the pilot's organization condoned the flyovers and routinely violated a number of regulations that prohibited such high-risk maneuvers (Hughes 1995).

The first accident represents the most obvious part of the airman's world—the physical environment. The physical flying environment includes everything from air density and aerodynamics to various types of weather phenomena. The second mishap represents the multilayered regulatory environment, which includes guidance for operating in our international, national, and local airspace systems. The final environment that flyers must be knowledgeable in is the organizational environment, in which policy guidance and organizational norms shape the way in which we fly. The third illustration above highlights the role of the organizational culture and its potential impact on a pilot's decision-making process.

Each area of the aviator's environments interacts in a major way with other areas of airmanship. For example, a thorough knowledge of the environment is important to understanding cues that your aircraft may be providing, and environmental knowledge feeds directly into situational awareness.

For purposes of initial discussion and analysis, it is best to separate the environmental factors to gain some insight on their individual importance. However, these factors do not occur in isolation in the real world, where all aspects of the environment—physical, regulatory, and organizational—interact with each other on a regular and continual basis. One example of how all three of these factors play upon each other can be seen in the example of an organizational culture that pressures pilots to bust minimums and regulations with regards to hazardous weather. This is exactly what was occurring during the 1970s with a small commuter airline operating on the coast of Maine. As you read the following case study, look closely for the interrelated effects of the physical, regulatory, and organizational environments.

Case study of interrelated environmental effects: Downeast Airlines

Downeast Airlines was a small commuter air carrier operating out of Rockland, Maine, a location notorious for its bad weather and sea fog. At first glance, it seemed as if it would be extremely difficult for a small company to make a go of it at this location, flying in an area where the weather was very often below takeoff and landing minimums for an FAR Part 135 carrier. (Air taxi and commuter airlines were operated under less restrictive Part 135 of the FARs, as opposed to the more stringent Part 121 for major carriers. Recent changes have been made to upgrade the safety restrictions on the smaller carriers.) But Downeast had beaten the odds, in no small part due to the single-minded approach of the owner, who routinely pressured pilots to break minimums, fly with overloaded airplanes, and accept mechanical defects (Nance 1986).[1] He "knew all about sea fog and approach minimums and had scant respect for any pilot of his who would cancel or divert a passenger-carrying, money-making flight because the actual cloud ceiling and visibility were slightly below the legal minimums." The owner made it clear that pilots who could not or would not push (or bust) the limits did not fit in his company. In short, the owner felt that pilots who wouldn't violate the regulations were cowards. This attitude permeated Downeast operations (Nance 1986).

There had been accidents. On August 19, 1971, a pilot had been pressured by the Downeast owner into diverting to Augusta, an

---

1. *Blind Trust* is an extremely insightful story of the multiple factors that led up to the "discovery" of Downeast's renegade operations, as well as other similar stories of life after deregulation of the airline industry. It is highly recommended for anyone wanting an inside view of the multiple organizational pressures for profit and how many were rooted out by dedicated NTSB investigators in the 1970s.

airport without a precision approach,[2] instead of Portland, because of the financial concerns with busing the passengers all the
way back up to Rockland. The captain complied, but after beginning to execute a missed approach at Augusta, he decided to try
and implement one of the scud-running techniques that Downeast
was becoming famous for. While attempting to establish himself in
visual conditions for a return to the airfield, he descended out of
the minimum safe altitude of 2000 feet and slammed into a fog-
shrouded hill at 520 feet, killing himself and two passengers.

There were two other weather-related accidents over the next few
years. In 1976, a pilot clipped a tree while flying well below minimums at Knox County Airport. The minimum altitude for the specific
portion of the approach where the accident occurred was 440 feet;
the tallest tree in the area was 90 feet. Draw your own conclusions.
According to John Nance in *Blind Trust*, this accident demonstrated
how far Downeast was prepared to go to circumvent the FARs.

> . . . *the Navajo was immediately taxied into the Downeast
> hangar, the hangar doors locked, and repairs commenced in
> secret. Even the other pilots (with the company) were not al-
> lowed to view the damage the next day. The accident was
> never legally or formally reported until it was discovered by
> the NTSB years later (Nance 1986).*

More than supervisory pressure affected the decisions of the pilots at
Downeast. Nance explains:

> *There was a more insidious force that perpetrated this attitude:
> peer pressure. When pilots who had no previous airline, com-
> muter, or military flight experience found that they could meet
> most of the schedules despite the fog, by sneaking around the
> minimums and the rules, they began to develop a sort of per-
> verse pride in their own abilities. A pilot who could fly an over-
> weight airplane, take off in less than legal weather, ignore
> mechanical problems in order to bring the airplane back to
> Rockland with revenue passengers . . . was more often than not
> proud of himself. Any pilot who came on board and couldn't
> do as well was less of a pilot—less of a man (Nance 1986).*

---

2. A precision approach, such as an instrument landing system (ILS) approach, offers
both course and glidepath information and brings the aircraft down to a lower
"decision height" in a more controlled manner than a nonprecision approach, such as
a TACAN or VOR, which offers only course guidance.

The result of this organizational approach (ignoring the environmental aspects of weather and regulations) eventually became a part of the corporate culture, instilling a twisted form of *esprit de corps* in the pilots caught up in a common misery. Conditions were tough on the coast of Maine, but they had found a way to hack it, the rules and regulations be damned. There was even an informal set of "Downeast minimums," or approach altitudes that were considerably lower than the FAA allowed, but routinely flown by these "intrepid" aviators. Multiple stories were told by passengers who recalled seeing trees brush by only inches below their aircraft and the night a loud thump was heard on the third attempt to get into Rockland. When the aircraft landed at Augusta, a large dent was found on the leading edge of the aircraft's right wing, but the pilot played it off as a bird strike from a seagull, in spite of the obvious fact that seagulls are too smart to fly in "pea-soup fog" (Nance 1986).

The straw that broke the camel's back for Downeast airlines came when their chief pilot, Jim Merryman, pushed one limit too many on a foggy night in May 1979. Flying with a weak copilot and a known mechanical defect into weather that was suspected to be below minimum, Merryman crashed the Twin Otter short of the runway, killing all but one on board the aircraft. Even then, the investigation would not have uncovered the multiple organizational problems with the company had it not been for one dedicated NTSB accident investigator, Dr. Alan Diehl, who stuck to the case for months, finally uncovering the multitude of issues that lay beneath the surface of the Downeast mishap.[3]

The sad saga of Downeast Airlines was clearly an extreme case, but it represents the interrelationships that exist between environmental factors. Thanks to improved human factors training, renegade organizations like Downeast have mostly gone the way of the dinosaur. But the drive to "get the job done" still exists today, and many experienced pilots tell stories about subtle organizational pressures that they have encountered. Hopefully, you will never be confronted by organizational or peer pressure to do anything less than the right thing, but an in-depth knowledge of all three environment factors—physical,

---

3. Dr. Alan Diehl, who would later become the USAF's leading civilian safety official, spent weeks interviewing former passengers and employees at Downeast, eventually piecing together the myriad human factors associated with the poor safety record at Downeast. These factors included falsified training reports and company pressure to routinely accept maintenance defects and to break minimums.

regulatory, and organizational—will help you *survive* if you do encounter such pressures and excel in the normal day-to-day culture of compliance.

A natural starting point for our discussion of environmental factors is the physical environment, the medium in which we ply our skills.

The physical environment

The physical environment is the medium in which we operate, and it influences almost everything we do in the aircraft. It includes some factors that remain relatively constant and others that change. This distinction is helpful in understanding the knowledge requirements relative to our environment, because while some elements can be learned once and counted on forever, others are naturally shifting and require updating on a regular basis.

Constants within our physical flying environment include the terrain, air density, and light. Obviously, we do not fly over the same type of terrain at all times, and we all realize that the characteristics of terrain change as you fly over it. But a valley that was in a certain location yesterday can be counted on to be in the same place tomorrow. Its location is predictable, but our perception of it might not be, especially if we are trying to interpret unfamiliar terrain off of a chart we have not adequately studied. Once you have knowledge of a "constant" environmental factor, you can rely on it not to change, at least not very quickly. Other constants include the characteristics of altitude or the light patterns caused by the time of day in your local area. Night flying offers a variety of challenges and rewards, and no discussion of the flight environment would be complete without a discussion of the factors involved with night operations.

Changing factors include nearly all forms of weather, from friendly winds to severe hazards such as ice and thunderstorms. Inadvertent flight from visual to instrument meteorological conditions is one of the two largest killers in general aviation today, and weather-related accidents are also high on the list in commercial and military operations. Weather affects nearly everything we do in flight, as well as on the ground.

Constant environmental factors: Atmosphere, terrain, and light

Three basic environmental factors can be approached as relative constants, meaning that once you have a firm grasp on the knowledge, it should become a permanent part of your mental kit bag. The role of the physical environment in basic aerodynamics, aspects of terrain, and the unique aspects of changing light conditions are all areas of knowledge that can be defined as relatively stable learning objectives.

Altitude and air density

Air is life—for you, the wings of your aircraft, and your engine. It would seem that telling an aviator about air would be roughly akin to telling fish about water. There is one major difference, however. Fish seldom drown themselves, whereas aviators often make fatally flawed decisions regarding atmospheric conditions, clearly demonstrating the need for a better understanding.

Hypoxic hypoxia, or lack of oxygen to the brain, kills some aviators every year. Although this condition is covered in more detail in Chapter 4, it bears another mention here under the discussion of the environment. Most aviators don't need to be told that there is less oxygen at altitude than at sea level. Most aviators are also aware of the regulations that require supplemental oxygen, either "readily available" or actually being used, above certain specified altitudes. Performance can be impaired at much lower altitudes, especially with flyers who are predisposed to hypoxia, such as smokers, those flying with the aftereffects of alcohol, or aviators who come from low altitude locations.

Symptoms of hypoxia are insidious, with the initial stages often resulting in a mild state of euphoria, masking the true nature of the problem. The NTSB and military accident files are loaded with stories about those who failed to give proper respect to this phenomenon. Some cases involve flyers who tempted fate by taking a known pressurization problem to altitude—a grievous and often fatal mistake. The following case study is unique. A hypoxic pilot who had gone beyond the "point of no return" with regards to being able to help himself was assisted to recovery by his wingman. This is especially poignant because it happened to a military pilot who had recently undergone his physiological training. This training included a demonstration in the altitude chamber, in which he was required to identify his personal hypoxia symptoms and demonstrate the recovery procedures to be used at the first indication of a problem. Had he not been flying with an astute flight leader on this day, he might not have survived to see another.

Case study: No air up there (Hughes 1995)

A flight of two A-10s departed for a weekend cross-country navigation training sortie. The pilots were planning to cruise at FL220, and the weather forecast was good. During the initial climbout, the flight was required to make several intermediate level-offs at the request of air traffic control. During one of these level-offs, the number-two pilot lowered his oxygen mask, which remained down during the ensuing climbout. Passing 13,000 feet, the flight lead noticed his number two drifting out of the prebriefed formation position. He casually advised his wingman to "close it up." There was no response. Lead again directed number two to close it up, this time more forcefully; still no response.

The flight leader suddenly realized that he was no longer leading anybody. He suspected that two had a physiological problem on his hands, one that was getting worse by the second. He quickly maneuvered to the "chase" position off his impaired wingman and began to give him stern directives in a final attempt to save his life.[4] Lead directed, "Go to 100-percent oxygen—hook up your mask." But the stricken pilot was now beginning to lose aircraft control. From his chase position, the flight lead began to issue flight instructions: "Lower your nose—roll left—level your wings!" There was a slow but positive response; the wingman was still able to comprehend some directives and subsequently managed to get the mask to his face. Within moments of resuming 100-percent oxygen, the pilot began to recover. An emergency was declared, and the flight returned to base, where medical attention was obtained and the aircraft systems thoroughly checked.

This case study illustrates the insidious and time-sensitive nature of hypoxia. It also points out several keys issues. Hypoxia can and often does occur at relatively low altitudes. In this case the symptoms were present by 13,000 msl. Secondly, the recovery steps, although simple, are usually beyond the capability of the impaired individual. This point is driven home by the fact that this individual had recent physiological

---

4. The chase position is a fluid formation position usually flown to the left or right and slightly behind another aircraft that may require assistance.

training in which he was required to demonstrate proficiency in personal hypoxia symptom recognition and recovery. Finally, the steps taken by the aircrew members following the incident were absolutely correct. Although they suspected the nature of the physiological incident, they recovered immediately to a suitable field to have both the pilot and the aircraft thoroughly checked out.

While the need for air is of obvious importance to the flyer, it is also critical to the aircraft. In fact, the loss of control can happen much more rapidly if a pilot fails to understand the critical nature of airflow to the airframe or the engine.

Airspeed is life. This is certainly one of the more repeated axioms of our hobby or profession, and justifiably so. Loss of aircraft control occurs more often from aerodynamic stall than from any other reason. No one intends to intentionally lose control of their aircraft, but typical reasons for inadvertent stall are failure to monitor airspeed and aircraft attitude, overaggressive maneuvering, or automation-induced problems. Although this book is not intended to teach basic aerodynamics, some general principles deal directly with environmental knowledge required for good airmanship.

All crewmembers must have an appreciation of inflight minimum control airspeed. Distraction is usually the culprit when we lose track of our airspeed. On crewed aircraft, this typically happens when one pilot has made a power decrease, either to adjust airspeed or to descend, and then becomes distracted or transfers aircraft control. Another common occurrence is forgetting that you have extended drag devices to decelerate or descend, which can result in a very uncomfortable feeling when you find yourself unable to level off even using full power.

A fundamental axiom of good airmanship is that if you are experiencing a controllability problem, accelerate. There are, of course, some exceptions to this rule, such as structural failure or unknown damage to the aircraft. Generally speaking, however, more air over your control surfaces means greater controllability. Getting more airflow might require a descent from altitude if the terrain permits, which is another good reason to have an accurate three-dimensional environmental picture in your mind at all times.

Controllability problems can occur for a variety of reasons, including fuel imbalances or flight control problems. A military C-21

Learjet was lost when its pilots allowed the aircraft to slow towards minimum inflight control airspeed and then made a turn into the "heavy" wing, which was caused by a fuel imbalance. Once the aircraft started to roll, the pilots did not have enough control authority to get it back.

Another part of the environmental control process is managing aircraft altitude. Altitude can equate to time or distance for the pilot and crew, but these must be analyzed carefully for several reasons. First, let's look at altitude as *time*. In many circumstances higher altitude can be a friend, such as creating conditions for better fuel efficiency to increase your endurance or taking advantage of favorable winds. It might also give you the ability to wait out a passing weather front or get technical assistance on the radio. It is very helpful to know your aircraft's best endurance altitude and airspeed to take advantage of it, and many pilots carry these charts in their "plastic brains" just in case they are needed on short notice. In addition to stretching the gas remaining in your fuel tanks, altitude can also be viewed as *distance* — which can be a good or a bad thing.

In many circumstances, especially in single-engine or ejection-seat aircraft, you will have a strong desire to determine where you will come down out of the sky if your aircraft suddenly stops running. In mountainous terrain or over water, distances from safe landing zones becomes critically important in an emergency. Therefore, two other bits of knowledge you will want to carry around with you are best range speed and altitude and best glide speed and ratio. Best range allows you to maximize your powered distance, and glide techniques take over after that.

In some conditions, more time or distance is *not* what is called for. The classic example is a cockpit fire, in which prolonged exposure to the heat or fumes might render you an unconscious passenger— a bad thing. This factor becomes increasingly important as we begin to see more and more composite materials used in aircraft manufacture that can actually fuel the fire or give off toxic fumes. We may no longer have the guarantee of an aluminum skin over a stout metal frame to protect us or to hold the aircraft together.

Many experienced pilots have the ability to constantly update their closest emergency landing site in their heads. If necessary, they are able to make an immediate emergency turn towards a suitable landing field and climb or descend as required. These decisions are automatic, preconditioned responses and are illustrative of an aviator who knows and manages the physical environment. This is good airmanship.

Terrain considerations

A decision to climb or descend must always be balanced against other environmental factors, such as terrain. Sir Isaac Newton taught us many important lessons, but perhaps none as significant to aviators as "what goes up, must come down." Ideally, we want to choose where the last half of that equation occurs. Knowledge of terrain characteristics in general, and the specifics of the terrain over which you intend to fly, are critical to safe flying, but they are two separate and distinct pieces of knowledge. The first is a generalized knowledge of how terrain affects the atmosphere, common terrain illusions, and the general cues and hazards necessary for safe flying in mountainous terrain. This study and preparation can be accomplished at any time once you are familiar with general aerodynamic theory. Once learned, this information should remain in your airmanship bag forever.

The second type of terrain knowledge—the detailed knowledge of specific terrain features in the area which you intend to fly—is mission- and area-specific. It requires daily study during the mission planning and preparation phase. Smart pilots prepare not only for their intended route of flight, but for any possible route of flight that might be required because of weather, emergency, or simply a change of plans in flight. The great aircrew lounge in the sky is inhabited by hundreds, and perhaps thousands, of former pilots who did not take the time to understand the terrain over which they flew. Unless we want to join them for a premature round at their next happy hour, we should learn from their mistakes.

Terrain is inextricably tied to atmospheric phenomena, such as fog, updrafts, downdrafts, and turbulence. While we can usually see the terrain, we are often surprised when we encounter the totally predictable updrafts, downdrafts, or turbulence. From the earliest days of flight, aviators have been perplexed by changes in the atmosphere caused by the terrain.

Case study: A mysterious force

Igor Sikorsky, one of the pioneers in aviation, had been successful in getting his S-2 aircraft to fly. He had made three successful flights in the small "pusher" aircraft, each one longer than the previous effort.

On June 30, 1910, he was going to attempt more than just a straight flight across a field, "having in mind a flight which would bring me back to the point of departure" (Aymar 1990). The planned route of flight would take him beyond the boundaries of the small field within which he had made all of his flights to date, over a small swampy area in a ravine at the end of the field. Sikorsky explained what occurred after a successful takeoff and a gentle climb to about 25 feet of altitude:

I crossed the field, flying a straight line and approaching the boundary, and started to turn towards the ravine. The smoke of burning castor oil and even drops of it were thrown by the propeller blast on my hands and face. The little plane was sailing through the air, gradually gaining altitude and giving the most delightful feeling of flying. I crossed the border of the ravine and then was some sixty to eighty feet high over the swamp. . . . My joy, however, was of short duration. Being busy with my first turn in the air and with the new sensation, I did not notice that the swamp below had started to move up toward the plane. Instinctively,[5] I pulled on the stick. The ship's descent was slowed for a moment, but it then became much worse, and the next moment the S-2 crashed on the opposite slope of the ravine. I climbed out from the debris of the plane with only scratches and bruises, but the S-2 . . . was a complete wreck.

Igor Sikorsky was more perplexed than upset; this crash for no apparent reason was quite a mystery. To an analytical mind like Sikorsky's, it was a mystery that needed to be solved. It took him a year to figure it out. His explanation hints not only at the nature of terrain effects on atmospheric conditions, but provides insights into the mind of one of the world's great airmen, who saw the need to understand *why* things happened to him in flight so that he could improve his next effort. Igor Sikorsky understood the essence of airmanship—increased knowledge and continuous improvement.

The exact reason for that disheartening crash was not clearly established and understood until a year later. The S-2, with the 25-horsepower engine and a homemade propeller, had barely enough power to stay in the air, and flew only a few miles (per

---

5. Sikorsky's choice of the word *instinctively* is interesting, because he had less than eight minutes of flying experience at this time.

hour) above its minimum speed. There was not a single in-strument on board. In fact, speed indicators for aviation were not yet in use. Therefore, I had no way of determining the loss of speed, except by feeling and experience, and of that, I did not have much because my total time in the air was then not over eight minutes. The S-2 had just enough power to main-tain horizontal flight. The turn required a little more power, but the chief trouble was created by a sort of air pocket, which existed frequently above the cool swamp in the ravine. Later on, I crossed it many times with more powerful and efficient machines, and . . . I could often feel the down pull . . . (but) for the little S-2, it was enough to cause a loss of altitude. By pulling the stick I made matters worse, stalled, as we call it now, and came down abruptly (Aymar 1990).

Although most all of us fly aircraft that have more than enough power to handle the updrafts and downdrafts caused by terrain, this phenomenon typically occurs to us on final approach, where we are configured for landing and have the power back, inten-tionally flying close to our stall speed in preparation for landing. By understanding the phenomena, we can anticipate up- and downdrafts and apply smooth pitch and power corrections to maintain a stable glide path.

An increased knowledge of terrain effects on performance did not completely cure the safety problem. Pilots continue to fly them-selves and their aircraft into conditions from which they cannot re-cover, especially in mountainous terrain. One common mistake that often results in tragedy is made by pilots who fly their aircraft into rising terrain, only to find out that there is no performance ca-pability to climb over it, and no room left to maneuver around it. Sometimes this error in judgment is caused by overestimating of aircraft performance, other times by weather, and sometimes by simply misreading the map. The following case study from the files of the NTSB dramatically illustrates the dangers involved with mountain flying.

Case studies: No way out (NTSB 1993, 1994a)

The recently certified 29-year-old commercial pilot was doing all he could to develop his aviation knowledge and skills. Although he only had 229 hours of pilot in command time (only four hours in the last 90 days and only *three* hours in type), he was actively attempting to learn more about flying. In fact, during the first week of October 1993, he and two student passengers were attending a junior college flying team orientation session, which included a conference on safety and flight evaluation. At the completion of the conference, several aircraft departed Kernville, California, for La Verne, a route that would require the aircraft to cross a mountain range that contained several narrow passes. The aircraft took several different routes on their way back to La Verne, as there were three canyon routes that headed toward their destination. On the sectional chart, the route with the lowest terrain was the only one that showed a road depicted through the canyon, when in fact there were multiple roads into many of the canyons along the mountain range. Some went all the way through the range towards La Verne, others did not.

At 2:15 Pacific Daylight Time, the mishap pilot took off in a Piper PA-32-300 and followed a road into a canyon that was just north of the preferred route. The canyon gradually narrowed and the road and canyon eventually ended with an abrupt rising terrain on three sides. With no capability to climb over the sheer cliff in front of him and no room to maneuver, the aircraft impacted the mountainside approximately 2000 feet below the crest.

This short case study illustrates two important points about environmental knowledge. First, it is imperative to clearly understand the nature of the terrain as it relates to the sectional chart. Second, the risk of flying a relatively underpowered aircraft into mountainous terrain is high and demands careful preflight navigation and performance calculations.

A similar incident occurred in 1992 and sheds additional light on what it is like in the final moments of an encounter with rising terrain and inadequate aircraft performance. The 19-year-old pilot was trying to build hours towards completing the requirements to take his commercial pilot's examination. He and a friend took off on May 5 in the early afternoon from Bermuda Dunes, California, in a Cessna 152. While sightseeing, the pilot elected to fly into a canyon surrounded by rising terrain in the vicinity of Pinyon Pines. Unknown to either of the aircraft's occupants, their fate was already sealed the moment they made the turn into the canyon. The passenger was immediately uncomfortable and after several stressed-out moments, he told the pilot, "Let's get out of here," to which the now-overwhelmed pilot replied, "I'm trying." In a last-ditch attempt to avoid the terrain, the pilot lost control of the aircraft and impacted the mountainside. The passenger survived, although he was seriously injured. A discouraging footnote to this tragedy was uncovered during the toxicological tests on the pilot. He tested positive for tetrahydrocannabinol carboxylic acid (THC), indicating a possible cause for the poor judgment.

Lost horizons: Understanding light considerations

The eye provides about 90 percent of our cues for orientation, with only about 10 percent coming from other sources, such as the vestibular (inner ear) system (Reinhart 1993). It should come as no surprise, then, that light levels in our environment are extremely important for pilots to understand.

First and foremost in this discussion is the need for us to be honest about our vision. Although many of us can pass our flight physical, we squint harder each year to read line 9 on the chart to keep that magical 20/20 rating. As the years progress, we find new and creative ways to complete this annual task. My own method is to stop reading anything three days before the eye exam, eat lots of carrots, drink two cups of strong coffee on my way into the examiner's office, and carry a rabbit's foot. But when I fly, I wear my glasses. The *last* thing I want is to miss that off-white Cessna descending towards me out of a gray cloud cover or to misread the altimeter in turbulent conditions at night. Beyond the need to overcome our vanity and put on the glasses, there are several key pieces of information about light conditions that we need in our airmanship kit bag.

Glare is both a friend and a foe to pilots. The glare off of a wing or windscreen flash is often the first visual cue that identifies conflicting traffic. For this reason, "polarized lenses" are definitely not a good idea for routine wear in flight.[6] However, during the landing phase, low level, or other see-and-avoid operations, glare can be a significant hazard, a point that was made shockingly apparent to a cropduster in September 1995.

6. Although polarized glasses can prevent traffic identification based on seeing a flash, many pilots carry a pair with them in case they absolutely need to reduce sun glare for safe operations. Like a bottle of sinus decongestant that most pilots carry with them in case of a sinus block, it is not a bad idea to have a good pair of polarized sunglasses in your bag if you are going to be landing to the west late in the afternoon.

Case study: Lost it in the sun

The 42-year-old cropduster pilot had more than 3500 hours and was finishing up a day's work late in the afternoon on a cotton field near Loop, Texas. His Rockwell International S-2 (not the same as the Sikorsky model described earlier) was somewhat high-tech for a crop duster, complete with an on-board SAT-LOC Global Positioning System (GPS). To lay the aerial spray down precisely on the rows of cotton, the pilot had to skim only a few feet above the ground, and, on one of his westbound legs, with the setting sun glaring directly in his face, the pilot was a little late to notice a sign sticking up out of the middle of the field directly in front of the aircraft. He quickly banked right to avoid the sign, but the glare hid a far more severe threat—a 115,000-volt power line. His left wing hit the power line, and the pilot lost control and crashed but amazingly survived.

Glare from the sun is not the only visual problem we encounter due to light conditions. Some combinations of light and terrain features can also result in a loss of reliable visual cues that are required to maintain spatial orientation.

Aviators must be on the alert for the potential trap set up by two separate sets of physical circumstances, each of which has claimed lives. The first is the illusion that is created by flying over smooth water, which causes a reflection much like the terrain surrounding it. If care is not taken to establish a "composite cross-check" using both instruments and outside visual cues, a pilot may soon end up wondering which side is really "up." From the files of the NTSB comes a story of two friends who set out to enjoy the California scenery but ended up falling prey to its beauty.

Case study: Mirror image (NTSB 1994b)

On October 1, 1993, the pilot and a friend set out from Klamath Falls, Oregon, in a Cessna 182 to fly down to Clear Lake Reservoir in northern California. It was a clear day with VMC conditions, and the pilot was experienced, skilled, and proficient, having flown more than 50 hours in the past 90 days. Once over the reservoir, the pilot and his passenger took turns doing wingovers and tight turns. During the postaccident interview, the pilot stated that they decided to do one more 360-degree turn over the water and then fly back to Klamath Falls. During the turn over the glassy water, a combination of the water, sun, haze, and background terrain caused an optical illusion, making it difficult to maintain reference with the horizon.

The aircraft crashed into the water about halfway through the turn. The pilot survived but was unable to rescue his friend, who went down with the sinking aircraft.

A pilot can also lose track of the horizon when flying over snow on cloudy days or in conditions of blowing dust. I have spent many hours flying high speed low level in my Rockwell B-1 over the flat-lands of Wyoming and Montana, where a combination of dirty snow and gray clouds create conditions in which the horizon can be lost or an illusion of a false horizon is caused by a line in the clouds. Worse, a line on the ground can create a dangerous condition in which there are inadequate references for safe visual flight, espe-cially at high speed. In spite of having adequate ceiling and visibil-ity to legally fly low level, I have often had to climb and fly IFR altitudes due to this obscuration.

Any of these circumstances can call for the pilot to make a judgment call based on the need to continue visual flight. There are few firm guidelines here, only the pilot's sense of his or her ability to main-tain situational awareness under the given conditions. The natural tendency under VMC conditions is to say "I can hack it" and to press on, but awareness of these hazards might be just the key you need to unlock the door out of this trap. Balance the benefits against the risks, and make a judgment based on a thorough knowledge of your environment and sound principles of airmanship.

Night flying

Night operations are unique, and not just because of reduced visual cues. Night operations provide another opportunity to see the inter-related factors of airmanship, encompassing individual preparation factors such as proficiency and fatigue, both of which usually play a larger role at night. Knowledge requirements for safe and successful night ops include, among other factors, an understanding of night vi-sion, fatigue, and circadian rhythms. Since many of these subjects are covered in other areas of this book, they are discussed only briefly here.

Most flyers know that it is useful to "dark adapt" prior to night fly-ing, but it is important to understand *why* if we are to do it effec-tively and not lose the benefits of dark adaptation once we have accomplished it. During daylight conditions, it takes the human eye about 10 seconds to adjust to changes in light intensity. At night, however, it takes as long as 45 minutes for the eye to adjust. There are several reasons for these differences, having to do with different chemical exchanges and types of vision. For our purposes, it is sufficient to understand that it takes nearly an hour to dark adapt, and all of that can be washed out by just a few moments of bright light.

At night your 20/20 visual acuity is likely to drop to near 20/200. Color vision is nearly totally lost, unless the source of the light is strong enough to cause your eyes to react as if it were daylight (Reinhart 1993). This is significant to pilots who rely on colors for many warning systems and aircraft identification features, such as position lights. Because of the physical design of the eye, it is best to use off-center vision by looking slightly away from the object you wish to see clearly. In addition, it is helpful to use a red lens on your flashlight if possible, as the red light will not wash out your night vision as rapidly as normal white light. Care should be taken here, however, as red light tends to obscure other red lights, such as warning indicators in the cockpit, and is less able to effectively illuminate cockpit instruments. It is probably best to use low-intensity white light for instrument panels, as it provides the best clarity of vision, and night flying primarily relies on instruments for orientation.

Effective night operations also require a knowledge of circadian rhythm and fatigue. Circadian rhythm is the body's natural cycle of alertness. It is driven by habit patterns and exposure to sunlight, among other factors. It is important to flyers because it means that we have predictable peaks and valleys in our performance. An analysis of normal (i.e., day worker) circadian rhythms shows that we tend to be most prone to error after midnight, with the error peak between 4 a.m. and 6 a.m., with another performance dip seen in the early afternoon. Fatigue compounds this trend.

Many experts cite fatigue as the single greatest safety hazard in professional aviation today. Yet in spite of its criticality, many flyers don't fully comprehend its nature—a dangerous oversight. Even the causes of fatigue are not fully understood. Although a lack of restful sleep is certainly a major player, other causes, such as dehydration, noise and vibration, caffeine, hypoglycemia, boredom, stress, illness, and over-the-counter medications also play significant roles in an aviator's fatigue level. Fatigue increases reaction time and has been known to cause short-term memory loss, changes in attitude, loss of initiative, depression, and expanded tolerance for error (Reinhart 1993). These areas are discussed in greater detail in Chapter 4, but they clearly establish night flying as a challenge that must be approached with a full understanding of the multiple risks factors involved.

Knowledge of the physical environment is made more challenging by the presence of changing conditions. While the effects of air density, terrain, and light patterns are relatively constant and predictable, other factors in our physical environment, most notably weather, are equally important and far less predictable.

Weather

It is not the intention of this book to conduct a course on meteorology, but no discussion of environmental factors is complete without addressing the challenge posed to aviators by changing weather conditions. According to Robert Buck, the author of *Weather Flying*, weather can only bother us in a few basic ways. It can prevent us from seeing, it bounces us around, and it can reduce our aircraft's performance to a serious degree (Buck 1978). Given the right circumstances, any one of these items can absolutely ruin our day. The bottom line on weather flying is that our decisions must be made based on a solid understanding of meteorology, updated for current conditions and flavored with a healthy dose of skepticism for the precision and capability of weather forecasters. Few challenges in aviation stretch the aircrew's capabilities as much as flying an instrument approach to weather minimums. Success requires a blending of all aspects of airmanship, including discipline, skill, proficiency, knowledge of self, team, aircraft, environment, and risk. In short, weather can challenge us to be complete airmen.

Basic meteorology

Aviators need to possess certain fundamental pieces of information related to weather formation. We need to understand these critical items because we are sometimes called on to be our own weather forecasters in flight, especially as it relates to localized phenomena such as fog, windshear, and terrain-induced up-and downdrafts. Flyers should also thoroughly understand weather hazards such as windshear, thunderstorms, turbulence, and icing. We should have a grasp of the relationship between the dewpoint and the temperature and understand that when the two come together, moisture comes out of the air. As already mentioned, we should understand how terrain affects the atmosphere.

Although Bob Dylan, the ageless folk singer, laments, "You don't need a weatherman to know which way the wind blows," aviators must rely heavily on experts for most weather information. Therefore, a key to your success is understanding what the meteorologist is telling you.

Meteorology has its own language and symbology, and it can be very difficult for the uninitiated to understand without a translator or adequate training. All aviators should thoroughly understand how to read weather charts, including all of the terminology that goes with them, such as isobars, winds aloft, frontal boundaries, inversions, etc. Many excellent resources are available for learning this information on your own, but perhaps the best method is to ask the weather briefer. These underappreciated and much-maligned professionals are usually more than happy to explain the source and meaning of the tools that they use for forecasting and briefing. For all of their charts, graphs, and direct lines to the National Weather Center, these people are not Nostradamus, and they cannot predict the future with 100-percent accuracy. It is also good to remember that while you are approaching the outer marker with deteriorating weather and less than an hour of holding fuel, your forecaster will likely be home eating dinner at groundspeed zero. In short, they can afford complete confidence in their predictions; we do not have that luxury.

Even a pilot with an advanced degree in meteorology (perhaps *especially* a pilot with an advanced degree in meteorology) leaves him or herself an "out" or an alternate course of action if the weather should suddenly change. Along these lines, it is good to remember a couple of simple equations. First, bad weather sucks fuel from your tanks. Whether you are required to hold until weather passes, divert to an alternate airfield, or simply circumnavigate midwestern summer thunderstorms, weather lowers the fuel in your tanks. The natural corollary to this equation is that fuel equals options. Less fuel equals less options. Unless you are fortunate enough to fly a military aircraft that is air-refuelable, you had better keep a diversion range summary chart and a list of alternates close at hand when weather could be a factor, which is nearly always.

Robert Buck, a former TWA captain who has flown the Atlantic more than 2000 times, distills his years of weather flying experience into a few simple steps. He states:

Weather and weather flying go like this: Check what it is and what it is going to be, and then watch closely en route to see what it's really doing. These are the steps:

1. Get the big picture (forecast).

2. Digest the forecasts.

3. Check what it's been doing.

4. Get current information in flight and watch through the windshield.

5. Mix it all together for flight management.

6. On the ground, after flight, look back and see what happened (Buck 1978).

All of this presupposes a thorough understanding of weather theory and a healthy respect for the consequences of an unintended encounter. The following case studies clearly illustrate the need for advance preparation and the hazards of overconfidence. In each case, the pilots were skilled and experienced, but the rapidly changing weather conditions created situations that were more than they could handle.

7-1 *Aviators haven't always had the luxury of satellite pictures or professional meteorologists, but they have always understood the need for understanding their physical environment. Wilbur Wright checks the wind speed and direction before flight.*

Case studies: Lack of respect (Hughes 1995, NTSB 1994)

The mishap sortie was a single-ship aircraft change out at a distant alert base. The filed route of flight was direct to the alert base with an intermediate stop at a military operating area (MOA) for air work. During the initial contact with center, the pilot requested a change to his clearance, indicating stops at two intermediate bases for practice approaches. He was informed by the center controller that only the VOR approach was available at one of his requested bases due to other equipment being NOTAMed out of service, something the pilot should have known if he had checked the NOTAMs prior to his departure. The pilot changed his request to stop at only one other base prior to continuing to his destination. Following completion of his MOA work, he proceeded to the intermediate field for his approaches, despite being advised of deteriorating weather at his final destination. The pilot coordinated for an ILS low approach to be followed by some VFR "sightseeing." As the pilot let down towards the intermediate location, he was informed that the weather at this location was also rapidly deteriorating. During the approach, the pilot seemed rushed, made a number of procedural errors, missed the approach, and was asked for his intentions by the controlling agency. The pilot, now clearly in the clouds, asked for a turn to remain VFR. The aircraft impacted a hill approximately 2220 feet msl while still on centerline for the runway after failing to execute the published missed-approach procedure.

The mishap pilot (MP) was 36 years old, an experienced F-15 pilot with more than 2880 total flying hours and 867 hours in the Eagle. He was current and qualified for the mishap mission, which was a relatively undemanding one for a seasoned fighter pilot. The pilot changed his original flight plan to practice approaches at an intermediate base but failed to check NOTAMs or weather as required. Flight planning was definitely a factor in this mishap. He changed his planned route of flight without coordinating with operations and indicated an intent to fly VFR in violation of command and Air Force directives. He exhibited an air of complacency by not reviewing a relatively complex approach he was obviously not familiar with and by electing not to continue immediately to his planned destination when he was advised of deteriorating weather there. In an attempt to fly the approach anyway, he became task saturated, was unaware of his close proximity to rising terrain, and slammed into a hillside.

Complacency, inadequate planning, discipline, lack of situational awareness, overconfidence, and a feeling of invulnerability all combined on this day in an error chain that led to this pilot's death and the destruction of his aircraft. This pilot failed to appreciate or respect what changing weather can do to an aviator's day. Although this pilot clearly had other airmanship problems, his failure to understand and respect what was happening with the weather was his final undoing.

A similar lack of respect was shown by a commercial airline captain in April 1977, resulting in the loss of 70 lives.

Just before 4 o'clock in the afternoon, Southern Airways Flight 242 took off from Muscle Shoals, Alabama, for a flight to Atlanta with an intermediate stop in Huntsville. There were reported and forecast thunderstorms throughout the region, but the DC-9 had a weather radar and the captain had picked his way around Georgia thunderstorms before. After contacting departure control and being cleared to 17,000 feet, the crew requested and received clearance to proceed direct to the Rome VOR. Moments later, according to the cockpit voice recorder (CVR), the captain commented to the first officer, who was flying the aircraft, "Well, the radar is full of it; take your pick."

The air traffic controller told Flight 242 that his radar scope was showing heavy precipitation and that the "echoes," the radar returns showing heavy weather, were about 5 nautical miles ahead of the flight. Flight 242 responded, "OK . . . we're in the rain right now . . . it doesn't look much heavier than what we're in, does it?" This question seemed to be an attempt to lead the controller towards a negative answer, but the controller replied, "It appears to be a little bit heavier than what you are in right now." For the next several minutes the crew discussed the radar picture that they were seeing in the cockpit and decided that, based on that information, the best course of action was to continue straight ahead.

After switching frequencies to contact Memphis center, the crew was advised that there was a SIGMET active for most of the southeastern United States, and that the crew should monitor VOR broadcasts within 150 miles of the SIGMET area. The weather was getting rougher and the turbulence was starting to bounce the DC-9 around a bit. Thirty-eight minutes after takeoff, the CVR recorded the captain's voice stating, "Here we go . . . hold 'em, cowboy."

For the next 10 minutes, the crew discussed options for working through the weather, aided by several commercial airliners that were giving pilot reports (PIREPs) on the weather up ahead. At one point the captain stated, "Looks heavy, nothing going through that." Six seconds later, he said "See that?" to which the first officer responded, "That's a hole, isn't it?" At this point, the crew apparently decided to attempt to penetrate the line of thunderstorms that they were rapidly approaching. The crew decided to disconnect the autopilot and slow to 285 knots—a clear indication that they knew they were planning for a rough ride. This ill-fated decision to continue sealed their fate.

Approximately 50 minutes after takeoff, Atlanta center cleared Flight 242 to descend and maintain 14,000 feet as part of the en route descent into Huntsville. About the same time, heavy rain and hail sounds were recorded on the CVR. During the next minute, Atlanta center made four transmissions to Flight 242, trying to amend their altitude clearance to 15,000 feet; none of the transmissions were acknowledged. Finally, Flight 242 came up on frequency and told Atlanta center to "standby." The CVR simultaneously recorded the first officer saying, "Got it, got it back Bill, got it back."

After the pilots had a moment to catch their breath and regain their composure, they reported to Atlanta center, "OK . . . we just got our windshield busted and . . . we'll try to get it back up to 15, we're at 14." This call was followed shortly by "Our left engine just cut out." Moments later the news got worse. After a garbled transmission from the cockpit of the stricken DC-9, which required a "say again" from Atlanta center, Flight 242 responded clearly, "Standby, we just lost both engines." The jet engines of the DC-9 had ingested so much water and ice that they had simply drowned out. For the next five minutes the crew tried everything they knew to restart the drowned engines, but to no avail. For a few hopeful moments it appeared that they might be able to glide into Dobbins Air Force Base, but they fell short by several miles, crashing into the Georgia countryside.

One of the 24 survivors of Flight 242 was a commercial pilot who was seated just in front of the left engine intake. He reported that the flight was routine until the aircraft encountered severe turbulence followed by very heavy precipitation, a lightning strike on the left wingtip, and hail. In a nutshell, the crew of Flight 242 was overconfident and lacked the appropriate respect for severe weather. They were convinced of the capability of their airborne radar and air traffic control's ground radar to steer them clear of any hazards. Based on this faulty assumption, they flew smack dab into a thunderstorm and killed 70 people.

There are several common denominators in these two tragedies. Both aircraft were flown by highly experienced pilots. Both were advised of hazardous and deteriorating weather conditions, and both proceeded on assumed risk, trusting in their capabilities to beat the weather. Both were dead wrong.

Although we did not use a general aviation example for this segment, unintentional flight into instrument conditions is one of the two major causes of fatalities in general aviation, the other being controlled flight into terrain. Weather should be studied, understood, and respected, but not feared. Although weather is often unpredictable, its patterns are logical and its hazards should be understood by all who fly. Like all other aspects of airmanship, your abilities to deal with the challenges posed by changing weather should be assessed prior to any encounter. This assessment and decision should be based on the whole airmanship equation.

Regulatory environment

Many regulations should be written in red ink, because their origins can be traced to the blood of fallen aviators. However, many aviators feel that flying has become overregulated and that our governing bodies overreact to every mishap. Others disagree, citing the need for more stringent restrictions to protect innocent passengers from undisciplined flyers.[7] Regardless of your opinion on the subject, we are bound by the regulatory environment in which we fly, and other flyers base their actions on the expectation of our compliance. Failure to know and follow regulatory guidance is a clear indication of poor discipline, the cornerstone of airmanship. Yet some aviators still take the approach that there are too many rules, and they simply can't expend the time or effort to learn them all. Their airmanship problems go much deeper than a lack of regulatory knowledge, and the foundational issue of personal discipline should be addressed before going any further into the flying business.

---

7. More regulations seldom fix the problem of undisciplined flyers; it just gives them more rules to ignore and tends to inhibit others who likely don't need the additional guidance in the first place. This tendency to place restrictions on rule-abiding flyers because of the actions of an undisciplined few who are unlikely to follow new guidance in any form is one of the principal irritants of the conscientious flyer.

It is also critical for aviators to understand the most important regulation of all, the one that explains that regulatory guidance is not a substitute for good airmanship in an emergency. Each of us who flies should have Federal Aviation Regulation Part 91, Section 91.3, Responsibility and authority of the pilot in command, etched permanently in our minds. Paragraphs a and b state:

(a) The pilot in command of an aircraft is directly responsible for, and is the final authority as to, the operation of that aircraft.

(b) In an in-flight emergency requiring immediate action, the pilot in command may deviate from any rule of this part to the extent required to meet that emergency.

These words and others that echo the same sentiments that are contained in nearly every military and commercial subset of these regulations, clearly place the burden of responsibility at the feet of the pilot in command.

Types of regulations

The regulatory environment is defined by three basic types of rules: air traffic control regulations, regulations related to aviator certification and procedures, and regulations related to aircraft and equipment. While some types of regulations should be committed to memory, others require only familiarization. Your own type of operations defines your needs and approach to learning regulatory knowledge.

The problem of nonadherance to rules and requirements is significant, as evidenced by the data contained in Fig. 7-2. An analysis of incidents *reported* to the NASA Aircraft Safety Reporting System (ASRS) demonstrates the magnitude of the problem of noncompliance. Keep in mind that these incidents only represent those that were self-reported, and include few, if any, military incidents, as the ASRS primarily services commercial and general aviation.

The data suggests that violations of ATC clearances represent the largest block of regulatory noncompliance, so let's begin with an overview of ATC regulations.

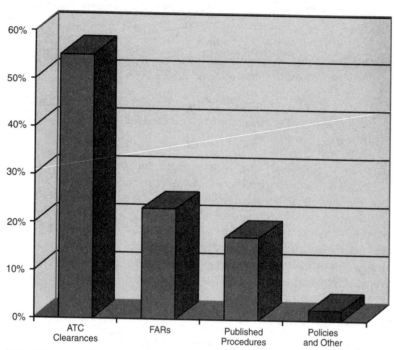

7-2 *Noncompliance is a severe problem in aviation. Although many of these self-reported incidents were unintentional, the failure to comply still indicates less-than-professional airmanship.*

Air traffic control regulations

Air traffic control regulations are designed to keep more than one aircraft from occupying the same space at the same time. Because of the great expansion in aviation, the airspace systems have to be complex enough to handle the multitude of different aircraft types and capabilities and yet simple enough for flyers to understand. A logical way to explore this environment is to work from big to small, so let's begin our discussion with the international airspace system.

The International Civil Aviation Organization (ICAO) establishes rules and procedures for international air operations, often referred to as ICAO procedures. If you are operating in the international environment, as international commercial or military aircraft often do, you had better have a clear understanding of the rules and procedures involved. Most commercial and military operations also have a familiarization period and a few "over the shoulder" flights for new crewmembers in this environment. ICAO procedures can become somewhat confusing and occasionally intimidating, but you also have to contend with individual differences between countries. Most countries have their own equivalent to the FAA, and nearly all have some idiosyncrasies that pilots should be prepared for. For example, the first time I flew into the United Kingdom, I was diverted for "purple airspace" and was asked if I would accept a "talkdown to a roller." (Purple airspace is the term used to describe protected airspace for travel by members of the royal family or other high-ranking dignitaries. A talkdown to a roller is a precision radar approach terminating with a touch and go back to approach control.) I thought for a moment that I might have slipped through a tear in the time-space continuum and was flying on another planet, but after a few more rides, I became comfortable with the differences in terminology. Even these differences in terminology are minor compared to important procedural differences, such as using European QFE altimeter settings instead of the American standard QNH. Those who operate in the international system understand fully how dangerous and embarrassing it can be to find your knowledge lacking in these critical areas while descending for your first approach in a foreign country.

The next layer of required regulatory knowledge is the national airspace system, which is regulated in the United States by the Federal Aviation Regulations, or FARs. For aviators to get around in the skies in an orderly fashion, we need some basic rules of the road. The FARs divide this up under Subpart B of Section 91: Flight Rules General. After introductory sections (91.101 to 91.144) that cover many important topics, such as basic right-of-way rules, aircraft speed, and altimeter settings, the focus shifts to visual flight rules (VFR) and instrument flight rules (IFR). Of course, there is also a set of rules that describe when to use each of these sets of rules.

Visual flight rules are covered in Part 91 of the FARs, between sections 91.145 and 91.159. They are designed to provide safe and orderly traffic flow, list equipment and flight-planning requirements, and establish altitude minimums to protect and minimize the nuisance of overflying aircraft to people on the ground. VFR operations are generally governed by see-and-avoid procedures and give the right of way to the least maneuverable aircraft. Even if our own flying activities are primarily conducted under IFR guidelines, airmen need to have a firm grasp on VFR procedures as well, as we often find ourselves operating under VFR guidelines and always share the sky with others who fly exclusively VFR.

Instrument flight rules go into greater depth than the visual rules because of the increased complexity and danger involved with instrument flight. Instrument flight rules are listed between sections 91.161 and 91.193 of the FARs and are designed to provide guidance for equipment, flight planning, and a variety of altitude, course guidance, and communications requirements.

Lower-echelon regulations

Other layers of regulations coexist with the FARs to provide guidance for special groups of flyers, such as the military. Air Force Instructions (AFI), NATOPs, and Army regulations are all examples. It is imperative for military flyers to understand that military regulations do not exempt them from operating by FAR standards. There are certain military waivers, and a great deal of the military regulations directly overlap the FARs, but Uncle Sam's aircrews must know and operate under the FARs unless there is a specific waiver or exception. If there is a weak area of knowledge in military pilots and aircrew members, the FARs may well be it.

On the flip side of this coin is the need for commercial, corporate, and general aviation pilots to understand the special needs and operating practices of the military. If you are going to be flying near a military base or operating area, get a briefing from the military safety officer on what to expect from the military's operations. They are always more than happy to provide this information, because a greater mutual understanding of operating practices increases the margin of safety for all.

Other organizations maintain an additional set of regulations for their own operating purposes. For example, the members of the Northern California Soaring Association have a set of rules to govern their flight activities. Although the general flying population are not bound by these rules, if you are going to be flying in northern California, it might be a good idea to familiarize yourself with their operating procedures.

The final level of regulatory guidance is situation-, location-, or mission-specific guidance. For all operators, these can take the form of Notices to Airmen (NOTAMs), or, in the military setting, they could be rules of engagement (ROE) or special mission instructions (SPINs). Unlike the rest of the regulatory environment, these instructions may well change on a flight-by-flight basis, so currency and in-depth understanding is critical.

Military implications are especially important here, because they go well beyond the safety of the pilot and crew. A misunderstood rule of engagement can, and has, resulted in international incidents and national embarrassment, jeopardizing the credibility of all military airpower. A sign on the wall of the mission briefing room at RAF Mildenhall, a military base in the United Kingdom, sums up the mature airman's approach to changing guidance: "If you have any unanswered questions as you leave this room, you are not prepared to fly."

Learning regulations

The first step in getting a handle on regulatory knowledge is finding out what you need to know. Whenever you check out in a new aircraft, new position, company, squadron, etc., you need to ask the senior instructor for a complete rundown on all applicable regulations, procedures, and policies. After that, double-check his or her list against a couple of other "old heads" to make sure nothing was left out. If in doubt, visit your local flight examiner office and get it straight from them.

Once you know what you need to know, the next step is to obtain *current* copies for your personal use. Your training organization should be able to procure these for you, but the FAA regulations can all be obtained through direct mail to the FAA or off of the Internet, using the keywords "FAA" or "FARs." The Internet versions come with a disclaimer as to their currency. You might not get a completely up-to-date set. Make certain that you have a way of finding out when these regulations change, and develop a system for updating your publications.

The second step is to briefly overview all of the regulations to familiarize yourself with their content and then set up a prioritized plan of study—and stick to it. Ask questions as they come up, and resolve any discrepancies that you might encounter between regulations and policy or procedure. For example, many military aircraft routinely exceed the FAA airspeed restriction below 10,000 feet. What gives? The military has many specific waivers to FAA regulations for training purposes, something that is good for all flyers who operate near a military base or training area to understand. The FAA has many handy study guides to their regulations that you can get free of charge from the local FAA office or off the Internet. Most other organizations and companies have similar aids to help new crewmembers learn the required knowledge. Take advantage of these. Finally, you need to assess your knowledge, hopefully before the flight examiner or mishap board does. Again, several practice examinations are available that can identify any weak areas that you might have.

Don't be discouraged by the multitude and layers of regulatory guidance. Start small. It is better to understand a little than to misunderstand a lot. Prioritize your study and stick to it. You will feel more confident in yourself and others will see and admire your mature and professional approach to airmanship. Also, as a part of the airspace system, you have a right and responsibility to provide input to changing regulations. Although many aviators are unaware of this opportunity, nearly all new regulations considered by the FAA have a period of time set aside for public comment, and these comments often greatly affect the final outcome of the rule or regulation. (Occasionally, Congress mandates a change or a new regulation is needed as soon as possible for safety reasons. In these cases there is often no opportunity for public comment.)

Regulatory competence and expertise is perhaps the least enjoyable and least glamorous aspect of airmanship. As such, it is the easiest discriminator of professionalism. Serious flyers take the time and make the effort to fully comprehend the regulatory environment. They wear their regulatory expertise like a badge of honor, and rightfully so. It is an absolutely indispensable part of total airmanship. Like many tough learning tasks, the best way around is to go straight through it. Once you have the basic knowledge base, it just takes a little refresher now and then to keep it tuned up. So roll up your sleeves and get started.

The physical and regulatory environments are only two-thirds of the picture. Perhaps the closest environment to the aviator on a day-to-day basis, and the one often perceived as the greatest threat, is the organizational environment.

Organizational environment

"You can't expect aviators to consistently practice something in the air that they don't see modeled by their supervisors or management on the ground," said John Lauber, a former NTSB member and vice president for safety and compliance at Delta Airlines (Lauber 1996). He went on to say that the final key to institutionalizing sound flight practices is to use the same principles of shared decision making, advocacy, and assertiveness within the organization on the ground that you want modeled in the air. This approach is certainly logical for all organizations, civilian or military, and sits at the heart of what has come to be called the "organizational environment."

Organizational environment or corporate culture, as it is often called, is a relatively recent idea that applies some of the concepts and methods from cultural anthropology and business to modern organizations. Northwest Airlines teaches its new instructors about the organizational environment with the following analogy:

A good analogy that some of you may remember from high school or college biology is the experiment of growing bacteria by exposing a shallow dish filled with a nutrient to the air. Soon after the exposure, colonies of bacteria become visible to the unaided eye. The growing medium, depending on its type, could either impede or stimulate growth. It could be nourishing or toxic or various degrees in between. Corporate cultures can do a similar thing with the people that become exposed to their influence, that is, either help or hamper them over time (Northwest 1993).

This approach seems to be reasonable for two reasons. First, we spend far more time on the ground than we do in the air. Second, ambitious flyers in the commercial or military sectors will naturally gravitate to the positions held by senior levels of the organization in an attempt to stay competitive for advancement.

So how do we analyze our own organizational environment? A good place to begin is to take a look at who are the "heroes, villains, storytellers, and wise persons" (Northwest 1993). By looking at these individuals, we can tell what the organization *really* values, as opposed to reading some public relations mission statement posted on a wall at headquarters.

It is interesting to note that formal organizational mission statements are typically displayed near the front entrance of an organizational headquarters, where visitors from the outside can read it. Perhaps more appropriate locations would be the snack bars and rest rooms or on each piece of correspondence sent to employees or subordinates, if the message is really designed for them.

Let's use the case study on Downeast Airlines from earlier in the chapter to illustrate these points. According to former NTSB accident investigator Dr. Alan Diehl, who uncovered the truth about the pressures inside Downeast, the organization outwardly assured its passengers that it was a completely safe operation and no unnecessary risks were being taken. Yet internally, the organization had a completely different look. Diehl explains: "The heroes in this company were those who could bust the limits and get away with it, to maximize the profit margins of the company. The villains were those who would not." Downeast was full of stories about pilots who brought in the passengers under extreme and illegal conditions. The wise persons were the sage old pilots who taught the newcomers the "tricks of the trade" for sea-fog flying. Although Downeast is certainly an extreme example, other organizations exhibit their norms less dramatically, but the tragic results can be just as devastating, as the following military C-130 example illustrates.

Case study: To condone is to encourage (Hughes 1995)

A C-130 aircraft impacted the terrain in a remote mountainous area on U.S. Forest Service land. The aircraft was destroyed, and 10 crewmembers and a survival specialist were fatally injured. This mishap occurred when the effects of three different factors combined at a critical time: the man, the machine, and the environment.

The events leading up to this mishap began several months prior, when a flight engineer from the mishap crew's wing submitted a written report concerning the mishap pilot pushing the aircraft too far, not taking inputs from the crew, and handling the aircraft aggressively during critical phases of flight. On a subsequent occasion, two students (both aircraft commanders) complained about this instructor pilot's seeming disregard for Air Force and Military Airlift Command, or MAC (now called the Air Mobility Command, or AMC), directives, other crewmembers' input, and weather hazards. In response to these grievances, the squadron operations officer (DO) verbally counseled the mishap pilot. The DO had been close friends of the mishap pilot for the past 14 years, but he told the pilot not to push so hard and advised the squadron commander of his actions. After the second incident, stan/eval flight examiners administered a no-notice flight evaluation, which the IP passed without difficulty. Most crewmembers felt that the IP did not alter his conduct in any way as a result of the counseling. Since the feedback from the senior leadership was limited, crewmembers might have felt that further complaining was futile.

Interviews with other members of the wing almost always contained such words as "aggressive, pushed the aircraft, pushed the crew, not receptive to inputs, extremely confident, mission hacker" and one report, "dangerous." He was considered an excellent stick-and-rudder pilot but overbearing as both an aircraft commander and as an instructor. His closest friend described him as someone you either liked or stayed away from. While this worked for social contact, as an aircrew member in this organization, you flew with whom you were scheduled to fly.

On the morning of the mishap, the navigator expressed concern to the IP and others about flying the western low-level routes due to forecasted weather. Both he and the IP had flown in the planned region before during periods of gusty winds and were aware of the potential for severe turbulence. The IP later ignored a SIGMET broadcasted over center frequency reporting, not just forecasting, severe turbulence for the entire area of the planned mission.

What the crewmembers could not have known on this fateful morning was that the aircraft had a structural flaw, a tiny stress crack in the wing that had resulted from a series of repairs and flight stresses. In the form of a microscopic crack, Murphy waited patiently for a day when an overaggressive pilot or severe turbulence would allow him to play his deadly hand. On this morning, he had both.

The mission was scheduled for a personnel drop followed by two low-level routes and 1 hour and 30 minutes in the local traffic pattern. After takeoff, wind conditions prevented the initial personnel drop and required a change of plans. At 1701Z, the aircraft entered a low-level route. At 1703Z, center broadcast SIGMET Hotel Alpha, valid until 2036Z. The SIGMET advised of occasional severe turbulence below 15,000 feet with very strong up- and downdrafts reported by numerous aircraft. At 1822Z, a local resident observed the aircraft near the base of a mountain in a steep bank estimated to be

90 degrees or greater, with the lower wing appearing to be shorter than the upper wing. Immediately thereafter, he observed a large fireball and black smoke. The aircraft impacted near centerline for the intended route of flight. At the time of the mishap, the prevailing winds were from 240 degrees at 30 knots gusting to 40 knots. Some gusts had been reported as high as 55 knots in the area. Vertical gust velocities of 1000 to 3000 feet per minute were calculated to be present at the time of the mishap.

Unlike the Downeast organizational environment, the pressures to perform in this case were internally generated and not a result of the flying unit's "personality." However, by failing to intervene in a forceful manner, the supervision of the unit had unofficially condoned the actions of the mission-hacking IP, at least in the eyes of many others in the unit. No formal documentation of reprimands existed as a result of the written complaints against the pilot. In short, the organization was permissive, and this unwillingness to discipline may have been seen as a license to continue "normal" operations by a pilot who unknowingly played into Murphy's hand.

Assessing your organization

Organizational practices should be carefully analyzed by each flyer as he or she joins a new group, whether that group is a commercial company, a military unit, or just a local general aviation flying club. Look carefully at who the heroes and villains are and seek out the storytellers and wise persons in the group. Brigadier general-turned-actor Jimmy Stewart recommended that we look at attitudes as well and drew on his experience as a bomb group commander in World War II to give this advice:

> *As a commander, the best you can do is observe attitudes. Call it what you will, but a pilot's attitude toward the mission, his commanders, his enemy, and his aircraft are an important indicator of what to expect. The attitude he displays is not only indicative of what to expect of him, it is also indicative of what to expect of the younger, less-experienced aviators in the group. Young pilots will do what they observe, not necessarily what they are told (Stewart 1995).*

Find out what the real values of your organization are. If you find that the norms of the organization are not consistent with good airmanship, you have a few options.

You can succumb to the temptation to practice poor airmanship. In reality, it might be the easiest path to take. You won't make any waves and might even advance within the organization. However, the risks and unprofessionalism involved with this choice make even the best outcome in this scenario a hollow one. If you yourself are fortunate enough not to have a mishap or foul up a mission, the lessons and attitudes that you perpetuate might well influence someone who does. If this occurs, you will have many sleepless nights asking yourself if you could have done anything to prevent the accident or if you contributed to it.

You can accept the organization as it is and attempt to practice good individual airmanship within it. This choice is often difficult because you might encounter considerable peer pressure or pressure from superiors to "fit in." It might also put you at a competitive disadvantage if you are unwilling to go as far as others in the organization.

You can leave the organization. This option also has its disadvantages. In the military, it might not be possible at all. If you are flying commercially, it means sacrificing employment, which is always a difficult decision to make unless you are independently wealthy. Even if you are involved with a local flying club, it might be the only one that fits your needs relative to distance and expense.

Finally, you can become an agent for positive change. By consistent word and deed, you can take a leadership role in moving your organization in a positive direction. Talk the talk. Keep airmanship issues on the front burner. Ask probing questions when you see poor airmanship on display. Furthermore, make it apparent that you recognize poor airmanship when you see it. It is likely that poor airmanship practices go unnoticed by those who have become used to seeing and practicing them. Walk the walk. Make it apparent that you are a complete airman, disciplined, skilled and proficient, with a thorough knowledge of yourself, team, aircraft, environment, and risk. Stay situationally aware and use good judgment. Work harder than the rest and set a new standard of excellence, one that is based on good airmanship, not shortcuts. By establishing yourself as a role model, you might advance much faster than by trying to fit in with poor airmanship practices. Be yourself and stay approachable; don't set yourself up for ridicule with a "holier than thou" attitude of superiority, even if it's true. Sometimes good airmanship speaks loudest without words.

A final word on the environment

Knowledge of the environment is by far the most difficult of the pillars of knowledge to master, mostly due to the sheer magnitude of the task and scope of information. Begin with a solid understanding of the physical environment and make flight planning a methodical exercise that completely covers all of the environmental factors you might encounter in flight.

The multilayered regulatory environment is of critical importance to the professional and physical longevity of all aviators. One sure way to find yourself at the bottom of the organizational pecking order is to get caught violating a regulation, either knowingly or out of ignorance. Develop a familiarity with all of the regulations, and then establish a disciplined plan of study to master the important ones. Ask questions about policies, special instructions, and rules (especially ROEs) that you don't understand. Like the sign at Mildenhall says, if you have any unanswered questions about the mission, you are not prepared to fly.

Find out about your organization. Seek out the true values and work to improve them if necessary. If the organization is too bad, like the example of Downeast, find another place to work or play. It's better than having them scrape you out of the bottom of a smokin' hole.

Finally, be at home in these environments. Swim like a fish. Once you have established a solid working knowledge of the physical, regulatory, and organizational environments, your confidence will increase, and you will be seen as a true professional. Whether you realize it or not, your positive example of airmanship will be emulated by others, and you will become a positive force for cultural change.

References

Aymar, Brandt, ed. 1990. The first flights of Igor Sikorsky. *Men in the Air.* New York: Crown.

Buck, Robert. 1978. *Weather Flying.* New York: Macmillan.

Hughes Training Inc. 1995. *Aircrew Coordination Training (ACT) Workbook*, draft case studies, provided on disk by Dave Wilson, Hughes Training Inc.

Hughes Training Inc. 1995. *CRM Workbook.* Reprinted by permission.

Lauber, John. 1996. Telephone interview with the vice president for safety and compliance, Delta Airlines. January 26.

Nance, John J. 1986. *Blind Trust.* New York: William and Morrow Co.

National Transportation Safety Board (NTSB). 1978. Report NTSB-AAR-78-3. January 26.

————. 1982. Report NTSB-AAR-82-10. August 24.

————. 1993. Report brief. Accident ID #LAX94FA005, NTSB File #1730.

————. 1994a. Report brief. Accident ID #LAX92FA202, NTSB File #2764.

————. 1994b. Report brief. Accident ID #LAX94LA003, NTSB File #1740.

Northwest Airlines. 1993. *Instructor Seminar Facilitator Guide.* Reprinted by permission.

Reinhart, Richard. 1993. *Fit to Fly: A Pilot's Guide to Health and Safety.* Blue Ridge Summit, Pa.: TAB/McGraw-Hill.

Stewart, J. 1995. A little bit of all of us. *Torch Magazine* (safety magazine of Air Education Training Command). May.

8

Know your risk

with Jim Quick,
director of operational risk management,
USAF Safety Center

First reckon, then risk.
Helmuth Von Moltke ("The Elder")

A personal view from Stephen Coonts

In my novels, Jake Grafton and other characters comment in numerous places on safety, what it is, and how to achieve it. Novels are written to entertain, but I found that you can't write about military aviation in a realistic manner and not touch on the role of chance, or luck, and professionalism. I just separated the two terms, and yet, I am not sure they are completely separate. One way of looking at it is what my father, a naval officer during World War II, used to tell me: "You make your own luck." I think in one sense he was right. That is the kernel of truth Lieutenant Colonel Haldane states in *The Intruders*: "This thing we call luck is merely professionalism and attention to detail, it's your awareness of everything that is going on around you, it's how well you know and understand your airplane and your own limitations. Luck is the sum total of your abilities as an aviator. If you think your luck is running low, you'd better get busy and make some more. Work harder. Pay more attention. Study your NATOPS (regulations) more. Do better preflights."

That's partly true—you'll certainly minimize your problems, but there's a limit to how much luck you can make. In *The Red Horseman*, Toad Tarkington muses, "A little dollop of carelessness could cause you to crash, burn and die. Sometimes even without the carelessness you crashed, burned and died—at a level too deep for philosophers, luck was involved."

In *The Intruders*, Jake wrestles with the whole concept of luck. People tell him he is lucky to have so narrowly escaped disaster, yet he feels he is unlucky that he got so close to the edge. Luck is like a banana peel, a slippery proposition. Are we unlucky because we had an accident, or lucky that it wasn't worse? Clearly, the perspective from which we view an event has a huge effect on its psychological import to us. This is the point that one of the characters in *The Intruders* makes to Jake, referring to investments: "There's no such thing as bad news. Whether an event is good or bad depends on where you've got your money."

Mathematicians tell us probability theory predicts everything, and it probably does on a macro scale. Yet humans don't live on that scale. For example, statisticians might tell us that it is probable that the fleet will experience one cold cat shot this year. We all breathe a sigh of relief—only one. Yet the pilot it happens to will come face-to-face with absolute catastrophe, a disaster of the first order of magnitude. One cold cat shot a year in the fleet is a statistic, but one cold cat shot happening to you is a major event in your life, perhaps the major event, a crisis you may not survive.

Even though the probability of a mishap is low, you'd think people would be reluctant to gamble with their lives. But people are addicted to it. They play the lottery, bet on sports. They go to extraordinary lengths to meet interesting specimens of the opposite sex because the hoped-for rewards justify the tremendous known risks. Success at a risk-free endeavor is impossible —everyone intuitively understands that. Risk makes life worth living; life itself is a gamble. Random chance rules our lives. What you are trying to say is this: Most people try to minimize the negative effect of random chance on them, or, said another way, they want to be the dealer. In aviation, we know how to do that: Know NATOPS, keep emergency procedures fresh and ready to use, stay situationally alert, be mentally and physically ready. If you are, you'll have the tools to make the best of whatever situation random chance throws at you. You'll be lucky.

I've never thought much of the old saw, "I'd rather be lucky than good." I think the good are lucky. Not the morally good, but the professionally good. There is just no substitute for sound, thorough preparation to avoid or cope with foreseeable misfortune. People who drive straddling the centerline can get around a few curves, but sooner or later they're going to meet a Kenworth coming the other way. That's not just predictable, it's inevitable.

Stephen Coonts is the author of several best-selling novels, including Flight of the Intruder, Under Siege, Final Flight, The Minotaur, and, most recently, War in the Air: True Accounts.

An old French proverb states "He who risks nothing, gets nothing." It is fitting, therefore, that the final pillar of knowledge in the airmanship model is knowledge of risk. It is appropriate because you must understand the other elements of airmanship, i.e., discipline, skill, proficiency, yourself, aircraft, team, and environment, to understand, assess, and manage risk, and you must manage risk to grow in airmanship. This chapter looks at risk by first asking the question, "Why risk at all?" Once the need for taking some risk in aviation has been established, we will define risk, identify sources of risk in aviation, attempt to determine what is acceptable risk, and conclude with several suggestions for organizational and individual risk management.

The reason for risk

Where there is no risk, there is usually no opportunity. In an aviation context, we constantly make decisions that affect our effectiveness, efficiency, and safety. Daily, we make decisions that involve risk in our most routine endeavors. When was the last time you remember driving to work? If you do, it was probably because something happened out of the ordinary, and the odds are that the event that brought you out of your trance placed you at higher than normal risk. Is there risk on your drive to work? Certainly. It's probably the most risky thing you do all day, but you are in control of that risk to a degree. Do you wear your seat belt? Do you stop at red lights, go on green? If you took no risk, you would most likely spend your entire life at home. Simply stated, risk is required to get things done.

Let's look into this phenomenon a little deeper. Using the driving example, do you always pay full attention to the task at hand, or do you channelize your attention on other things, like changing the radio or cassette or unwrapping a Big Mac? The odds are that you do, and those tasks deprive you of total situational awareness and elevate your personal risk beyond that which is required to accomplish "the mission." But these risks are deemed acceptable because, in our minds, the benefits outweigh the costs. If a benefit is to be gained, we take small risks in stride. We know that we have the capability to handle them, so we proceed about our business, operating on an assumed-risk basis.

But what about the risk that we can't control—that piece of metal in the road up ahead that is about to cause a blowout or the slowly disintegrating engine part that will leave you stranded on the mountain pass ahead? How do we plan and account for *unknowns*? The same types of risks happen in an airplane, but the consequences are usually far more severe.

As aviators, we intuitively understand that we must plan for the unknown as well as addressing known risks. Methods for accomplishing this delicate balance are discussed later in this chapter. Managing risk through personal awareness and process improvement makes a lot of sense. It is also relatively easy to do if we take the time to make it a part of our regular flight habit pattern.

In an era of diminishing national assets, shrinking organizational and military budgets, and increased public interest in aviation safety, aviators must ask themselves, is the cost of doing business too high, too low, or just about right? If hundreds of lost lives and crashed airplanes per year is just about right, then I suggest we're not too bad off. If the cost of doing business as usual is too high, as I suspect it is, then something has to change.

A partial answer might lie in the way we make decisions based on risk. Figure 8-1 illustrates the impact risk has on opportunity. The faster you go, or the more risk you take, the more opportunity there is for gain, but gain comes at a price—elevated risk. Conversely, as we slow down, or become more risk aversive, opportunity diminishes. The key to managing risk is to arrive at a state we'll call "risk neutral." Risk neutral is where risk and opportunity mesh, or, said another way, where the risk being accepted meets the requirements of the flight.

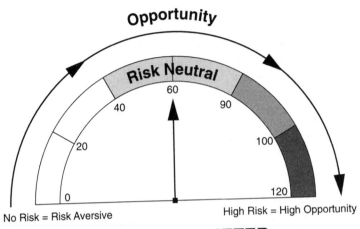

RISK SPEEDOMETER

8-1 *The risk speedometer illustrates the relationship between risk and opportunity. The goal of risk management is to find a "risk neutral" position, where the need to take risk is in balance with the potential benefits.* ASRS 1995

We should continually strive for a risk-neutral position, but it is only achieved by lending definition and proportion to risk. Aviators need to make decisions with all possible information. Without defining risk, such decisions are a gamble. With risk definition and control, we have accomplished *risk management*. But even effective risk management can be like a good night in Las Vegas; things may be looking great, your luck is running at an all-time high—and then things change. In the sky, unlike in Las Vegas, your life literally rests on the next roll of the dice.

The appropriate role of risk has been misunderstood for too long. For example, for decades in the military, commanders have said that "safety is paramount." This simply is not so. If safety were placed ahead of all other things—paramount, then we'd *never* take risks. We'd never fly airplanes, work on them, drive vehicles, or work at a military mission. Likewise, a commercial company would carry no passengers, and general aviation pilots would wash and polish their aircraft and wonder what it would be like to fly them but they wouldn't be in business very long. Safety can never become paramount. In fact, nothing in aviation is ever really "paramount" unless we are talking about the kind of show where military members are required to forfeit their lives for the mission. The mission is important, but we seldom seek to achieve it at all costs. The typical flight operation, military, commercial, or general aviation, when done properly, will be successful *and* safe. Whatever it is you are seeking to accomplish on your individual flight, even if it is just sightseeing during the autumn color season—you must determine ahead of time how much you are willing to risk to achieve those objectives.

To make this determination we must first come to grips on the true meaning of this thing we call "risk."

Risk-taking

Risk is the probability and severity of a loss linked to a hazard. Aviators manage risk by controlling, avoiding, reducing, spreading, or transferring it. Each of these techniques is discussed in detail later in this chapter. To get a better handle on risk, we also need to understand what is meant by the term "hazard." A hazard is any event that can overtly or covertly intrude between you and the job at hand. A hazard has the potential for causing injury, damage to equipment or property, or mission degradation. Because of the scope of risk, identifying hazards can be difficult.

Detection of hazards and the application of measurement to the level of risk they represent is often referred to as risk assessment. Once we have identified and assessed the risk we are willing to take, we crank our engine(s) and fly. In short, we become organized and logical risk takers. If we approach risk with a logical process, we call it risk management. If we approach it without a process, we call it gambling. Ask yourself which side of the coin the following F-111 crew was on. Were they managing risk or testing fate as reckless gamblers?

Case study: Pushing too hard

The scheduled mission was a night, three-ship, flight lead upgrade sortie for a young F-111 pilot and his weapon systems officer (WSO). One aircraft aborted prior to takeoff, but otherwise the mission went as briefed until the night range-work. The plan was to use terrain-following radar (TFR) hooked to the autopilot for the bomb runs, which were planned for 400 feet agl passes. The terrain-following system hugs the nap of the earth at a prescribed altitude and significantly reduces the work load of the crew as they deliver their weapons. The mission proceeded normally until the mishap crew had TFR system problems (the aircraft had history of TFR problems) and called off "dry" on the second pass, meaning that they did not release their weapons. On the third pass, the range control officer (RCO) observed the aircraft descending below 400 feet and transmitted, "Pull up, pull up, pull up!" but the crew did not respond and flew the Aardvark into the ground. There was no attempt to eject and both crewmembers were fatalities.

The mishap pilot was a 28-year-old captain with 1333 total hours and 385 in the F-111. The mishap WSO was a 27-year-old captain with 919 hours total with 800 hours in the F-111. The pilot's history was characterized by frequent lapses in judgment on the ground and in the air. He was universally characterized as argumentative, stubborn, and hardheaded and felt that every mission should be flown as a "combat mission." He was known to approach each training situation as if he could come under enemy attack at any moment. The mishap WSO was described as the opposite in personality. He was considered quiet and passive but still a competent crewmember.

The accident investigation revealed that the crew elected to hand-fly the last pass at 400 feet agl in direct violation of directives and exhibiting extremely poor discipline. The fact that the WSO had only computed bombing ballistics for 400 feet might have influenced their decision. This fact notwithstanding, the regulations require passes no lower than the minimum en route altitude (MEA) without the use of the TFR system. In spite of the lack of TFR, the crew apparently did not use other available aids, such as the altimeter, radar altimeter, and vertical velocity indicator (VVI) to maintain their situational awareness during the bomb run. It is most likely that the pilot was looking outside for visual cues on the range and did not perceive the shallow descent to ground impact. The WSO was probably looking at the radar scope in an attempt to get the bombs on target. They both missed important cues to the impending disaster.

The findings were that lack of discipline, poor judgment, and lost situational awareness were causal factors in this mishap. Hand-flying below MEA was a conscious decision to violate directives and indicated a breakdown in discipline by both crewmembers. It showed extremely poor judgment and a breakdown of airmanship. Why would highly trained and otherwise professional aviators pursue this course of action?

This mishap involved many different hazards, including the diverse aircrew personalities, the malfunctioning terrain-following system, the willingness on the part of both crewmembers to violate known directives, and finally, lost situational awareness that led to flying a good aircraft into the ground. Had the F-111 crew thought about the possible consequences of the multiple risk factors they faced and weighed them against the value of another bomb run on a routine training mission, they undoubtedly would have chosen a different course of action. But they didn't and paid the ultimate price for failing to do so. They most likely got caught up in the moment and didn't have time to sort it out. Based on an "I think I can" attitude, they gambled and lost. It is highly likely that in their concentrated efforts to get the bombs on target, they did not take the time to fully understand or assess the risks involved with their course of action. It is certain that they did not manage the risk well. If they had a risk-management process at all, it failed miserably when they needed it most.

Hazard identification

Perhaps part of the problem with the F-111 crew was that they did not adequately identify the multiple sources of risk that they faced ahead of time. Let's see if we can name them all: night operations, low-altitude operations, malfunctioning TFR system, overaggressive pilot, nonassertive WSO—did we miss any? Of course, it's much easier at groundspeed zero, using 20/20 hindsight, than it is at 540 knots and 400 feet off the desert floor. But that's the whole point, isn't it?

A hazard can be identified from many sources. One of the best is simply talking to the buddies you fly with. The safety types call this talk "a group process with functional experts directly from a common environment." In our case, we are most likely talking about other aviators with common experience. If this process is used in a positive way, these discussions can be productively funneled into hazard identification. "What if?" every situation as far as you can go. Ask questions until there are no more questions to ask. What should come out of this process is a picture that not only identifies common risk factors but also provides possible methods for addressing these challenges. What would a group of

F-111 crewmembers suggest as possible courses of action when hypothesizing about a malfunctioning aircraft on a routine night training mission? You can bet that hand-flying a night bomb delivery at anything less than the minimum en route altitude would be dismissed as far too risky. "Hangar flying" is uniquely and ideally suited for hazard identification.

A second source for hazard identification is the local safety representative. The following is a partial list of assets the safety rep has at his or her disposal, along with some suggestions for possible use.

- *Mishap reports* come from a variety of sources. They can be found within the organization, from the NTSB, and various military sources. Obviously, a "missionized" or specific hazard identification is the best, and often your safety representative would be happy to help you construct a complete list based on your individual aircraft and mission type.

- *Military inspector general (IG) reports.* The IG, a military inspection team, claims that "we're only here to help you," and they do, by providing important feedback and written documentation on military hazards and local procedures to combat them. Everybody in the military gets a visit, and all reports are official records and are archived, so they are easily assessible.

- *Mishap and incident databases.* The Department of Transportation and various military safety agencies retain extensive databases. All information is not releasable to the general public, but many reports and statistics are. One of the most useful and informative databases is the NASA Aviation Safety Reporting System (ASRS), which has literally thousands of incident reports available to anyone who asks. They can even be purchased on CD-ROM with an easy-to-use search engine to identify common areas of hazards for your particular type of aircraft or location. The ASRS CD-ROM can be purchased through AeroKnowledge, Inc., 2425 Pennington Road, Trenton, NJ 08638; (609) 737-9288.

- *Surveys.* Safety experts often use surveys to identify hazards, and you can too. Design your own questionnaires. Target an audience and ask some very simple questions related to such topics as "What will our next accident be? Who will have it? What will cause it?

When will it happen?" Don't ask questions like "Does the boss support safety?" The answer is always the same and does nothing for the improvement process.

- *Quality tools.* Although some people will likely reject this option out of hand, the quality process can be useful in identifying hazards. Teams develop scenarios, flow charts, logic diagrams, causal tree analysis, brainstorming, and job hazard analysis. But if you hate the quality movement, skip this one rather than scrapping hazard identification all together.

The goal of all these tools is simply to develop a hazard inventory, as specifically as possible, that highlights hazards related to particular flight scenarios. But to accomplish this objective in a thorough fashion, we must first understand where these hazards come from—the sources of risk. So far we have identified *why* we must risk, *what* risk is, and *how* to go about identifying it. The next logical step is to look into *where* the various sources of risk reside.

Sources of risk to airmanship

Risk can be found in any part of airmanship. Although dozens of risk assessment models can be used, let's first look to the airmanship model as a guide. The model works well for identifying risks, as well as attributes to successful flight operations. Lets begin with risks to flight discipline.

Risks to discipline

Discipline is the bedrock foundation of airmanship, and risks to discipline can come in many forms. There are, of course, the "hot dogs" and show-offs who violate flight discipline for the thrill or recognition. These unprofessional aviators are also a risk to the rest of us who share the sky with them, reason enough to keep a watchful eye out at all times. A second point is important to make here. Even if you yourself are not an undisciplined flyer in the sense we have just described, it is not wise to associate too closely with those who are. Your association might well be the wind that fills the show-off's sails, and by not taking a strong stance against any breeches of discipline, you may be playing an enabling role for the less-disciplined flyer.

Simple laziness and lack of attention to detail are more insidious risks to flight discipline. Consider the following ASRS report, which illustrates two cases of failed checklist discipline, which resulted in an embarrassing and potentially hazardous situation.

Case study: Self-induced emergency

During cruise flight at FL310, with the aircraft on autopilot, the captain noticed that only one inverter was powering the ac electrical bus. The other inverter was off line, inadvertently missed on the taxi-out checklist. Noticing the anomaly, the captain reached up to the overhead panel to turn on the other inverter but inadvertently shut off the good inverter, thereby shutting off all ac power to the aircraft, including the autopilot gyro INS, altimeters, air data computer, and other primary flight instruments. The inverters were both brought online quickly, with complete recovery of all systems. In the interim, however, the captain allowed the aircraft to descend approximately 350 feet before all systems were recovered, which also resulted in an overspeed condition, further complicating the recovery. The controller, who likely saw an automatic altitude alert from the excursion, asked our altitude just as we were releveling at FL310. We naturally replied "level at 310." No further comment was received from ATC.

Several factors contributed to this sequence of events. First, the pilot flying (PF) answered the challenge on a challenge-and-reply checklist item during the taxi-out checklist without actually checking the position of the ac electrical system switches. By paying lip service to a challenge-and-response item, he negated the effectiveness of the checklist and set the stage for future problems. When the failure occurred, the pilot didn't continue to fly the aircraft or maintain altitude with reference to the standby attitude indicator, which continued to function after ac power was lost. He failed to follow the first and most basic step of any airborne emergency—fly the aircraft. Corrective actions would include better adherence to checklist discipline and including in the prebriefing who flies the aircraft and who deals with the emergency procedures. In this case, the emergency was completely self-engineered (NASA 1994).

This short case study illustrates the less-talked-about side of discipline, inattention to detail. It demonstrates that even a simple error can have compounding effects. Although the crew on this aircraft did not get violated by the FAA or have a midair collision, the stage was set for either to occur. There are other threats to discipline as well.

Perhaps the single largest risk to discipline are the temptations caused by the lack of supervision or oversight in flying operations. Aviators seldom have anyone looking over their shoulder and, if they do, it may be as a partner in crime. If you add one more hazard to this flexible environment, perhaps a family member or a friend who wants to see you fly, you have a potentially explosive mix that can and often does lead to a sudden loss of judgment. The mishap databases (and cemeteries) are filled with tales of otherwise solid aviators who wanted to do something special for a friend or family member. It is worth mentioning again that the two most dangerous words in aviation are "watch this!"

As you assess the risk to discipline prior to a flight, ask yourself the following two questions. Are any circumstances present that might tempt me to deviate from the proven path of regulatory and procedural compliance? Secondly, how faithful have I been to my checklist lately? If you can answer each one of these questions honestly, you will have identified the majority of risk factors related to your personal discipline.

Risks to skill and proficiency

Every flight has the potential to overwhelm even the best of aviators. Keeping your finger on the pulse of mission requirements, as well as tracking your personal skill and proficiency level, is the only way to keep from taking unwise and unnecessary chances. There are two primary inputs to this equation. The first are the demands, or possible demands, of the mission. The second are your capabilities: skill and proficiency. If demands exceed capabilities, you are out of luck. Chapter 3 discusses the techniques for maintaining skill and proficiency in some detail, and Chapter 4 discusses self-assessment. Both are necessary to understand and evaluate a given situation or decision that involves risk. Personal risk identification requires a thorough understanding of these two areas and adds another step to identifying hazardous areas of the mission.

Personal safety windows

Every flight involves particular segments that have greater or lesser risk. For example, a typical general aviation mission might involve relatively high risk during takeoff, approach, and landing; moderate risk during airwork such as practice stalls and point turns; and relatively low risk during cruise. A military mission might involve high risk during takeoff, landing, and low-level flight, and perhaps moderate risk during air refueling. What ever your mission type, it is relatively simple to draw a picture or graph of the relative risk across a timeline of the mission. A military risk line for a typical B-1B training mission is illustrated at Fig. 8-2.

The risk line indicated in the chart should be about the same for everyone who flies this training profile. That is to say, this line should indicate generic risk and known hazards that would face anyone flying this particular profile. To truly personalize the risk assessment and locate the areas that are most hazardous for *you*, an additional step must be taken. After an honest appraisal of our capabilities in the same events as we just analyzed for relative risk, we must plot a line of our *capabilities* in these areas. For example, I may have flown several low levels recently but not air-refueled for more than two weeks, and my last approach and landing was nothing to brag about. My skill and proficiency in low level would be high, but perhaps my confidence is not as strong in the air refueling or traffic pattern areas. Figure 8-2 illustrates a "perceived capabilities" line over the same mission profile. The key to personalized risk assessment is to look for the areas where the lines come the closest together. If the capabilities line ever dips *below* the requirements line, it is time to consider getting an instructor to accompany you on the flight, or leave the mission for someone more proficient or skilled. If neither of these options are acceptable, extreme caution should be taken during that critical phase of flight. Unless the flight mission is absolutely required, and you are the only one who can fly it, it may be far wiser and mature to exercise discretion and decline the flight.

The risks associated with inadequate skill and proficiency are simply the inverse of the skill-building techniques outlined in Chapter 3. Lack of a personal training focus, inadequate resources, and conflicting demands on your attention or time are all obvious risks. But a more insidious risk source is a lack of understanding of the

Personal Risk Analysis

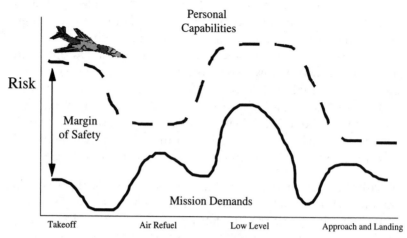

Mission Segment

8-2 *Personal risks can be assessed by simply comparing your capabilities to mission demands. The area between the lines can be seen as a margin of safety. In the case of this aviator, the match of capabilities and demands identifies air refueling and approach and landing as the areas of highest risk. This identification of risk might lead to increased vigilance and a more conservative approach during these phases of flight.*

mission demands. If you don't fully comprehend what the flight can throw at you, you are unlikely to be adequately skilled or prepared to handle whatever may come.

Risks to knowledge

There are several risks to acquiring and keeping a thorough knowledge base in aviation. The first roadblock may be our own attitude towards learning. Many aviators love to fly but hate to study—a sad but true fact. While we may be willing to put in the time and effort required to get through a training or checkout program, once that phase is completed, we are all too glad to put the books away and get around to the "real" business of flying. This is a huge mistake. Professional aviators understand that the "real" business of flying is only in small part a physical task. Airmanship growth takes place between the ears, and while flying experience is certainly important, it is only a portion of a well-balanced airmanship diet.

A second risk to knowledge is the rapidly changing technical orders, flight manuals, regulations, policies, and procedures. In an aviation organization, to stand still is to fall behind. We must constantly keep abreast of changes to maintain knowledge currency. Perhaps "keeping abreast" is not even the optimal method. Find out what changes are being considered and why. Join the debate on prospective changes in the early stages, and you will be much more comfortable with the change than if you are merely a "receiver" when the official change takes place.

Not all changes involve policy or regulations. New research in areas such as aviation human factors, cockpit/crew resource management, ergonomics, flight physiology, and others are constantly being presented and published at conferences and in various professional journals or magazines. Airmanship is developed each and every time you take a few moments to feed the knowledge base across all areas of airmanship.

Risks to situational awareness and judgment

Any crack in the airmanship structure puts SA and judgment at risk. Poor discipline, degraded skills or proficiency, or lack of knowledge all have a negative effect on the two capstones of airmanship. But beyond the elements of the airmanship model, physiology and simple distraction are risk factors that must be accounted for.

Physiological concerns constitute a large risk to situational awareness and can also impair judgment significantly. These concerns are discussed in detail in Chapter 4 but are worth another mention here. Consider the physiological impacts on SA and judgment in the following case study from a commercial pilot, found in the files of the ASRS.

Case study: Tired, hungry, and complacent (NASA 1994)

I had filed IFR to Hayward [California] due to stratus layer over the [San Francisco] Bay area. I intended to make the VOR approach to Hayward, then cancel IFR when I got below the clouds and proceed to San Carlos. After departing Reno, the flight was normal until I got into the vicinity of Livermore, when Bay Approach advised that current weather would probably not permit flight across the Bay under visual flight rules (VFR). I stated my intention to land at Hayward and wait until safe flight across the Bay could occur.

I received clearance for the Hayward VOR-A approach. When I broke out of clouds at 900 ft msl and made visual contact with a rotating beacon, I proceeded toward that beacon, made visual contact with the runway, and landed. After rollout, I contacted Approach Control to cancel my IFR clearance. Approach asked where I was (a bad sign). I stated that I was on the ground at Hayward, but they advised that I was actually on the ground at Oakland. I have listed below the specific errors that I now recognize as having led to this incident.

I firmly believe that the underlying cause of this incident was pilot fatigue, caused by an upset of my circadian rhythm due to lack of sleep, improper nutrition during the 18-hour period prior to the incident, and high stress during the 10-hour period prior to the incident. This fatigue led to errors of poor judgment about my condition to fly. I had originally planned to sleep a few hours before the flight. However, upon arrival at the FBO in Reno, I felt OK, so I reasoned that I would make the flight, then sleep when I got home.

A second factor was a poor review of approach. I had flown the approach numerous times in the past and thought I was familiar enough with it, so my review was not as thorough as it should have been. I was behind the airplane throughout the approach. Specifically, I was off course, fast, and started the timer late, which led to a failure to execute missed approach. I failed to factor in the "late start" of the timer. The timer expired just as I spotted the beacon. I should have realized that having started the timer late and being fast that I would be far beyond the airport when the timer expired, but I had an improper mindset. I was not mentally prepared to execute a missed approach, and assumed that I would break out, see the airport, and land—which is exactly what I did. (Only one problem—wrong airport.)

This incident made me keenly aware of what fatigue can do to a pilot. I now know that staying awake is not enough. Fatigue makes the mind lazy and a target for a complacent attitude. I knew I was fatigued, but (ironically) I believe my fatigue made me incorrectly think that I was okay to fly. I will take more care to prevent myself from being in situations where I might be flying in such a fatigued state. Additionally, this incident serves a good lesson of what complacency can do to flying safety. My complacency brought on a mindset that I would make the approach and land like I had done so many times before. Therefore, I will use this to recognize times that I could succumb to complacency and take new measures to prevent it.

This case study shows how even a familiar flight can become hazardous when your body and mind are not well nourished and rested. The next case study shows how the same factors can have a profound impact on even the most mundane pilot tasks, such as taxiing on a familiar airfield.

Case study: Lost in the dark (NASA 1994)

This was day three of a three-day (night) trip, all night flying, getting to bed very late. The first night I changed hotel rooms three times due to noise. Day two was not much better for rest or food. I had a new first officer (FO) who had just returned from three-year furlough (nonflying), and this added to the stress level. The last two legs of the trip were very short (25 minutes each way) Phoenix-Tucson-Phoenix. Although I was fatigued, I thought the short turn could be safely flown.

While we were taxiing out, our progress was blocked by another aircraft on an intersecting runway. The international ramp was empty, so I taxied across it and began to taxi on what I thought was the centerline of the new taxiway. There were blue raised taxi lights on left side of taxiway but none on the right at this point. The FO had misplaced his departure plate, so I was glancing at mine and taxiing at the same time. I was actually on the double yellow stripe (the right-hand edge) instead of the single center stripe. The wide and empty international ramp on my right disappeared as we approached the taxi bridge. Now blue (flush-mounted taxi) lights appeared very close to the aircraft at the 1 o'clock position about 30 feet ahead. For a second I thought they were taxiway centerline markings since they were obviously flush with the pavement.

I recognized the error prior to taxiing onto the bridge and the blue taxi lights and promptly got back on the true centerline (narrowly avoiding a taxi incident). Fatigue is subtle and can be deadly when combined with poor nutrition. The next time, I won't fly when I am this tired . . .

Both of these case studies illustrate professional commercial pilots who succumbed to the physiological hazards of fatigue and poor nutrition. Their lessons should be heeded by all who fly.

Beyond the risks to airmanship that are human-centered, which we have discussed, there are other significant sources of risk, including the aircraft, organization, and environment.

Aircraft-related risks

Four areas of potential risk are related to the aircraft, all of which must be included in a thorough risk assessment: design, maintenance, servicing, and technical data. Of course, the individual aviator's perception about all of these areas also plays a large role. Let's take a look at each individually.

The design, or "user friendliness," often referred to as the ergonomics of an aircraft, can be a critical source of risk for the aviator, especially if he or she is new to the aircraft. Nearly all aircraft have built-in design hazards of one type or another. It may be a speed brake handle that is easily caught with a cuff or similarly shaped and colocated switches that accomplish radically different functions. Given a high-workload environment, such as the one an F-16 pilot was faced with in the following case study, ergonomic hazards can lead, even with a highly qualified fighter pilot, down the road to disaster.

Case study: Wrong switch (Hughes 1995)

The mishap pilot was a 26-year-old captain with 930 total flying hours and 677 hours in the F-16. On a clear day, while flying a VFR straight-in approach and completing checklist procedures for a possible gear-up landing, the mishap pilot inadvertently moved the fuel master switch to the closed position. The main fuel shutoff valve closed, the engine flamed out, the mishap pilot was unable to restart it, and he ejected successfully.

Several human factors led this pilot toward his critical error. First, the mishap pilot was aware of the mishap aircraft's recent history of gear-indication problems. He combined this information with three distinct "klunks" he heard during gear extension and wingman observance of what appeared to be all three gears extended normally to conclude that the red light in the gear handle and lack of a green nose-gear light were false indications of unsafe gear. While reviewing applicable emergency checklists, the mishap pilot delayed completing the "landing gear up" emergency procedure checklist, because he considered the problem to be only an indicator problem. The mishap pilot began a VFR straight-in approach and was then directed to open the air refueling door (depressurizing the external fuel tanks) and activate tank inerting (reducing internal fuel tank pressures) in accordance with the "landing gear up" checklist procedure as a precaution in case the nose gear collapsed upon landing. To allow concentration on flying the VFR straight-in, the pilot attempted to active both switches without looking at them. He properly opened the air refueling door and inadvertently closed the *fuel master switch* instead of activating the tank inerting switch next to it. Oops.

The official cause of the mishap was the mishap pilot's inadvertent moving of the fuel master switch to off due to habit pattern interference. The mishap pilot's delay in completing applicable portions of the checklist was not causal but did place him in a position in which he had to accomplish the checklist while simultaneously flying the straight-in approach. The pilot's decision to activate the required switches without visual confirmation was a technique to avoid task saturation while continuing the approach. In retrospect, it was obviously a bad call. Had he been aware of the potential ergonomic hazards associated with the close proximity of the two switches, you can bet your next month's flight pay that he would have visually confirmed the operation. A second point of emphasis is that this pilot put himself unnecessarily at risk by deciding to conduct two operations simultaneously—the visual approach and the emergency checklist. An unwise work load management strategy combined with an ergonomic design risk proved disastrous on this day, and a fully functional F-16 fell out of the sky in large part due to poor risk analysis.

Maintenance factors

Even the best designed and equipped aircraft can become a death trap if it is poorly maintained, and maintenance factors must be considered when we assess aircraft risk. But this can be a difficult task for many pilots and other flight crewmembers, who are often not mechanically inclined or trained to make difficult systems assessments. Four areas *can* be looked into by the novice: documentation, time, parts, and servicing.

All aircraft are required to maintain detailed maintenance records. As operators, we must be fully conversant with what the various parts of this documentation mean to us. There will likely still remain some areas that remain mystical and shrouded in secrecy by the "brotherhood of the wrench." Ask questions until you receive satisfactory answers. At a minimum, we must be able to determine if the aircraft meets current inspection requirements, has been preflighted, and what the recent maintenance discrepancies have been.

The second area of concern for the operator should be the airframe time, usually measured in flight hours. This is not just a "total hour" concern but can indicate other high-risk events in an aircraft's life cycle. For example, we should be suspicious of an aircraft that has not flown recently. Why hasn't anyone flown it lately? Was it down for repairs? If so, what were they? Is there a problem with the aircraft that has caused others to refuse to fly it?

Likewise, care should be taken when flying a brand-new aircraft or one that has just returned from a "phase inspection," teardown, or overhaul. These are the times when the aircraft has been visited by many specialists and "knucklebusters," all who attempt to do a good job, but the risk of error is increased by the many operations that are required during these inspections. The flights that immediately follow these inspections are when maintenance-related failures are most likely to occur.

Closely related to documentation are aircraft parts and servicing. Any replaced or repaired part should be viewed as the source of a potential hazard, and most pilots in command will go as far as making these mandatory briefing items to the rest of the crew. Additionally, aircraft servicing should be carefully checked, not just to see that it was done, but to ensure that it was done correctly. Keep in mind that the servicing specialists are often the most junior and least skilled members of the maintenance team, as well as often being the most overworked and time-stressed. This combination can lead to problems of under- or overservicing or failure to close up servicing access caps, doors, and panels.

Technical data and the flight manual

The final source of aircraft-related risk comes under the heading of technical data. Specifically, the operator needs to be aware of what information is available, what the information means, and how it can be used in the operational context. Many sources of technical information are available beyond the flight manual, but it is certainly the place to start when looking for risk. The *warnings, cautions,* and *notes,* are specifically designed to highlight important risk-related topics and can form the basis of a good initial study plan. The following definitions explain how these three areas are helpful.

A *warning* is information that, if not followed, can result in personal injury or loss of life. An example of a warning from a single-engine flight manual states

> **WARNING:** *The cockpit fuel flow gauge is a direct readout instrument. Damage to the gauge can result in fuel entering the cockpit.*

This is a pretty clear risk indicator. *Cautions* are just as directive, but the risk is to equipment rather than personal injury, as the following example illustrates.

> **CAUTION:** *Do not check rudder deflection when the aircraft is stationary. The nosegear steering system cannot be deactivated and this action will result in scrubbing of the nosegear tire and possible damage to the steering mechanism.*

A *note* is an item in the technical manual that the writers feel requires additional emphasis. Notes can also point towards hazard and risk identification. Consider the possible risk associated with the following note about an aircraft hydraulic system.

> **NOTE:** *Care should be exercised when turning and stopping the aircraft simultaneously, as the brakes and nosewheel steering system share a common reservoir.*

The obvious but unstated corollary to this note is that if the common hydraulic system were overtaxed or underserviced, you may not have reliable braking available to you during a turn, which is a good thing to know in advance.

Each of these aircraft-related risks make a direct linkage between our knowledge of the aircraft and our situational awareness and judgment during operations. Organizational risk factors are equally important for a flyer to understand.

Organization-based risks

The job of the organization is to clearly communicate, with actions, words, and examples, what is expected of its members. If the relationship between the aviator and the organization is a formal employee-employer arrangement, then the organization also has other responsibilities. It should provide training, resources, and motivation to pursue and reach individual goals that are in line with the overall goals of the organization. Unfortunately, all organizations do not operate this way. In many cases, the goals of the organization and the individual conflict, and when this occurs, risk is born. Consider the plight of the following aircrew.

Case study: Organizational risk (NASA 1994)

After being held on call for nearly 12 hours, and after requesting time off for crew rest, and after the captain, first officer, and myself individually indicated it would be unsafe to operate a 6-hour flight from Philadelphia to Oakland, a company supervisor directed us to operate the flight. In the interest of safety to everyone, we refused and now our jobs are in jeopardy.

After deadheading in to Philadelphia, we were scheduled for a hotel pickup to the airport. The first officer and I had checked out of the hotel and as the captain was checking out, he received a message which said "stay put, your trip is late, please call Philly gate." We were kept on call with no release time, no reschedule time or any other guidance for when or if we would depart for nearly 12 hours. Many calls were made to Philly Ops, many attempts were made to call scheduling, plus each of us checked for messages frequently. At no

*time were we able to find out what time our flight would op-
erate, or what time to plan on if it did operate.*

*After nearly 12 hours of waiting . . . the captain again at-
tempted to call scheduling. It took several minutes; scheduling
indicated they wanted us to depart the hotel at a time that
would not give us adequate crew rest. The captain indicated
we had been up for many hours waiting to be called and we
were quite tired. He then conferred with the first officer and
myself. We expressed that we were fatigued, and to fly a six-
hour trip at this time of day would be unsafe. The captain told
scheduling we needed some sleep before making the flight and
requested about eight hours of rest. The captain was then
handed over to a company supervisor and repeated the information. After several minutes, the captain gave the phone to
the first officer, who shortly thereafter handed the phone to me.
The person on the phone identified himself as an assistant
manager or assistant chief pilot. With no discussion of what
was going on, he said "I am directing you to operate this flight
and I want to know if you will take it." At first I didn't know
what to say. My first thought was, how can I take a flight without a captain and a first officer? I told him we had been on
call all day, I had been waiting around in my uniform all day,
and in my opinion it would be unsafe to fly. He asked me
nothing about why I felt it would be unsafe. He said, "we have
reviewed your rest and we believe you are rested. I am directing you to operate the flight. Will you take it?"*

*After overcoming my disbelief that someone in a leadership
position within our organization would direct a crew to operate an unsafe flight, I told him no. He then said, "you all
are released from duty; we'll get back to you later."*

As a result of their actions, the entire crew was taken off the flying
schedule for two weeks while the company was making up its mind
on what to do next. After a hearing with management and the pilot's
union, each crewmember was given a "letter of warning" to be
placed in his personnel jacket. The future ramifications on the aircrew members' careers is indeterminable.

This case study highlights the types of risks that can crop up when organizational and individual interests conflict. Most airmen will clearly
come down on the side of the aircrew, while many management types

may feel otherwise. Only through a thorough understanding of the organizational priorities can these risk sources be identified and managed.

Environmental risks

The final source of risk are risks associated with the environment. Although this small section cannot possibly identify all environmental risks that an aviator may encounter, the following questions can act as a guide to identifying individual risk factors in your environment.

What are the potential weather hazards that I may encounter on my flight? Consideration should be given to clouds and instrument meteorological conditions (IMC), visibility, fog, winds (both aloft and at your destination and possible alternates), windshear, microbursts, thunderstorms, turbulence, icing, temperature, braking conditions, etc. Each of these potential hazards must be balanced against your aircraft's instrumentation, and equipment, such as anti-icing and navigational equipment, as well as your available fuel and personal proficiency and qualifications. Beyond the "soft" environmental hazards, you must ask yourself a few questions about the terrain.

What types of terrain might I have to fly over, either on my primary route or any potential modification thereof? Is my aircraft equipped for mountain flying? Do I have the appropriate survival equipment on board to handle a water landing or a crash landing in a remote area? What about communication and navigation equipment? Are they sufficient to cover the route I intend to fly on this flight?

If we peel away one more layer of the onion, we see that we must look at airspace considerations. What are the airspace considerations for my flight? Will I be operating under IFR or VFR guidance, or perhaps both? What is the minimum equipment needed to operate this aircraft in the planned environment? Are there any restricted areas or no-fly zones? Will I be transiting any high-density traffic areas that require special handling, such as terminal control areas (TCA), airport radar service areas (ARSA), transition areas (TA), or control zones (CZ)? Have I looked at the IFR and/or VFR Supplements or spoken to someone familiar with the areas I intend to fly into? Where is the highest potential for conflicts and midair collision? These and other questions can help you identify the sources of risk associated with the airspace, leaving a few more areas to touch upon for a relatively complete assessment, beginning with the organization.

Several risk factors can originate inside an organization. Are there any organizational pressures that might cause a conflict with good judgment? How important is this flight? Do you sense any hidden agendas associated with this flight? If you encounter any of the above, how do you plan to handle it? How far will you go before saying—and meaning—"no?"

Once we have identified the risk to our flight, we must decide how much risk we are willing to accept and how we plan to manage it.

Measuring risk

Risk management in aviation cannot be reduced to simple equations. Some elements of society have developed risk management into a science, but aviation is not yet one of them. For example, the insurance industry would cease to exist as a business if it were ineffective at risk management. The difference is that the insurance industry can spread risk over huge numbers, and individual anomalies become insignificant as they "disappear" into the statistical abyss of the whole. For aviators, however, an individual anomaly is very significant indeed, especially if it happens to you. While we should be very skeptical about claims of mathematical exactitude in aviation risk management, we can use some valuable principles to help us understand how we can begin to measure and manage risk.

The insurance industry uses a straightforward risk equation that looks like this (Jensen 1995):

$$Risk = probability\ of\ loss \times cost\ of\ loss \times length\ of\ exposure$$

or

$$R = PCE$$

This approach appears useful for aviators as well. As our knowledge of risk sources increases, we become attuned to the relative risk of various activities. While we may not have the "P" figured out to the fifth decimal point, like the insurance statisticians, we don't really need this type of pointless precision. For example, we know that if we fly into a thunderstorm, we are far more likely to get hurt than not, that the costs are very high, and that exposure time in a thunderstorm greatly increases our risk. This leads us to an operating strategy of 1) avoiding thunderstorms in every case if at all possible and 2) getting out of them as quickly as possible if we encounter them inadvertently. While this example is pretty straightforward, others can become less clear-cut and require continuous assessment of the risk/benefit trade-off. Consider the following case study using the probability × cost × exposure process.

Case study: Courting disaster

On December 26, 1989, United Express Flight 2415 (Sundance 415), a British Aerospace BA-3101 Jetstream, crashed approximately 400 feet short of runway 21R at Tri-Cities Airport at Pasco, Washington. The airplane crashed while executing an instrument landing system (ILS) approach at approximately 10:30 p.m. Pacific Standard Time—just 29 minutes after takeoff. The sequence of events leading up to the crash show a blatant disregard for a known risk—icing. There were multiple opportunities to address the problem on the ground and in flight, but the pilot accepted the risk at each decision point. Using the various aspects of the airmanship model, analyze the pilot's actions and try to determine what factors might have had an impact on his decision making and judgment.

Sundance 415 arrived in Yakima, Washington, from Seattle at 8:45 p.m. and reported no mechanical difficulties. As the crew prepared for the next leg of the flight, the company station agent at Yakima reported that she had seen the first officer of Sundance 415 and another company first officer, who was commuting to Pasco on the flight, "knocking" ice off of the wing leading edge surfaces. She also observed that ice was sliding off the airplane. The agent asked the captain if he wanted to deice, mentioning a new glycol-dispensing apparatus that was available. The captain told the agent that she should check with him later. She checked back with the captain later and was told that he did not want deicing. The agent asked about at least deicing the tail area, where the pilots could not reach. The captain declined again and walked over to company operations to update the weather forecast. Upon his return, the company agent offered once again to coordinate deicing, stating that the equipment was "fired up" and once again inquiring about the condition of the aircraft's tail. The captain declined twice more and loaded four passengers and the two first officers for his flight to Pasco. It is noteworthy that Sundance 415 was the only aircraft at Yakima (out of six) that did not get deiced the evening of December 26.

Just prior to takeoff, the tower controller called Sundance 415 with this observation, " . . . one thing I forgot to let you know is that there have been numerous reports of light to moderate mixed icing between the tops and the bases [of the cloud deck that was stable at around 1000 agl] and that's between eighteen [hundred] and four thousand feet." The first officer of Sundance 415 replied, "ah thanks, . . . we did experience a little of that coming in ourselves." In spite of the repeated warnings, at approximately one minute after 10 p.m., Sundance made a safe takeoff and climbed uneventfully to 11,000 feet. The company agent probably breathed a sigh of relief when she saw the Jetstream climb safely away from the runway. Safely airborne, the crew called in to Seattle Center, at which time Sundance 415 received clearance to Pasco.

Fourteen minutes later, the Seattle controller began to issue clearances for the en route descent: "Sundance 415, descend at pilots' discretion, maintain 6000, the Pasco altimeter is 30.27." Subsequently, the crew received several vectors to align the aircraft for the ILS runway 21R approach at Pasco, and some confusion began to develop between Seattle center and the crew. At 10:26, the controller advised the crew " . . . five miles north of DUNEZ (approach fix), turn right heading 180, maintain 3000 until established on the localizer, and you're cleared for the straight-in ILS runway 21R approach." This vector would bring Sundance 415 onto the final approach course tighter than usual, adding more stress to an already deteriorating situation. After receiving the clearance, the first officer responded, "Okay, you were, uh, partially broken up, uh, for Sundance 415, can you repeat that?" The controller repeated the clearance, and the first officer acknowledged. One minute later, the controller advised the crew " . . . radar service is terminated, frequency change approved, have a nice day." The crew responded that Sundance 415 was "switching to tower" for landing clearance.

Approximately one minute after this radio call, the first officer advised, "Seattle Center, Sundance 415 is, uh, doing a missed approach out of Pasco, we'd like, uh, vectors for another one, please." After some communication confusion, the crew apparently changed its mind about the missed approach and attempted to rejoin the final approach course and glidepath. The first officer advised center, "Okay, we just had a couple of flags on our instruments, everything appears to be all right now, we're going to continue with the approach, Sundance 415." A few moments later the crew contacted Pasco tower, stating "Sundance 415 is on short final for runway 21R now."

The local controller at the tower observed the aircraft at an altitude that he called "higher than normal" with a rate of descent that appeared greater than he was used to seeing. Shortly thereafter, he witnessed the aircraft nose over and crash short of the runway, killing all six aboard the aircraft.

The NTSB determined that there were several contributing factors to this disaster, but the primary cause was the crew's decision to continue an unstabilized approach. Two factors that led to the unstabilized approach were ice accumulations on the surfaces of the aircraft, which degraded aerodynamic performance, and short vectors to the final approach course, which put the aircraft inside the outer marker at the course intercept point. Perhaps under optimum conditions the pilot would have been able to fly the ice-covered aircraft safely into Pasco, but the combination of short vectors, distraction from the "flags" on the instruments, and the poor aerodynamic performance of the aircraft was too much. The continued exposure to a risk (ice) with a high potential for disaster was exacerbated by the pilot's decision not to have the aircraft deiced before takeoff. As with most aviation tragedies, a combination of events led to the final accident, but even a little risk management could have avoided it.

Controlling risk

Before you control risk, you have to understand that there are some alternatives to risk. It can be accepted, reduced, avoided, spread, or transferred. The trick to managing risk is in knowing how much we are accepting. Until you exercise a process that identifies and assesses risk, you don't know what you have. Consequently, you can't control what you don't know.

First of all, risk can be reduced. When risk is fully understood, its component parts can be manipulated to affect a reduction in risk. For example, in the previous case study, the flight crew could have reduced risk by deicing the Jetstream before flight, by refusing to accept short vectors to final approach, or by executing the missed approach when they began to get behind the aircraft.

Risk can also be avoided. For example, the crew of Sundance 415 could have elected not to fly into icing conditions at all, thereby avoiding the high risk altogether. If we wish, we can run a totally risk-aversive operation. The only problem with risk aversion is that the outcomes become limited. Risk aversion has its place as assets become limited, but conversely, "doing more with less" is the antithesis of this concept. We must decrease risk or manage it more effectively to "do more with less" and not pay the price of decreased safety. But some risk is always inevitable in flight operations.

Another method of controlling risk is to spread it around, thereby decreasing the risk to each individual. We commonly spread risk out by either increasing the exposed population or by lengthening the time span in which there is risk exposure. An example might be a training situation in which two student pilots require multiple emergency procedure patterns. The IP could train them both in one sortie, risking the potential for fatigue or complacency, or the students could fly on different sorties, perhaps with different instructors, thereby spreading the risk.

Risk transference is a little more difficult to apply and is commonly used in industry by underwriting risk through insurance policies. Pilots transfer risk to instruments or automation. In adverse weather conditions, a pilot entrusts navigation to cockpit instrumentation and automation. Care must be taken when transferring risk. Aviators are taught not to place their entire trust in a single instrument or autopilot function, but to back up what the system is saying or doing with other, preferably manual, means as well. Another example of risk transference deals with student fighter pilots who fail to progress during training. These pilots are often reevaluated and sent to other aircraft, thereby transferring the risk to a lower-risk flying environment, perhaps in a crewed aircraft.

Another risk-control tool is to "engineer" the risk out of the system. An example of the effectiveness of this approach can be seen in Japanese industry, where they have almost no ground mishaps due to the depth of engineering controls built into their industry. The Japanese have literally engineered hazards out of the workplace. Before a tool or machine is ever placed in production, the tool design engineers and *users* conduct a risk-management process to get rid of hazards. Management often pays through the nose for this type of process control, but in the end, the process benefits tremendously in quality production without injuries. In aviation we have seen many successful efforts toward this end, including the Ground Proximity Warning System (GPWS) and Traffic Alert Collision Avoidance System (TCAS). Many flyers have the opportunity to suggest, test, and use various types of engineering controls to help reduce risk. But don't fall into the trap of believing that automation is *the* answer—or even most of the answer—to risk management and control. No computer can ever replace the wondrous mechanism that is the human mind, with all of its knowledge, courage, fears, and intuition. Many a flyer has experienced the feeling that "something's not right" only to look around and catch a "could have been" disaster. No computer will ever duplicate that capability. But engineering solutions do have their place, as seen in this case study.

Case study: Using automation to reduce risk (Prime 1995)

In August 1994, a small piece of sophisticated computerized risk reduction paid for itself many times over. According to Air Force records, two B-52s from Barksdale Air Force Base—call sign Storm 01/02—nearly collided with a Saudi Arabian 747 filled with passengers over the Mediterranean Sea during a record-setting around-the-world flight. Through an error by air traffic controllers, both the Storm flight and Saudi 113 were assigned FL280 and were flying reciprocal headings on a collision course about 150 miles from the Egyptian coast.

While it was cruising quietly during hours of darkness, the Saudi 747's TCAS system suddenly went off. Although both flight crews could "see" the other aircraft on radar, neither had any indication that there was an altitude conflict. The Saudi pilots had the presence of mind to quickly turn on their landing lights, which allowed the B-52s to see the oncoming traffic and take evasive action just moments before the aircraft would have collided. The bombers broke hard right, and the jumbo jet climbed hard and turned right, missing the American military jets by less than 300 feet.

Brigadier General Peyton Cole, the commander of the 2nd Bomb Wing at Barksdale, was the aircraft commander of Storm 01 and had just ended a stint at the controls when he "felt the aircraft bank to the right . . . I just about vaulted into the IP seat. I heard the Saudi pilot trying to raise Cairo radio to report the incident." Currently, most military aircraft are not equipped with the TCAS system and only the quick recognition and reaction of the Saudi pilots averted a disaster of huge proportions.

This case study illustrates that while automation can certainly be helpful in reducing risk, it still requires human interpretation and decision making. The TCAS system may have alerted the Saudi pilot, but if both aircrews had not taken the appropriate actions as a result of the alert, a disaster could still have occurred.

Another successful engineering fix is the Ground Collision Avoidance System in the A-10 aircraft. There were 31 controlled-flight-into-terrain (CFIT) A-10 mishaps before money was allocated to lessen the cost of "doing business" in the low-altitude environment. Now, A-10 pilots are warned before they hit the ground, and CFIT mishaps have been greatly reduced. There are some system limitations, but this modification has definitely reduced risk, improved safety, and enhanced combat capability.

After engineering, the best way to control risk is to guard or control the environment around us. This is another way of saying limit exposure through establishing and adhering to risk-management guidelines. On factory floors you can see yellow and black lines that define areas where hard hats and safety spectacles are mandatory. Likewise, aircrews have safe separation criteria when dropping weapons, specific cloud clearance requirements when operating under visual flight rules, and landing minima during instrument approaches. Creating or changing procedures to limit or minimize exposure that is consistent with flight requirements is useful. The only problem with this approach is that often the limits are unrealistic and get bigger when they don't work the first time. These control measures are not successful when they artificially establish criteria (such as limiting minimum altitudes in low-level flight training or establishing a "bubble" around another aircraft in an air-to-air training scenario). They often fail because they attempt to limit realistic training and because they are perceived to be, and are, in fact, artificial. Many operators feel too constrained by rules, the rationale for which cannot be adequately and logically explained. As a result, these rules are often ignored.

Summary of risk management

Three rules must be understood to make sound decisions related to risk:

1. *Do not accept unnecessary risk.* We accept risk all the time without knowing it. The trick is identifying and exposing

risk, then breaking it down into component parts, and managing the parts. Risk properly managed is acceptable.

2. *Make risk decisions at the appropriate level.* Who in your organization should accept risk? The risk taker or the manager/supervisor/commander who accepts the mission? The answer might be "all of the above," depending on the situation. The real trick is assigning risk accountability at the lowest level, with appropriate levels of review. On an individual level, however, the buck stops with you.

3. *Accept risk when benefits outweigh costs.* Obviously, this is related to the previous two rules. We accept a lot of risk, but knowing its dimension is critical when one decides what the benefits are. Ergo, you must know what the benefits are to balance the risk. If you do it wrong, the mission becomes either too risk-aversive, which jeopardizes mission accomplishment, or too risky, which jeopardizes safety. Neither is desirable, and both can seriously degrade your performance.

Aviation will never be completely free of risk, so the ability of an airman to understand and analyze multiple risk factors, and then act on them using skill and good judgment, will always be one of the principle bench marks of expertise in aviation.

References

Hughes Training Inc. 1995. *Aircrew Coordination Workbook.* Abilene, Tx.: Hughes Training Inc.

Jensen, R. S. 1995. *Pilot Judgment and Crew Resource Management.* Aldershot, UK: Avebury.

NASA ASRS. 1994. Accession Number 134927, The AeroKnowledge. ASRS CD-ROM

————. 1994. Accession Number 245988, The AeroKnowledge. ASRS CD-ROM

————. 1994. Accession Number 261766, The AeroKnowledge. ASRS CD-ROM

————. 1994. Accession Number 297539, The Aeroknowledge. ASRS CD-ROM

Prime, J. A. 1995. 1994 BAFB mission's near collision revealed. *The Shreveport Times.* December 21.

9

Situational awareness

Are they shooting at us?
F-111 weapon systems officer during Desert Storm

A personal view from John W. Huston, Major General, USAF (Retired)

More than 300 years ago the great English poet John Donne put teamwork in a perspective that is as relevant and important today as it was when he wrote "No man is an island." The history of military operations is dominated by the necessity for teamwork whether we are discussing the phalanxes of ancient Greece and Alexander the Great or maintaining the avionics on an F-16. Teamwork is an essential part of good situational awareness (SA), which, in turn, is essential to success in aviation.

Those of us who flew in bombers in World War II quickly became believers in the necessity of maintaining good SA through teamwork. Headed eastbound in the early morning, our German fighter adversaries attacked out of that bright sunlight, and it took the diligent team efforts of the entire crew to identify the enemy Messerschmitt or Focke-Wulfs. Eight of the team crewmembers manned guns while the pilot and copilot kept the Flying Fortress in the tight formation that maximized the bomb group's team awareness and protection. The remainder of the mission, whether on the bomb run, returning home triumphant from success over the target, or a hazardous landing in a severely damaged plane with wounded crewmen, put our concept of teamwork to the ultimate test. While over enemy territory, our "little friends" (fighter escorts) reflected teamwork at its best; their wingman concept provided much of our welcome protection necessary for the long missions over Europe.

My most vivid memory of teamwork and precise situational awareness involved ditching in the North Sea in August of 1944. As the prospect of an unscheduled dunking seemed likely, all members of the team worked to lighten the aircraft to prevent a water landing. The radio operator, using data furnished by the navigator and coordinating with the pilots and engineer concerning our altitude, position, fuel reserve, and consumption, initiated contact with the British Air Sea Rescue center, which then began to plot our changing positions. Continued plotting of our declining altitude and their estimate of where we would land in the water resulted in a boat being dispatched towards our expected impact point. All of the crewmembers except the two pilots secured the loose items at their stations, checked our Mae Wests, and assumed the proper positions, as we had been trained, to the radio room of the B-17. After a terrific impact with the waters of the North Sea, we quickly exited, pleased to see the plane's life rafts deployed. All of us carefully made our way into these rubber saviors and ten minutes later (but what seemed much longer), we were picked up by the air rescue boat that the British had sent to our predicted landing location.

Returning to our home base later that same day, we pondered the vast network of team players that had been part of our mission and rescue. Oftentimes forgotten by the publicity accorded those of us who flew was the tremendous network of team players who played such a necessary role before the lead airplane charged down the runway. Each bomb and fighter group was supported by a vast team network, encompassing all the talents necessary to support a small city. These coordinated skills ranged from jeep drivers to armorers, maintenance personnel, cooks, meteorologists, and parachute packers, among a host of others. Not to be ignored was the Allied teamwork rescue effort that had plucked us from the waters, coordinated through the various American and British agencies, most of which, other than the rescue boats, were unknown to us.

Little has changed concerning teamwork in the 50 years since my experiences as a 19-year-old officer. Although we have made a quantum leap forward in many areas of science and technology, the need for a cooperative team effort and good situational awareness by all remains constant no matter what our job or assignment.

John W. Huston is the former chief historian of the United States Air Force.

Approximately six minutes after entering the mountainous low-level route, the B-1 bomber began a slow left turn into rising terrain, stabilizing at nearly 50 degrees of bank. Although the crew had flown this route many times before, there was no moon this night, and, in the absolute darkness, they may not have been aware of the proximity of the rising sheer terrain to the north. As the aircraft exceeded radar altimeter bank-angle limits, the terrain-following (TF) mode of the autopilot commanded a fail-safe fly-up. This fail-safe 2.4-G pull up was designed into the terrain-following system to prevent the autopilot from flying an unsuspecting aircrew into the ground should failure of a component occur or if a flight limit was exceeded. The system was functioning as designed this night and would have cleared the terrain if the crew had allowed the fly-up to continue as normal procedure dictated. For reasons unknown, one or both of the pilots took action that terminated the fly-up, rapidly rolled towards level flight, and began a shallow descent. The aircraft impacted the terrain with a force of greater than 200 Gs at nearly 550 miles per hour (USAF 1992).

What went on inside the cockpit during the last 30 seconds of this tragic example is anybody's guess. What can be said, with considerable certainty, is that a lack of situational awareness (SA) contributed to the disaster (Fig. 9-1).

Situational awareness, also known as *situation awareness* or just *SA*, is a filter through which every aviator's action must be viewed. It can be considered as a "feeder fix" for judgment and decision making. With it, a flyer can construct an accurate mental model of reality and even project potential courses of action against likely future scenarios. In this manner, one can fly efficiently, taking

9-1 *Lost situational awareness often has tragic results, as evidenced by this crash site. A B-1B aircrew flew a fully functional bomber into this ridge line at nearly 600 miles per hour on a moonless night.*
USAF 110-14 Accident Investigation photo

advantage of opportunities and avoiding pitfalls. Without it, we misperceive important cues, or worse, proceed blindly into the unknown, often with disastrous outcomes like the one just outlined.

Situational awareness is an extremely complex and involved phenomenon. There are many different academic definitions for it, and flipping through the psychological journals on SA can result in a severe headache in relatively short order for nonacademic flyers. In this chapter, we attempt to distill the complex issue of SA down to its essence and answer five essential questions:

1. What is SA?

2. How important is it?

3. How do we recognize if we are losing SA?

4. What immediate actions should be taken in the event of lost SA?

5. Finally, what can we do to improve individual situational awareness?

Situational awareness defined

In a nutshell, SA is the *accurate* perception of what is going on with you (and your crew, wingman, lead, student, instructor, etc.), the aircraft, and the surrounding world, both now and in the near future. Another way of saying the same thing is when perception matches reality, and you are able to act upon it in a timely and rational manner, you are situationally aware. Dr. Mica Endsley, a noted SA researcher, defines it as "the perception of the elements in the environment within a volume of time and space, the comprehension of their meaning, and the projection of their status in the near future" (Endsley 1989). Crew Training International, one of several companies that trains military aviators about situational awareness, teaches fighter pilots with the diagram in Fig. 9-2, which hints at the temporal aspect of SA.

9-2 *Situational awareness is a four-dimensional phenomenon, encompassing both space and time. Good SA means analyzing the past and present to help prepare for the future. This process is essential to safety, effectiveness, and efficiency in flight.*
Crew Training International

It is critical for aviators to understand the role of time in the SA equation. Most pilots intuitively understand the need to "stay ahead of the aircraft," or else things can get quickly out of hand. To stay ahead of the aircraft, an aviator has to be able to accurately perceive elements of a current situation, comprehend their meaning, and project their implications on likely future scenarios. Endsley refers to these requirements as the three levels of situational awareness. Let's look briefly at each.

Levels of situational awareness

In the first level of situational awareness, an aviator simply must be able to perceive a relevant cue, a warning light for example, and recognize it for what it is. But even at this most basic level of situational awareness, many variables can cause a breakdown in the SA process. Some of these are controllable, others are not. Errors in level-one SA are often referred to as "input errors" (Sventek 1994). Input errors include *incorrect data*, such as a bad oil pressure gauge or a faulty warning light. A second input error is *misinterpreted information*. It has been suggested that the B-1 crash that was outlined at the beginning of this chapter might have been the result of a misinterpreted cue.[1] The B-1 has a tendency to "shudder" at high angles of attack, such as was encountered during the automatic fly-up maneuver moments before impact. This shudder could have been misinterpreted as a stall indication. If so, the actions taken by the pilot to disconnect the autopilot, roll wings level and push over, make perfect sense—except for the proximity of the terrain. The failure of the crew to account for the terrain is an example of the third type of input error, *misprioritized information*. It is indicated by a flyer who fails to see the relative importance of a cue or event. One of the classic human failure accidents in history occurred when the airline captain of a DC-8 failed to heed the warnings of his flight engineer and ran out of gas while circling the destination airport in VFR conditions—a clear case of misprioritized information. This case study is analyzed in detail later in this chapter.

The enemies of level-one SA are channelized attention, distraction, and task saturation, all of which prevent important information

1. The author was an instructor pilot and check airman in the B-1B at the time of the accident in question. Many hours of speculation between experienced aviators resulted in several theories of what occurred. It is my belief that the pilot in control at the time of the accident believed that the aircraft was about to depart controlled flight and took the actions he did as a form of stall recovery, completely forgetting the proximity of the rising terrain on his left.

from being observed or acted upon in a timely manner. Suggestions for dealing with these attention problems are discussed later in this chapter.

Level-two SA is when the aviator puts the observed event into the big picture. It attaches meaning and significance to the observation. A pilot in a single-engine aircraft flying over mountainous terrain views a low oil-pressure indication much differently than a multi-engine crew flying in the local traffic pattern. The ability to attach the appropriate significance to an event is crucial to successful flying and dependent on several factors, including the airmanship pillars of self-knowledge, systems knowledge, and environmental knowledge. This level of SA requires a *gestalt*, or holistic interpretation of events (Endsley 1989). Many new flyers have difficulty developing situational awareness at this level or beyond, because they have few experiential patterns on which to place new information of events. Simulation, "hangar flying," and aviator cross talks can all help to build this capability in lieu of actual experience.

It has been said that great broken-field runners in football have the capability to see holes in the defense before they appear—merely by sensing the movements and intentions of both blockers and defenders. Endsley refers to this type of anticipatory capability in flight as level-three SA (Endsley 1989). An aviator who accurately perceives and assimilates an indication or event into the current situation must look into the future to "see" the implications of the event on future courses of action. It is often a difficult aspect for many flyers because it requires innovation and changing the existing plan—actions many flyers try to avoid. Nonetheless, SA is useless if it cannot be converted into action, and that often requires a pilot to modify courses of action by looking at the new reality in terms of what it means for the remainder of the flight.

Let's use the example of the solo airman with the low oil pressure to summarize our understanding of the three levels of situational awareness. While scanning the cockpit over a cup of coffee, the pilot notices a low oil-pressure indication on his only engine. This demonstrates level-one SA. His knowledge of his current location, over mountainous terrain, and the knowledge of the engine's oil requirements attach importance to the indication and illustrates the second level of situational awareness. Finally, level-three SA is represented by the pilot projecting the event or cue into multiple futures. The pilot must quickly decide if his best course of action is to climb to obtain gliding capability to a suitable landing airfield while he still has an engine, but he must balance this course of action against the additional strain on the engine that the climb will require. This scenario also demonstrates how a flyer's situational awareness influences judgment and decision making.

Although academic definitions and explanations lay the groundwork for a good understanding of situational awareness, the experiences and opinions of expert flyers are equally, if not more, insightful. Let's take a look at what military aviators who stake their lives on accurate and consistent SA have to say about it.

I know it when I see it: Eagle drivers list components of good SA (Wagg 1993)

What do 173 F-15 pilots have in common? Answers to this question might range from big watches to a dubious degree of credibility regarding war stories. In this case however, it is a commonly held view of what constitutes "good" situational awareness. In a study conducted by the Armstrong Laboratory in conjunction with the Air Force, operational F-15 pilots were asked for written definitions of critical components of SA. The following seven areas surfaced. By analyzing our own traits against those compiled in this study, we may better view our own SA abilities and work on strengthening weak areas. F-15 pilots considered these elements as critical to good SA:

1. *Building a composite image of the entire situation in three dimensions.* How good are you at viewing your situation in 3-D? Can you disassociate yourself enough, but not too much, to view the big picture in your mind's eye? Fighter pilots say this is critical to good SA.

2. *Assimilation of information from multiple sources.* If your idea of fun is listening to multiple radio frequencies, a student (or instructor) trying to tell you why he or she did what he or she just did, and figuring out a new bingo fuel based on the SIGMET you just received, then stop reading and move on to number three. If not, think about how much information you can realistically handle and develop strategies for prioritizing.

3. *Knowing spatial position and geometric relationships.* Although this element is directly related to the first element,

the Eagle drivers agreed it was important enough to merit individual consideration. Questions like *Where is the other aircraft? How fast am I closing? What is his relative vector?* all relate to your ability to think geometrically. How well do you score here?

4. *Periodically updating the current dynamic situation.* In a dynamic environment, to stand still is to fall behind. How often have you felt that things were happening too fast?

5. *Prioritizing information and actions.* No one can do it all, so what stays in and what drops out? What do we do first, second, never? Thorough preparation and comprehensive briefings can answer many of these questions at groundspeed zero.

6. *Making quality and timely decisions.* This element is the real utility and essence of SA. If you accomplish 1 through 5 but cannot make a decision, even the best situational awareness is useless.

7. *Projecting the current situation into the future.* Modern aircraft move fast. If you are not anticipating, predicting, and basing actions on upcoming events, you will fly in a reactive mode. When was the last time you felt you were behind the aircraft?

Equally as important as what these Eagle drivers *said* about situational awareness was what they *did* on simulated missions, as recorded by other pilots and expert observers. The study found several links between other areas of airmanship and SA, including "flight leadership and resource management (team), decision making (judgment), communication (skills), knowledge of tactics, and knowledge of weapons capabilities (aircraft)" (Wagg 1993). These findings clearly demonstrate the value of a systems approach to airmanship, in which each segment of the whole directly affects capabilities in other areas. It is safe to say that based on all available data, no one can "specialize" in situational awareness. It must be approached holistically, balancing development across all areas of airmanship.

Because situational awareness is so complex, it is often neglected by individual flyers seeking self-improvement, who see it as tied exclusively to experience. But concrete steps can be taken to improve situational awareness almost immediately by any flyer if the importance of SA is understood.

How important is SA?

Case study and mishap analysis from all areas of aviation tell us that situational awareness is essential to everyone who flies. While SA is very important to corporate, commercial, and general aviation flyers, military operations provide unique insights into the nature and importance of situational awareness. There are many reasons for this, but it is primarily due to the multiple mission demands placed on young aviators, the competitive nature of combat training and operations, and the data captured as a result. In an analysis of all U.S. Air Force mishaps from 1989 through 1995, channelized attention resulting in lost situational awareness tied with decision making as the single largest contributing factor cited in mishap reports (Magnuson 1995).

Mishap data is not the only indicator. In 1983, the Air Force conducted an Operational Utility Evaluation (OUE) of the new Advanced Medium Range Air to Air Missile (AMRAAM) at the McDonnell Douglas flight simulator in St. Louis. The simulations used line fighter pilots to achieve the best possible test parameters for the weapon evaluation, but what the test discovered about the pilots was perhaps more important than what they discovered about the missile. "While the main objective of the study was to determine how the AMRAAM performance compared to that of the AIM-7M, one of the major fallouts was a striking correlation between the flight's SA and the mission outcome" (VEDA 1988). At the test outbriefing, it was stated that the situational awareness of the pilot was more important than any other factor—"including the type of ordnance or type of jet being simulated" (VEDA 1988).

Another study of air-to-air combat kills estimated that 90% of all aircraft that were shot down never even saw their attacker. While the importance of situational awareness cannot be overemphasized in the military environment, perhaps the most dramatic and tragic examples of the importance of situational awareness come from the commercial aviation arena.

Case study: United Airlines Flight 173, DC-8 at Portland, Oregon[2]

On December 28, 1978, a McDonnell Douglass DC-8 ran out of gas while circling in clear weather within 20 miles of the Portland International Airport. The details of this case study clearly illustrate two major points about situational awareness. First, distraction is

2. This case study is paraphrased and quoted from NTSB 1979.

the enemy of situational awareness, and second, "group SA" is really limited to the lowest common denominator on the aircrew, especially if he or she is the pilot at the controls. As the case study progresses, note how the captain appeared to lose track of time—what psychologists call temporal distortion. Further note the captain's loss of level-two and -three SA. Although the fuel condition was brought to his attention on several occasions by both the first officer and the flight engineer, he was not able to attach appropriate significance to the fuel state, and further, he was clearly unable to project the future implications of his true emergency, low fuel. Finally, note how, in the final moments, both the captain and the first officer lose positional SA and accept an assigned heading away from the landing airport with only minutes of fuel remaining on board. There are several other lessons to be learned from this case study, including the need for appropriate assertiveness and the vital role of nonpilot crewmembers. As you read this case study, ask yourself when the crew should have first realized the criticality of low fuel, and see how many distractions are pulling on the aircrew's attention as they fly themselves into a disaster.

Flight 173 departed Denver for Portland at 2:47 p.m. with 189 persons onboard, including six infants and eight crewmembers. Many passengers were undoubtedly returning home from a holiday excursion to see family or friends. The planned flight time from Denver to Portland was two hours and 26 minutes. Family members awaiting their arrival in Portland would expect them at the gate at 5:13 p.m. The planned fuel for arrival at Portland was somewhere in the vicinity of 14,800 pounds. (When the aircraft left Denver, it had 46,700 pounds of fuel on board; the planned fuel consumption was 31,900 pounds. During the accident investigation, the captain stated that he was very close to his predicted fuel for the entire flight to Portland ". . . or there would have been some discussion of it.") This fuel reserve met all regulatory and company contingency fuel requirements and would provide well over an hour of holding fuel with a clean configuration. At approximately 5:06 p.m., Flight 173 called Portland approach to request descent and approach clearance. They were instructed to maintain heading and expect a visual approach to runway 28. The flightcrew responded "we have the field in sight."

As the DC-8 descended through 8000 feet, the first officer called for the flaps to be extended to 15 degrees and then asked for the landing gear to be lowered. As the captain complied with both requests, the crew and passengers noticed a distinct "thump," and the aircraft yawed slightly to the right. The captain stated "it was noticeably unusual and . . . it [the landing gear] seemed to go down more rapidly." A post-accident analysis of the landing gear by accident investigators from the NTSB revealed that the retractor strut on the aircraft's right landing gear had broken sometime during the flight. Although this caused the right gear to extend more rapidly, once it was down, the landing gear was securely locked in place. It is possible however, that either the rapidly extending gear or the earlier malfunction caused the landing gear indicator lights in the cockpit to malfunction. The captain stated "we got the nose gear green light but no other lights."[3] To confirm the status of the gear, the second officer visually checked the landing gear position indicators on the top of the wings and reported that they all indicated down and locked.

At 5:12 p.m., one minute before the scheduled landing time, Portland approach control instructed Flight 173 to contact tower for landing clearance. The flight crew responded "negative, we'll stay with you. We'll stay at five (thousand feet). We'll maintain about 170 knots. We got a gear problem. We'll let you know." This was the first indication to anyone on the ground that the aircraft was experiencing difficulties. Approach control instructed Flight 173 to "turn left heading one zero zero and I'll just orbit you out there until you get your problem (fixed)."

For the next 23 minutes, the aircrew worked through the emergency and precautionary actions. These included double-checking the gear position indicators. During this period, the first flight attendant came forward, and the captain discussed the situation with her. He told her that after they ran a "few more checks," he would let her know what he intended to do.

At about 5:38 the flight contacted the United Airlines Systems Line Maintenance Control Center in San Francisco. According to recordings, the captain explained to company dispatch that the landing gear had malfunctioned and explained what the flight crew had done to assure that the landing gear was fully extended. He reported about 7000 lbs of fuel on board and stated his intention to hold for another 15 to 20 minutes. At this point in time, the aircraft had used approximately 7800 lbs of fuel—half of the initial reserve—since

---

3. Accident investigators speculate that the rapid extension may have caused a gear indicator malfunction in the cockpit.

arriving in the local area at 5:12 p.m. It should have been clear that it would be dangerous, if not impossible, to hold for another 20 minutes. The increased fuel consumption of the landing gear down configuration had not been realized or accounted for. The personnel from United in San Francisco sought to clarify the captain's intentions. "Okay, United 173 . . . you estimate that you'll be making a landing at about five minutes past the hour. Is that OK?" The captain responded, "Yeah that's a good ballpark. I'm not going to hurry the girls. We got about a hundred sixty five people on board and we . . . want to . . . take our time and get everybody ready and we'll go. It's clear as a bell and no problem." Is it possible that the sight of the airfield passing so close by on each orbit gave the captain and crew a false sense of security? The aircraft continued to circle under the direction of Portland approach in a triangular pattern southeast of the airport at 5000 feet. The pattern kept the aircraft between 5 and 20 miles of the runway, depending on their location in the pattern.

At about 5:44 p.m., the cockpit voice recorder indicated a conversation between the captain and first flight attendant concerning passenger preparation, crash-landing procedures, and evacuation procedures. During this conversation, the captain neither designated a time limit to the flight attendant nor asked how long it would take to prepare the cabin. (This information comes from the captain's initial interview. He stated that he assumed 10 or 15 minutes would be reasonable and that some preparations could be made on the final approach to the airport.) Two minutes later, the first officer asked the flight engineer, "How much fuel we got . . .?" The flight engineer responded, "five thousand." Two minutes later, at 5:48, the first officer, now clearly concerned about the fuel state, asked the captain ". . . what's the fuel show now . . .?" The captain responded "five." At nearly the same time, the fuel feed pump lights began to blink, indicating that that the total usable fuel was indeed somewhere near 5000 pounds. (According to data received from the manufacturer, the inboard feed pumps lights on the DC-8 begin to "blink" at or near 5000 pounds of fuel remaining.) In only eight minutes, from 5:40:47 to 5:48:50, the aircraft had consumed 2000 pounds of fuel. It should have been obvious at this point that "another 15 or 20 minutes" of flight would be extremely hazardous, if not impossible.

At 5:50 the captain directed the flight engineer to "give us a current card on weight (for landing data purposes). Figure about another fifteen minutes." To this the first officer responded, "fifteen minutes?" The captain replied, "Yeah, gives us three or four thousand pounds." The flight engineer, now also gravely concerned, responded strongly to the captain's 15-minute estimate. "Not enough. Fifteen minutes is gonna really run us low on fuel here." The captain then instructed the flight engineer to contact the company representative in Portland and tell them that Flight 173 would land with about 4000 lbs of fuel. At this time the aircraft was about 17 miles from the airport, heading northeast (toward the runway).

At 5:55, the aircraft continued toward the airport, and the flight engineer reported "approach descent check is complete." One minute later the first officer asked "how much fuel you got now?" The flight engineer told him that there was 1000 pounds in each tank, for a total of 4000 pounds. At this critical juncture, the captain sent the flight engineer back to the cabin to "see how things were going." While the flight engineer was off station, the captain and first officer discussed several aspects of the emergency, including giving the flight attendants ample time to prepare. As the flight engineer came back to the flight deck, he reported that the cabin would be ready in "another two or three minutes." Simultaneously, Portland approach requested that the aircraft turn left to a heading of 195 degrees. At this time, Flight 173 was only 5 miles from the runway, configured for landing. But the first officer acknowledged and complied with the request to turn away from the field.

From 6:02 to 6:05 p.m., the crew proceeded outbound from the field while they coordinated with Portland approach and the cabin crew. At one point during the transmissions, approach control asked Flight 173 when the approach would begin. The captain responded "They've about finished in the cabin. I'd guess about another three, four, five minutes." The cockpit voice recorder revealed that the crew continued to discuss contingency plans for the approach, such as checking the landing gear warning horn as further evidence that the landing gear was fully down and locked and whether the automatic spoilers and antiskid would operate normally with the landing gear circuit breakers out.

At 6:06:40, the captain finally made his decision to begin the approach. Almost simultaneously, the first officer said, "I think you just lost number four [engine]," followed immediately by advice to the flight engineer " . . . (you) better get some cross feeds open there or something." The first officer repeated the observation two more times to the captain. "We're going to lose an engine. We're going to lose an engine." Following these alarming statements, the captain asked "Why?" The first officer responded with a single word: "fuel."

At this point Flight 173 was now 19 miles southwest of the airport and turning left. Following several moments of confusing conversation between the flight crew members as to the aircraft's fuel state, the captain called Portland approach and requested "clearance for an approach to (runway) two eight left, now." Portland immediately granted this *first* request for approach clearance.

From 6:07:27 until 6:09:16, the following were partial transcripts from inside the cockpit of Flight 173:

6:07:27

Flight engineer: "We're going to lose number three in a minute, too."

6:07:31

Flight engineer: "It's showing zero."

Captain: "You got a thousand pounds. You got to."

Flight engineer: (There was) five thousand pounds in there . . . but we lost it."

6:08:42

Captain: "You gotta keep 'em running."

Flight engineer: "Yes sir."

6:08:45

First officer: "Get this . . . on the ground."

Flight engineer: "Yeah. It's showing not very much more fuel."

6:09:16

Flight engineer: "We're down to one on the totalizer. Number two is empty."

6:10:47

The captain requested the distance from the airport. Portland approach replied, "I'd call it eighteen miles."

One and a half minutes later he made the same request, and Port-land replied, "twelve flying miles."

6:13:21

Flight engineer: "We've lost two engines, guys."

6:13:28

Captain: "They're all going. We can't make Troutdale." [Troutdale was a possible alternate landing field short of Portland International.]

First officer: "We can't make anything."

6:13:46

First officer: "Portland tower, United 173 Heavy, Mayday. We're—the engines are flaming out. We're going down. We are not going to be able to make the airport."

It was the last transmission from Flight 173.

About 6:15, the aircraft crashed into a wooded section of a popu-lated area of suburban Portland about 6 miles southeast of the air-port. Miraculously, only eight passengers and two crewmembers were killed in the crash, due in great part to the fact that no fuel re-mained to cause a fire during the crash landing.

This case study clearly illustrates the dangers of distraction. But look-ing at this tragedy through the lens of the airmanship model points to other areas of concern. There was no indication of any failure of flight discipline, skill, or proficiency. The entire crew was qualified in their respective positions; however, as we move up the airman-ship model to the pillars of knowledge, we can begin to understand how this seemingly impossible mistake could occur. The indications are that it goes beyond mere distraction.

Three pillars of airmanship knowledge apparently failed to support situational awareness and judgment on this fateful day in Portland: knowledge of aircraft, team, and environment. First, it appears as if the crew was not completely aware of the clues that the aircraft in-dications were providing. The crew had all of the information that it needed to make a clean decision about the landing gear malfunction almost immediately after the gear was lowered. The flight engineer's visual inspection showed that the gear was down and locked, and the captain had made the call early not to attempt to recycle the landing gear. Yet the captain continued to work landing gear indication issues. The fuel flow indications were also a clue available to all three crewmembers and should have indicated that the actual fuel reserve (in time) had been reduced by close to 50 percent. If this was not immediately apparent, it should have been realized when the first call was made to maintenance in San Francisco. After approximately 20 minutes of flight with the gear down, the aircraft had used approximately 7800 pounds of fuel—half of the initial reserve. It should have been clear that it would be dangerous, if not impossible, to hold for another 20 minutes. The failure of the crew to realize or act on these clues indicated not only a loss of situational awareness but possibly a weak pillar of aircraft knowledge.

The second pillar that failed to support good airmanship was knowledge of the team. As the event unfolded, it became apparent that the crew was functioning as individuals and not as a synergistic unit. It is apparent from the transcripts of the cockpit voice recorder that the first officer was concerned about the fuel state earlier than his crewmates, who were perhaps more distracted by their specific job responsibilities. Yet the first officer only hinted at the problem initially and failed his team by a lack of proactive action or assertiveness. The captain failed to provide adequate explanation of the situation or appropriate guidance to the cabin crew. This failure may have prevented the sense of urgency in preparing the cabin that the dwindling fuel state called for. The flight engineer was in the best position to see and analyze the fuel problem but either failed to do so or was apprehensive about bringing it to the attention of the captain in a forceful manner. Clearly, the pillar of team was weak or nonexistent in this case study.

There are also indications that the crew might not have had an adequate appreciation for the environment. It is possible that the proximity of the airport and the sight of the runway passing by on each trip around the holding pattern gave the captain and crew a false sense of security—as if distance from the field was not a real concern. Only five minutes before the engines began to flame out, the first officer accepted an outbound vector from the approach controller when the aircraft was only five miles from the field. This was the last opportunity to adequately address the rapidly deteriorating situation, and they missed it.

The flight of United 173 should be viewed as a failure of airmanship and not just airmen. The weak knowledge and understanding of aircraft, team, and environment, set the stage for the situation to overwhelm three highly skilled professional aviators with tragic consequences. The shortcomings in these three areas contributed to a loss of situational awareness, which in turn led to poor judgment and decision making. It should not come as a surprise that loss of situational awareness is even more prevalent in general aviation, but the underlying causes and preventative measures are much the same.

Case study: Cessna 210D, January 1, 1989 (NTSB 1989)

On New Year's Day 1989, John Wilkins (a pseudonym) and three passengers departed Fort Meyers, Florida, in a Cessna 210D, en route to Lansing, Illinois. The passengers most likely felt that they were in safe hands. Although John was not instrument rated, he was an experienced pilot with more than 2500 flying hours. After lifting off at 9:26 a.m., they made the first leg of the trip uneventfully to Tullahoma, Tennessee, where they stopped to refuel.

For reasons unknown, John elected not to get a weather update at Tullahoma. Perhaps he was pressing to get back before dark, or maybe he felt his initial weather briefing would suffice. Whatever the reason, he took off under visual flight rules en route to Illinois. Not too far into the flight, he began to encounter weather and made a decision to attempt a landing at Alvin, Illinois. Unknown to him, the weather had deteriorated severely. The weather at Alvin was reported as an indefinite obscured ceiling, one-quarter mile visibility with fog. In short, a pilot without an instrument rating was attempting to make a landing into conditions that were below minimums for many commercial and military aircraft and crews. Once he made the decision to attempt the approach, John had in effect sealed his fate. The aircraft crashed into a tree at 35 feet agl about 5 miles north of Vermillion County Airport, killing the pilot and one passenger and seriously injuring the other two. The accident investigation found no evidence of aircraft malfunction or part failure.

The NTSB listed the official cause of the accident as "continued flight by the noninstrument rated pilot into instrument meteorological conditions (IMC), and his failure to maintain sufficient altitude" (NTSB 1989). The airmanship model suggests that the lost situational awareness came from multiple failures. The first was a failure of flight discipline. The failure to update the weather forecast when flying during winter months in the Midwest is inexcusable, and the 54-year-old pilot had enough experience to know it. Further evidence of failed discipline was demonstrated by the decision of a non-instrument-rated pilot to fly into IMC conditions—the single most deadly sin in general aviation today. Further up the airmanship model, we see a failure of the pilot to understand his own limitations as well as those of his aircraft. Finally, we see a lack of respect for the environment and the changing weather conditions. Failed discipline and inadequate knowledge of self, aircraft, and environment led to lost SA and a fatally flawed decision.

Recognizing lost SA

It is essential that every aviator understand not only the definition and importance of situational awareness but also how to recognize and recover from lost SA. It is important to keep in mind as we analyze SA that it sits near the top of the airmanship model, and it varies as a result of many factors. It is a complex, multifaceted phenomenon, and no flyer maintains perfect SA. It is also important to realize that the situational awareness of the group is limited to that of the individual controlling the aircraft at the moment of truth. Therefore, it is important to be able to recognize lost SA in others. If he or she doesn't "have it" when it's needed, all of the combined SA from Jackie Cochran to Chuck Yeager won't save the day.

The role of the pilot is changing from one of merely a control manipulator to that of an information processor, flight or crew leader, and systems manager. A short review of existing literature on situational awareness points out three important points:

1. Lost SA can occur gradually or all at once.
2. Lost SA seriously degrades the ability to achieve flight objectives and is often the prime factor in accidents.
3. There are nearly always sufficient cues available to the crewmember for recognition and recovery from lost situational awareness.

The rate at which situational awareness is lost depends on several factors and can vary from a prolonged descent into lost SA to an almost instantaneous state of "cluelessness." A healthy respect for the rapid onset of this phenomenon means that a flyer should not routinely operate at the edge of his or the aircraft's operational envelope. Leave yourself a little time for error recognition and recovery. Don't build yourself into a situation in which you have to do *everything* right to escape. Leave multiple exits for every situation and maintain a watchful posture on yourself and your aircraft at all times. In short, when at all possible, fly defensively, even if it's only Murphy you're up against.

The case studies demonstrate the devastating consequences that can occur with a bout of lost situational awareness. But it doesn't always end up in twisted metal and fiery death. Sometimes it's just the mission that suffers. Take the case of an F-111 crew during the Gulf War. As they were returning from a bombing run in northern Iraq, the weapon systems officer (WSO) alerted the pilot to a surface-to-air missile (SAM) launch against their aircraft. After executing a high-G evasive maneuver and defeating the threat, the crew breathed a sigh of relief and continued on the new heading for several seconds before returning to their flight-planned route. During this self-congratulatory interlude, the aircraft flew right into the lethal threat ring of a *known* anti-aircraft artillery (AAA) emplacement. Although the aircraft and crew returned home safely that day, they admitted it was due more to the inaccuracy of the Iraqi gunners than anything the crew did to evade the AAA (Anonymous 1992). Sometimes it just pays to be lucky.

Unfortunately, the high-tech wizardry of modern aviation has not yet developed a "lost SA" warning system. So until they do, aviators must rely on self-assessment and teamwork to identify the problem before mission effectiveness or safety is compromised. Analysis of accident data and failed missions show that at least four cues to impending lost SA are usually available to the crewmember. Doug Schwartz, a human factors expert from FlightSafety International, pointed out, "These cues are not always black and white; judgment and discretion are required. For example, there may be a good reason for using an undocumented procedure, like a condition where there is no prescribed procedure" (Schwartz 1992). The presence of one or more of these cues, however, is adequate grounds for a conscious assessment of your current state of SA.

The following list illustrates several "pinches" that an aviator might feel as a precursor to lost situational awareness.

1. *Ambiguity or confusion:* A fuzzy feeling that you are missing something or that sixth sense feeling of uncertainty, often referred to as the *pinch*.

2. *Fixation:* Channelized, single-focused attention.

3. *Reduced frequency or poor communication:* When we start to lose SA, strained or difficulty in conducting communications outside of the cockpit is often one of the first symptoms. A related cue is the failure to react to incoming communications, often requiring a second transmission from those trying to contact you.

4. *Failure to meet targets:* If you suddenly find yourself 10 minutes early, you should ask yourself why and resolve the discrepancy.

5. *Reduced maneuvering:* Although not necessarily a bad thing when you are disoriented, reduced maneuvering at inappropriate times can be a cue to lost SA.

6. *Failure to stay ahead of the aircraft:* Good situational awareness keeps the crewmember well ahead of the aircraft, preparing and anticipating concerns in the immediate, intermediate, and long-term environments. If you suddenly find yourself exclusively reacting to immediate concerns, you may be losing situational awareness.

7. *Use of an undocumented procedure or violation of a minimum:* This problem may involve a simple checklist deviation, or it may be something much more serious, like violating weather avoidance criteria.

8. *Attempting to operate aircraft or weapon system outside of known limitations:* Lowering the landing gear or flaps above placard speed or attempting to release weapons outside of release parameters usually indicates lost situational awareness.

Although this list is by no means all-inclusive of symptoms of lost SA, it does point out that the skills for recognition can and should be learned. Through disciplined self-debriefing of minor cases of lost SA, this skill can be finely tuned. Try adding this simple step to your daily debriefing technique. After flying, look at the above list and ask yourself, "Did I have perfect SA today?" If the answer is no, look for the clues that were present to reinforce your future sensitivity to them. It's a simple self-improvement step—and it might save your life.

The next steps: Admitting and recovering from lost SA

Recognizing lost SA leads us to the next, often more difficult step of admitting it to yourself and others. Aviators are known for hiding weak areas. No one likes to admit a human frailty, but the risk is too great to keep lost SA a secret. Besides, your wingman, copilot, student, or instructor probably already knows. Just a short phrase like "Stay with me, I'm getting a little disoriented" will shift a teammate into a support role and safety monitor. Or in the military, it may require one of the standard phrases such as "terminate," "knock it off," or "time out."

But someone else may not always be there to pull you through an episode of lost SA, or maybe they are as bad off as you are. What then? There are several simple steps that experts recommend to reestablish equilibrium and give yourself a chance to "catch up." The age-old commandment to "fly the aircraft" can be broken down into a few useful, suggested techniques, the first of which should be in **BOLDFACE**, just like critical actions in an emergency procedures checklist. If you experience any symptom of lost situational awareness, consider taking the following steps:

1. **Get away from dirt, rocks, trees, and, metal! If in formation, increase separation. If flying low level, climb.** If you suspect your wingman (or lead) or other crewmember has lost SA, be assertive and question him or her if time is available. If not, direct the appropriate action. It is better to take a hit at the debrief than to answer questions for the accident investigation team or, worse yet, not be available to answer the questions.

2. *Stabilize the aircraft.* After removing yourself and others from immediate danger, stabilize conditions. By reducing the rate of change, you give your overtaxed mind a chance to get caught up.

3. *Buy time.* If flying under instrument flight rules (IFR), the old trick of requesting "present position to anywhere," gives you a moment to get your act together. Request a holding pattern instead of the en route descent; take another trip around if you are already there. Delay your range entry time if you need to. Remember, "better late than never" applies as much or more to aviation than any other endeavor.

4. *Seek information.* *Listen* to all inputs and sources of immediate information—visual, aural, seat of the pants, and intuition. Resolve any internal disagreements. Find out why you are early, late, or ahead/behind on the fuel curve. Ask ARTCC for verification of position, if required. In short, restore your confidence in yourself and your equipment.

5. *Learn from the experience.* Telling yourself "I don't want to go through that again" is not sufficient debrief. Analyze the episode. What led to the lost SA? What cues were present? How can you avoid it in the future?

Remember, you alone are the master of your inflight destiny. Only you really know when you are not performing or thinking at normal levels.

Prevention: Five keys to improving situational awareness

The goal of this chapter is to develop an understanding of situational awareness as it relates to the other areas of airmanship, as well as to provide practical suggestions for improving SA. In an ideal flying world, distractions would not exist, and all aviators would have solid foundations and pillars of airmanship. Unfortunately, such is not the case, and we must develop strategies for minimizing the mission and safety risks associated with lost situational awareness. The following planning guidelines are suggested for preventing an episode of lost SA.

1. *Define roles.* Avoid channelized attention by clearly outlining duties and responsibilities for everyone on the flight team. This is best done on the ground prior to flight.

2. *Manage distractions.* Establish and adhere to standard operating procedures. Do what is expected or brief otherwise. Acknowledge and verbalize breaks in routine or sequence.

3. *Reduce overload.* Recognize and admit to yourself and others when you're too busy to stay on top of it all. Delegate or reduce load to a safe tasking level.

4. *Avoid complacency.* When you're starting to feel bored or euphoric, that is when an aircraft bites the hardest.

5. *Test assumptions.* Take nothing for granted. Double-check data and question all hunches.

6. *Intervene.* Assertive questions are demanded when conditions threaten the safety of flight. Regardless of crew position, stay alert and mission-oriented.

Situational awareness feeds aviator decision making and judgment. As with the other aspects of airmanship, it cannot be viewed in isolation. We have seen how all foundations and pillars of airmanship can support or detract from situational awareness, as well as the disastrous consequences that can occur if it is lost at the wrong moment. Although complex, SA can and should be improved by individual effort. It is not an "innate" quality, as many would have us believe, but rather a higher-echelon piece of the airmanship whole, ours to improve to whatever degree we as individual flyers have the will and discipline to achieve.

References

Anonymous. 1992. Personal interview with F-111 WSO.

Endsley, Mica. 1989. Pilot situation awareness: The challenge for the training community. Paper presented at the Interservice/Industry Training Systems Conference, November 1989, Fort Worth, Tx.

Magnuson, Kent. 1995. Human factors in USAF mishaps, October 1, 1989 to March 1, 1995. Albuquerque, N.M.: USAF Safety Agency Life Sciences Division. December 1.

National Transportation Safety Board (NTSB). 1979. Aircraft accident report of United Airlines Flight 173, Portland, Oregon, December 28, 1978.

―――――. 1989. Accident ID number CHI89FA039, File number 1985.

Schwartz, Doug. 1992. Training for situational awareness. Presented at the Ohio State University Fifth International Symposium on Aviation Psychology, Columbus, Ohio.

Sventek, Jeff. 1994. Aircrew awareness and attention management briefing. San Antonio: Trinity University.

United States Air Force. 1992. Accident investigation report, B-1B mishap. Dyess AFB, Tx. November 22.

VEDA, Inc. 1988. AMRAAM OUE tactics analysis methodology briefing. Quoted in Stiffler, Don. 1988. Graduate level situational awareness, *USAF Fighter Weapons Review.* Summer:15.

Wagg, Wayne L. 1993. Preliminary report from the Situational Awareness Integration Team (SAINT) briefing slides. Mesa, Ariz.: Armstrong Lab.

10

Judgment and decision making

Judgment is not the knowledge of fundamental laws; it is knowing how to apply a knowledge of them.
Charles Gow

A personal view from Richard Jensen, Ph.D.

Judgment is important for a variety of reasons. One that is often overlooked is the way in which the FAA regulations and procedures are constructed. Unlike other regulations, which are designed to protect the weakest or least-skilled member of a group, the FAA designs its regulations to maximize the capabilities of the *best* pilots and aircraft. This is because the FAA has a dual purpose of promoting aviation while at the same time regulating its safety.

In the United States, the FAA testing program for certification of aircraft equipment and flight procedures uses the best pilots and equipment to test all operations that it regulates and designs regulatory guidance and procedures based on these data. Therefore, the FAA regulations are offered under the expectations that airmen will interpret them based on their own skill and aircraft capabilities, which may be considerably less than those who test-flew the procedures.

Although none of us like to think of ourselves as mediocre airmen, very few have the skill of the FAA test pilots. Therefore, when we make a decision, for example, to fly an approach to absolute minimums, we are asking ourselves to fly to the skill level of experienced test pilots with perfectly calibrated equipment.

253

Think about that the next time you decide to "push" a decision height or minimum descent altitude on an approach. You alone are responsible for assessing your individual, team, and aircraft's capabilities relative to the task at hand.

There is a certain truth to the popular belief that judgment is just common sense applied to decision making. However, for this belief to hold water, "sense" must imply a knowledge and intense awareness, realization, and in-depth understanding of all of the factors involved in the decision. Like the wise man once said, "common sense isn't always that common."

Richard Jensen is director of the Ohio State University Aviation Psychology Laboratory, editor of the International Journal of Aviation Psychology, *and author of* Pilot Judgment and Crew Resource Management.

Good judgment sits at the pinnacle of airmanship, and all other factors, including discipline, physical flying skills, knowledge, and situational awareness support good judgment. Good judgment, in turn, supports all three major goals of airmanship: safety, mission effectiveness, and efficiency. That is why good judgment is important, not only to individuals, but to organizations as well.

Judgment, like all other aspects of airmanship, must be viewed holistically, as an interactive part of the airmanship whole. Much of the research attempts to isolate pilot judgment from other elements of airmanship. While this approach is useful for controlling variables in the research environment, it is incomplete and therefore inappropriate for our purposes. Current reseach definitions lack the depth needed for a full understanding of the interdependency of the many aspects of airmanship associated with judgment. Judgment is demonstrated in almost everything an aviator does, from applying simple rules and procedures to routine situations to tackling and solving complex inflight emergencies. It even extends beyond the cockpit itself to include adequate mental and physical preparation for flight, briefings, debriefings, and critiques. It is even likely that individuals who exercise poor judgment in their day-to-day activities carry these traits with them into their aircraft. As the pinnacle of a complex human equation, understanding and improving judgment requires a solid foundation of airmanship knowledge and a process for study and improvement.

The scope of poor judgment

Poor judgment has a significant impact on all areas of aviation, including general, military, and commercial. The tragic results of poor judgment are in evidence in nearly every accident study conducted in recent years. In one five-year study of general aviation accidents, decisional errors accounted for more than 50 percent of fatal errors (Jensen and Benel 1977). See Fig. 10-1.

Percent of Fatal Accidents by Error Type

10-1 *Poor decision making is a deadly airmanship flaw, followed closely in lethality by mistakes of perception and physical motor errors. Airmanship can be viewed as a hedge against fatal errors when all areas are well-developed and balanced.* Diehl 1989

Similarly, a 10-year analysis of military human factors errors associated with mishaps identified three areas of poor judgment as causal factors more frequently than all other error types combined. The number-one cause was "selected wrong course of action," closely followed by "intentional failure to use accepted procedure" and "delay in taking necessary action" (Magnuson 1995). Each of these three areas indicate poor judgment and decision making on the part of the aircrew members.

Dr. Alan Diehl, a former NTSB and U.S. Air Force accident investigator, combines the findings of several studies to compare decisional errors across the spectrum of aviation (Fig. 10-2). His data show that judgment errors account for greater than one-half of all aircrew errors.

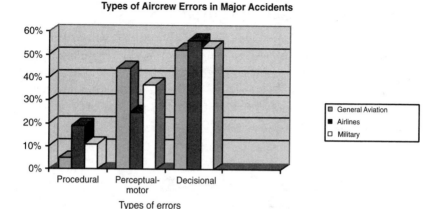

Types of Aircrew Errors in Major Accidents

10-2 *Decision-making errors cross all boundaries as the leading cause of major accidents.*

Nonstatistical data also suggest the severity of the decision-making problem. Recent advances in modern aviation technology can be seen as a condemnation of aviator decision making. The majority of human factors engineering advances in recent years have been directed toward taking more and more flying and decision making out of the hands of the fallible human being. Early examples included the autopilot and automated navigation systems to help prevent altitude and route deviations. More recently, the Ground Proximity Warning System (GPWS), Traffic Alert and Collision Avoidance System (TCAS), and various windshear alerting systems have come on line to assist pilots in determining when they are about to run into the ground, another aircraft, or an environmental hazard. These advances were prompted by accidents and incidents that were directly related to poor judgment and decision making.

On the training and operational fronts, checklists, policies, procedures, and regulations have become more directive. Organizations appear to feel that rigid standardization is a partial answer to poor aviator judgment. While both automation and standardization have led to a safer, more predictable flight environment, it may have inadvertently created a generation of flyers that rely too much on automation and checklists. These aviators may be unprepared or caught off guard when they are required to make tough decisions on their own.

Although most flyers welcome technological advances, neither automation nor rigid standardization is the answer to all problems in aviation decision making. There are several reasons why we must continue to strive for better judgment in our aviators. First of all, all aircraft are not automated. Secondly, automation has and will continue to fail periodically. Finally, increased automation has led to new types and levels of complacency, and many pilots feel that they are simply along for the ride. So judgment and decision making remain critical skills for pilots and other flight personnel in all sectors of aviation, perhaps more so than ever before.

The judgment problem demands our individual attention. While aviation psychologists and training organizations continue to address judgment and decision-making issues on a grander scale, the buck stops with each of us. Self-improvement begins with a basic understanding of how to approach this complicated and often confusing phenomenon we call pilot judgment.

Judgment: Art or science?

Flyers should view judgment as both a science and an art. It is emerging as a science due to excellent and extensive research done in fields such as aviation psychology. Modern research has revealed certain attributes, personality types, and patterns of thought that can lead to improved decision making, but there are multiple reasons why judgment must also be viewed as an art. First, aviation decision making cannot be reduced to "if-then" formulas that fit all situations. The complexity of the inflight environment produces far too many variables to account for in simple decision matrices. Second, there is an intuitive side to pilot judgment, a subconscious "sixth sense" that comes into play in many situations. It might take the form of a gut feeling that "something isn't right," like the hair standing up on the back of your neck as you prepare to take a certain course of action. Conversely, this sixth sense can lead to an inner sense of certainty of action that instills confidence in tough situations. Although little scientific research addresses this sixth-sense phenomena, many aviators say that they have used it often, and they know the value and validity of this internal approach.

Perhaps because individuals differ in their ability to recognize and react to various inflight challenges, judgment has taken on an almost mystical quality among aviators, leading some to believe that one either has it or doesn't. Nothing could be further from the truth. Decision making is an acquired skill and can be systematically developed through knowledge and practice.

This chapter does not suggests a cookie-cutter or checklist approach to all situations. However, judgment and decision making can and should be developed systematically. The first step is to possess or develop the prerequisite structures of airmanship discussed in earlier chapters, starting with uncompromising flight discipline, and adding skills and a broad knowledge base. The second step is to understand how aviators make decisions, as well as the types of decisions that are made. Finally, by applying the decision-making strategies suggested, aviators can build good-judgment habit patterns that will serve them well for a lifetime of flying.

To accomplish these objectives, this chapter defines judgment, overviews relevant research on aviation decision making, and ties in case studies illustrating important points. It concludes with strategies for improving individual judgment and decision making.

Judgment defined

Webster defines judgment as "the process of forming an opinion or evaluation by discerning or comparing" (*Webster's* 1990). For the purposes of aviation decision making, however, we must go beyond this basic definition. The Federal Aviation Administration uses a more detailed definition that hints at several aspects of the airmanship model:

> *Pilot judgment is the process of recognizing and analyzing all available information about oneself, the aircraft, the flying environment, and the purpose of the flight. This is followed by a rational evaluation of alternatives to implement a timely decision which assures safety. Pilot judgment thus involves one's attitudes towards risk-taking and one's ability to evaluate risks and make decisions based upon one's knowledge, skills, and experience. A judgmental decision always involves a problem or choice, an unknown element, and usually a time constraint and stress (FAA 1988).*

Even this expanded definition may not totally capture the complexity of judgment, because it fails to adequately incorporate the team into the process and leaves one with the feeling that judgment is exclusively an individual act. This would be the wrong message. In aviation, judgment is the process of comparing and evaluating courses of action—as identified by the individual flyer and the extended flight team. These courses of action are partially developed long before they are actually needed and involve preparation, communication, knowledge, and skill acquisition to make the most appropriate decision when the time comes. Good judgment tips the scales of chance and probability in your favor. The "appropriateness" of the decision is determined by the objectives of the flight and the flyer.

In most all cases, safety is the most important factor in good judgment and decision making. But flyers—especially commercial and military flyers—cannot always take the *most* conservative approach to all situations. In fact, in some military missions, safety may well be a secondary concern behind mission accomplishment, but these type missions are few and far between. The benefits of solid decision making and good judgment hold across the spectrum of aviation, whether you are a general aviation pilot trying to determine if you should chance a takeoff under deteriorating weather conditions or an F-15 driver weighing the pros and cons of a one-versus-four fight. As mystical as this process seems, there has been a great deal of excellent research accomplished in the past two decades, and several pieces of solid information and guidance have emerged to aid our understanding and self-improvement. The crucial first step towards consistent good judgment is to effectively assess each situation.

Step one: Assess the situation

An accurate assessment of a situation is the first and perhaps most important step in aviation decision making. Flyers must understand that there are often no hard and fast solutions to airborne problems, but that all problem solving begins with situation assessment. Judith Orasanu, an aviation psychologist with the NASA-Ames Research Center, breaks this assessment process into three distinct activities. First, an aviator must define the nature of the problem. Second, he or she must determine how much time is available to handle the problem. The final step is to assess the level of risk presented by the situation (Orasanu 1995). These three steps must take place before an aviator can effectively move to the next steps of decision making—creating alternatives and selecting a course of action.

The failure of the captain on United Flight 173 (Portland fuel exhaustion; see Chapter 9) to accurately assess the time available for

addressing his landing gear indication problem dramatically illustrates what can happen as a result of an incomplete situation assessment. In fact, researchers studying firefighters and tank commanders
have discovered the greatest difference between experts and
novices is the ability to accurately assess the situation (Klein 1993).
It is likely that this fact also holds true for flyers. If a single area
should be addressed first to improve aviator judgment, it must be improving situation assessment skills.

Assessment of a specific situation links overall situational awareness
(big picture) to a specific decision. That is to say, aviation decision
making involves the leadership and management functions of assessment within the context of the big picture. For example, your
overall situational awareness could indicate that an emergency recovery of a large aircraft is taking place at the field at which you intend to land. No one has directly informed you of this, but you have
pieced this together by listening to the radio chatter about the preparations for the incoming emergency. Specific problem assessment
would be your dwindling fuel state and deteriorating weather at
your alternate. Your decision is therefore based on multiple factors
from the big SA picture, as well as a specific assessment of your own
individual problem. You must now decide whether to increase your
speed to arrive at the airport before the emergency, which will
worsen your fuel state even more or proceed directly to the deteriorating alternate before the weather goes below minimums. This decision is tough, to be sure; to take action based on anything less than
knowledge of all factors could put you in an unrecoverable situation
if the dice roll the wrong way. By ensuring that a complete situation
assessment is accomplished, an aviator can begin to calculate the
odds of success or failure of a given course of action and then execute accordingly.

To summarize, when confronted with an airborne challenge, aviators should ask themselves three simple questions to assist in follow-
on decision making.

1. *What is the nature of the problem?* With this question, the
   flyer determines the nature of the problem and seeks
   confirming evidence from multiple sources, if available. This
   initial step is critical, because all following steps are based
   on an accurate identification of the problem at hand.

2. *How much time do I have to work the problem?* The answer
   to this question may be based on many factors, including,

weather, fuel, remaining hours of daylight, fatigue, and the nature of the problem itself. It determines the scheduling function during execution of the final plan—a step that is critically important in light of the temporal distortion that often accompanies aircraft emergencies.

3. *What are the risks associated with this problem?* This step is designed to help the aviator identify all of the possible negative outcomes associated with the situation and develop courses of action based on acceptable levels of risk. Here the flyers should ask themselves, "Is this problem life-threatening, mission-threatening, potentially embarrassing, etc.?" This step also should prompt "What if . . . ?" questions like, "What if they close the runway?" "What if I lose another engine?" or "How does this affect my go-around capability?"

These simple questions can assist in prioritizing the problem-solving effort versus other mission priorities and should become a standard part of each aviator's approach to abnormal situations. They can direct the use of specific checklists or procedures and serve as the most important input to the next step in aviation decision making—developing courses of action. However, this approach does not suggest that situation assessment is always a deliberate, thoughtful process. Pilots are occasionally required to make snap decisions, often with life-and-death consequences. Situation assessment is even more critical under these circumstances, as evidenced by the following case study.

Case study: Situation assessment (Hughes 1995)

A KC-135 aircraft crashed shortly after takeoff, due, in the final analysis, to misapplication of flight controls by the aircraft commander. A deeper analysis indicates that this mishap likely occurred due to poor situation assessment caused in part by the distraction of loss of water injection to the number-one engine and a nearly simultaneous crosswind gust that caused the aircraft to roll rapidly to the left. (The KC-135A used water augmentation to increase thrust on its engines. The water injection only lasted for approximately 160 seconds under normal circumstances. When water was "lost," it was a considerable reduction in thrust, which was especially critical during high gross weight takeoffs.) The pilot overcompensated for the crosswind gust and the aircraft rolled right, resulting in the right wingtip and number-four engine contacting the ground, rendering thrust insufficient for takeoff. At approximately the 1000-feet-remaining marker, the aircraft became airborne momentarily, entered a stall buffet, and settled to the ground shortly after passing the departure end. The aircraft commander kept the throttles advanced, attempting to gain speed for takeoff. At no time was the "crash landing after takeoff" procedure implemented. The aircraft continued on the ground until impacting a creek embankment, broke up approximately 3800 feet beyond the end of the runway, exploded and was engulfed in flames. All crewmembers and passengers were fatalities.

Prior to the takeoff roll, indications were present that there may have been some problems with the water injection system initiation, which is designed to add thrust to the engines during the critical takeoff phase. This could have caused the pilots to be preoccupied with thoughts of losing thrust on the heavy-weight takeoff, resulting in a distraction from the onset of the takeoff roll. Prior to brake release for takeoff, the aircraft commander advanced the throttles, retarded them to idle, and then advanced them again. This action is consistent with steps taken for a minor water augmentation problem. During takeoff roll, a malfunction occurred that degraded thrust, most likely a loss of water augmentation on the inboard engines. The aircraft commander became distracted with the malfunction and momentarily relaxed his crosswind controls. To avoid contact with the runway, the aircraft commander pulled up on the yoke and applied right aileron control inputs. The aircraft became prematurely airborne and rolled to the right, resulting in the number-four engine scraping the runway, leading to the failure of the engine and loss of water augmentation to the outboard engines. At this time the aircraft had insufficient thrust for sustained flight, and a mishap became unavoidable.

In all likelihood, the pilot channelized his attention on dealing with the loss of water augmentation during the takeoff roll and momentarily diverted his attention from the strong crosswinds. The forecasted winds and wind advisories given to the crew should have alerted them to the possibility of strong crosswind gusts and the need for constant vigilance to maintain lateral and directional control of the aircraft. With the takeoff clearance, the tower accurately reported the winds of 210 degrees at 16 knots; however, surface winds of 230 degrees (magnetic) at 16 gusting to 24 knots were recorded during the takeoff roll on runway 17. Although crosswinds may have contributed to the unstable takeoff, the limits for the aircraft were not exceeded. Witness statements indicated white smoke from the number-three engine, an indication of loss of water injection, occurred at a point on the runway after decision speed was obtained. Analysis of the takeoff roll concluded that failure of inboard water at this point was consistent with actual takeoff distance. Although it was a serious performance degradation, sufficient thrust was still available for a safe takeoff.

Shortly after the aircraft experienced the loss of water augmentation, it was observed to abruptly pitch up and start a roll to the right. The board concluded that the aircraft commander pulled back on the yoke and started the right roll to prevent the left wingtip or the number-one engine from contacting the runway. This overrotation caused a decrease in airspeed approaching initial buffet and loss of lateral control authority. This momentary distraction and lack of vigilance created a true emergency. Once the aircraft entered this thrust-deficient condition, the pilot would have had to do *everything* right to avoid a mishap.

In a nutshell, this pilot failed to identify the true nature of his most immediate and dangerous problem. The takeoff profile was such that there was little margin for degraded aircraft or pilot performance. Additionally, in the final stages of the mishap, optimum pilot technique and awareness was essential for any possibility of a successful recovery. It has been said that awareness implies a grasp of the overall situation, whereas attention involves concentration on the appropriate things at the right time. This mishap illustrates an example in which the pilot failed in the first of the three situation assessment steps. He did not accurately perceive the nature of the problem. It appears that he perceived a performance problem and misdirected his attention away from the more serious threat of aircraft controllability in the high crosswind environment. The entire sequence of disaster took only seconds. Judgment demands preparation.

Situation assessment includes the ability to handle multiple, simultaneous inputs and to determine what is important. In this mishap, the pilot was faced with a malfunction after decision speed occurring in a very demanding takeoff environment, i.e., heavy weight, 30-degree flaps, significant crosswinds. The fact that the malfunction occurred in the most critical phase of the takeoff is significant in that it reduced the time that the aircraft commander had to make an accurate decision. In time-critical emergencies like this one, steps two and three of situation assessment (time available and risk) must be addressed before the actual emergency occurs, for example, during pretakeoff briefings. Some actions must be "hard-wired" to ensure rapid and appropriate response. There will be little time with which to consciously process the problem after it occurs. These automatic responses can be aided by a "preassessment" of potential problems before critical phases of flight such as takeoff and landing, especially when high-risk factors, such as strong crosswinds or low visibility, are expected.

A second case study illustrates a situation in which the problem was clearly perceived, but a faulty analysis of time available and the relative risk led to a fatally flawed assessment of the situation. Although this example is also from military aviation, the pilot's poor decision to fly into instrument conditions and the end result of this decision is a sad refrain sung often in the general aviation sector as well.

Case study: Trying to hack it (Hughes 1995)

The mishap pilot was a 30-year-old first lieutenant with 553 hours total flying time and 258 hours in the F-16. He was the last man in a 14-ship of attackers and was flying his first major exercise mission as part of a 51-ship strike package opposed by 12 fighters and an array of simulated ground threats. After coming off his first attack run, the mishap pilot attempted to regain visual on his element lead while proceeding to the planned egress point. Low, layered clouds at the egress point forced the first 13 aircraft to deviate from the planned course to maintain clear of the clouds. Using minimum communications tactical procedures, preceding flight members, including the mishap pilot's element lead, chose not to call for a revised egress route/plan over the radio, which would have alerted and prepared the mishap pilot for the upcoming weather. The mishap pilot allowed his aircraft to enter instrument conditions in the vicinity of high terrain, then elected to press ahead in hopes of regaining visual contact instead of aborting the route to a prebriefed altitude that would maintain terrain clearance. The mishap aircraft impacted sharply rising terrain in controlled flight without an ejection attempt, and the pilot was fatally injured.

The mishap pilot became task-saturated while attempting to regain visual contact with his element leader, and, as a result, he failed to adequately assess the situation. He did not apply well-established, prebriefed weather route abort procedures to ensure terrain clearance. He misprioritized mission tasks ahead of maintaining clear of the clouds. Once in the weather, he did not effectively understand either the time available to make a safe decision or the relative risks involved with continued flight in instrument conditions. Based on this incomplete assessment of the situation, he flew into a mountain.

Peer pressure and a desire to hack his first major mission without having to resort to a nontactical peacetime training procedure (weather route abort) were also probable factors and may have induced the mishap pilot to accept what was clearly, in retrospect, an unacceptable risk given the high terrain in the vicinity of the egress route.

Some strategies for improving situation assessment

There are several suggested methods for improving one's situation assessment capability. To improve upon basic recognition of the problem, Orasanu suggests developing an established cockpit-scanning pattern or "cross-check" of all cockpit instruments, not just the ones of current interest. In this way, a flyer is better able to recognize deviations from "normal" indications, as well as establish a periodic timetable for monitoring. On a crew aircraft, this work load should be distributed, with the pilot not flying or other crewmembers sharing responsibility for monitoring. These crewmembers should be trained to verbalize deviations in a timely manner and with appropriate assertiveness (Orasanu 1995).

A second method to aid in situation assessment is to play "devil's advocate" with your situation. Playing devil's advocate means to "challenge interpretations of situations, especially before making decisions and taking actions that have severe consequences if wrong, such as shutting down an engine" (Orasanu 1995). This technique is also useful in assessing the potential future consequences of a given event. In some cases, it may be best to take a "worst case" approach as an added margin of safety and to plan for various contingencies that can be associated with an event or malfunction.

Finally, all events should be critiqued in detail once the aircraft and crew are safely on the ground. Ask yourself, "What was the first indication of trouble? How could I have improved my scan or cross-check to notice it? What lesson can I take from this situation?" Many flyers keep a personal journal to accompany their standard pilot's log. In this journal, they keep detailed records of lessons learned from each difficult situation encountered. Writing it down reinforces the experience and can aid in building internal models for dealing with similar situations in the future. The pilot's journal provides a record for relevant and reflective self-improvement.

Assessing the situation is only the first step in decision making. Once the problem is understood, complete with the potential future implications, a plan of action must be devised.

The next step: Determining alternatives

As aviators, we are required to make many different types of decisions, but all involve the choice between alternatives. Some are straightforward alternatives, such as abort or continue. Other decisions require the aviator to devise creative solutions to ill-defined problems. It is clear that some decisions require more mental effort than others, and this is based on the type of decision that must be made. J. Rasmussen, a noted researcher on decision making, has identified three major types of decisions faced by cockpit crews, each requiring a different level of effort. (Rasmussen actually breaks these decisions down into six components, but for the depth of this discussion the three major catagories will suffice. For a full explanation on these decision types, see Rasmussen 1993.) These decision types are rule-based decisions and knowledge-based decisions of both well-defined and ill-defined problems. If you understand the nature of these decision types, you are well on your way to improving your performance with each.

Rule-based decisions

Rule-based decisions involve applying known rules to given situations. Successful rule-based decision making presupposes a solid understanding and in-depth knowledge of the rules to be applied. This simplifies decision making but still requires the flyer to make certain "go-no go" decisions or to apply predetermined procedures after recognizing the situation as one to which a known rule applies. An example of the first type would be a backup or emergency oxygen system that is below the required pressure for flight. Regulations say fix it before flight, but depending on the perceived importance of the mission, this decision may become less clear-cut. In this case, the call to go or abort rests with the individual discipline of the pilot (see Chapter 2 for a full discussion on discipline).

A similar type of rule-based decision is the procedural decision. These types of decisions merely require that the aircrew member recognize a situation and apply known directives. The FARs require that if an aircraft loses any navigation equipment, the pilot is to notify the air traffic controller as soon as possible. Some pilots do and some don't, but the requirement exists, and a disciplined procedural decision maker will follow the directives every time unless extenuating circumstances dictate otherwise. As straightforward as this seems, a large number of accidents and incidents still occur as a result of poor rule-based decision making.

Case study: A simple rule not followed (NTSB 1989)

On February 9, 1989, a DC-9 cargo aircraft operating for Evergreen International Airlines departed Ogden, Utah, for San Antonio, Texas. Unknown to the pilots, the inspection door of the aft pressure bulkhead hatch had not been installed, so as the aircraft began its climb to en route cruise altitude, it failed to pressurize. The first officer was flying this leg and noticed the lack of pressurization during the climbout. In accordance with FAA regulations and common sense, he leveled the aircraft off at 16,000 msl to further investigate the problem while maintaining a safe level of oxygen in the cabin compartment. The captain countermanded this decision and ordered him to continue the climb to the assigned flight level, FL330. The first officer was very uncomfortable with the order, which he recognized as grossly exceeding regulatory guidance for an unpressurized cockpit. But rather than confront the captain, he complied and continued the climb.

The captain left the cockpit with a portable "on demand" oxygen walk-around bottle and went aft to see if he could fix the problem. These small oxygen bottles have a 15-minute supply of oxygen and are designed for emergency use, but they have been known to exhaust their supply in less than 10 minutes under conditions of high demand. After not hearing from the captain for several minutes, the first officer became concerned and attempted to signal him but received no response. Although reluctant to countermand the captain's orders, the first officer was now convinced that something was very wrong and began a series of descents, finally leveling off at 13,000 feet msl. The first officer was hoping that the lower altitude might bring the captain back forward if there was a physiological problem, but after 30 minutes of waiting, his fears were so great that he put the DC-9 on autopilot and left the cockpit to see if he could locate the captain.

He found him in the forward cargo area, unconscious with his foot caught in the netting of a cargo pallet. Although the captain was wearing the oxygen mask, the bottle's quantity gauge showed completely full, indicating little or no use. After attempting to revive the captain, the first officer returned to the cockpit and made an emergency landing at Lubbock, Texas. The captain was rushed to the hospital, where he was declared dead on arrival. Post mortem pathology exams determined the captain died of hypoxic hypoxia—oxygen starvation. The oxygen bottle was subsequently tested and found to be fully functional.

The NTSB lists several causes for this tragic accident, including improper maintenance procedures for failing to install or inspect the inspection hatch and improper use of the portable oxygen system. But the key event in this sequence, the decision most responsible for the death of the captain, was the failure to apply a simple and well-known regulation regarding altitude restrictions in an unpressurized cockpit. The first officer's lack of assertiveness certainly contributed to the disaster.

Although rule-based decisions appear to be easiest to apply in flight, they are often mishandled in classic cases of poor judgment. In these cases, there are usually underlying causes or missing pieces of airmanship, that lie at the root of the problem. In the case of the Evergreen crew, it was a lack of flight discipline on the part of the captain and a lack of crew coordination skill and assertiveness on the part of the first officer that set the stage for disaster.

Knowledge-based decisions

Knowledge-based decisions require solutions that go beyond mere procedural knowledge application. Rasmussen states that there are two types of knowledge-based decisions. The first are those decisions required to solve well-defined problems. These include selecting between options, such as determining which direction to go around a thunderstorm; and scheduling decisions, such as a decision to descend now or later. While knowledge-based decisions require more mental effort than rule-based decisions, they are not as demanding as the final category—solving ill-defined problems.

Solving ill-defined problems is the most challenging aspect of aviation decision making, because you are often unable to determine the actual nature of the problem—the first step in decision making. So what now? Orasanu describes the dilemma of ill-defined problems as follows:

> . . . ill-defined problems . . . may or may not be clarified in the process of dealing with them. Ill-defined problems are ones resulting from ambiguous cues that make it impossible (initially) to say what the problem is that needs fixing. . . . Two strategies may be used to cope with this type of situation: manage the situation as though it is an emergency without clearly defining the problem, or diagnose and define the problem, and work out a solution (Orasanu 1993).

Some airborne occurrences demand action, but the lack of information as to the nature of the problem can leave a crewmember without a clear course of action to pursue. If a crewmember hears an unusual noise or feels something "different" in the aircraft, a determination must be made as to the response, especially if the problem appears to be time-sensitive and life-threatening. In some cases, the problem becomes so frightening that the logical course of action is to land as soon as possible, if there is time. The following case study illustrates a self-induced "ill-defined problem" that nearly led to disaster on a routine military training flight. It shows an example of poor judgment based on a lack of checklist and procedural discipline. In addition to introducing the concept of the self-induced emergency, it highlights two important points for further discussion: the hazards of seizing on the first idea as the solution to a problem and the value of a team approach to problem solving with ill-defined problems. The case study is written in the first person by an active duty Air Force pilot who understandably preferred to remain anonymous.

Case study: The routine hop home

> It was a simple mistake, but it almost killed us. I was a copilot on a KC-135A returning from a short TDY at Wurtsmith AFB, Michigan. It was only a couple hundred miles back to K. I. Sawyer, so we took off with a light fuel load. I can't remember exactly what the fuel load was, but I do recall that "runway available" wasn't a problem for takeoff, and that we would still have plenty of gas at the (final approach) fix back at home.

While I was executing a textbook takeoff and climbout, the aircraft commander (AC) was working the checklist. I say "checklist" kind of tongue in cheek, because the truth was we had scarcely cracked the cover on the actual checklist for months. We had the standard items committed to memory and we were both pretty proficient, so we really didn't need it. Looking back, it almost seems like a cliché, but that's the way we felt. We were soon to learn that Murphy preys on the complacent.

One step in the checklist calls for the crew to "drain the reserves," a procedure which takes the fuel held in two small reserve tanks in the wing tips, and drains it into the aft body tank so that it can be burned by the engines. The flight manual also required a verbal cross-check of the action so that the rest of the crew could monitor the fuel panel—an important step on a flying gas tank with manual CG (center of gravity) control. It should be stated that the old fuel panel was an ergonomic nightmare designed by a sadistic Boeing engineer with a hangover in the 1940s. [The "old fuel panel" has been replaced by a newer, computer assisted FSAS (fuel savings and advisory system) panel. Although much prettier and possibly more user-friendly, it is no more immune to human error than its predecessor.] But I digress—back to the story.

When attempting to accomplish this seemingly simple fuel management function, the AC had inadvertently opened the drain valves on the right wing main tanks—a step that if left unchecked would result in a 27 thousand pound left to right fuel imbalance and an extreme aft center of gravity. This would create a very unstable aircraft at any speed, and one that was downright unflyable if you got too slow. Because there had been no verbal coordination as the checklist required, no one caught the mistake.

As I proceeded with the constant speed climbout, I began to notice a definite rolling tendency towards the left wing— which I counteracted with ever-increasing right aileron. I mentioned to the aircraft commander that I thought that the aircraft "felt funny" and that it was taking "a lot of right yoke to hold her level." He replied that I probably needed a little rudder trim and went about his business. After trying some rudder trim with precious little result, I voiced my concern again —using his first name to get his attention. "Hey Dan, something is definitely wrong with this airplane." At this point, he took the jet with the standard "I've got the aircraft." If I live to be a hundred years old, I will never forget the look that came over his face as we transferred aircraft control. While I had gradually added in rudder and aileron to counteract the rolling moment, he was unprepared for the severity of the condition. As I released my control inputs, the aircraft rolled rapidly left and the look on his face went from being perturbed by a wimpy, bothersome 2nd Lieutenant copilot —to the stark realization that we were only moments from departing controlled flight. He yelled across the cockpit, "Damn it Co! Why didn't you say something sooner? When did you first notice this?" Unsure of which question to tackle first, I started supplying what seemed like the most useful information. "It seems like it started right after flap retraction, but I'm not really sure." The AC replied, "Well the flap gauge says they are symmetrical, but I don't trust it. Let's start slowing to (flap) placard speed and milk them down to see if we can fix this."

Unaware of the true nature of the ever-worsening problem, we had seized upon the flaps as the most likely source of our problem and began to slow a marginally controllable aircraft towards its inflight minimum control airspeed. In a nutshell, we were barking up the wrong tree and about to make the situation much worse.

It didn't take us long to realize that this was a stupid plan. As the flaps started down, Dan neared the stops on the control surfaces with no noticeable improvement in the control of the aircraft. Nearly simultaneously, it dawned on both of us that we were about to kill ourselves. Dan slapped the flap control handle to the fully retracted position, shoved the nose over, and firewalled the throttles, severely overboosting the venerable Pratt & Whitneys. We needed airspeed in a hurry, and that was something the old steam jet was reluctant to provide. [Although officially dubbed the "Stratotanker," the KC-135A had a variety of affectionate nicknames, including the "waterwagon" and "steamjet," which referred to the use of water injection to aid in takeoff performance.] But the turbine blades held together, and we traded altitude for airspeed.

Leveling off at 500 feet above Lake Huron, we notified departure control we were experiencing problems and needed a block of altitude from the surface to 5000. They complied.

At this point the Boom Operator, a 20-year-old from the hills of Pennsylvania and the only enlisted man on the crew, came forward. He was apparently attempting to ascertain for himself what was going on—being unable to tell from the incoherent interphone babble of the pilots. As he quickly surveyed the situation, he asked "Pilot, why are you guys draining the main tanks on the right wing? Do we have a leak or something?"

After giving each other a brief "I can't believe how stupid we are" look, Dan and I quickly stopped the fuel migration and set about to apply the time-honored aviation tradition of hiding the evidence of our mistake. Through a combination of high-speed, low-altitude, high-drag flight, we managed to burn all of the fuel out of the left (heavy) wing and enough out of the aft body tank to place the aircraft barely within CG limits for an on-time landing. Once on the ground, we swore an oath to each other not to breathe a word of our incompetence until such a time when we had all received our final promotions—a day long since past.

In retrospect, I am left with five lessons from this ugly episode. The first thing I learned was that adrenaline release occurs in two stages—a minor dose at the onset of serious difficulty, and the full load when terror sets in. Second, a self-induced emergency on a routine hop home can kill you just as dead as any catastrophic disintegration of multiple engines that we practice in the simulator. Third, checklist discipline and crew communication can keep you out of trouble when individual complacency sets in. If we had used the checklist and/or notified the crew as required by the flight manual, we may well have averted the problem. Fourth, don't ever forget your basic aerodynamics. If you are experiencing a controllability problem, accelerate, stabilize the aircraft, and don't change the configuration until you know what's going on. Finally, don't seize upon the first possible solution to your problem as the definitive answer. Keep your cross-check going.

In my nearly two decades of flying, I have never been closer to death. It is still difficult to believe how fast it all happened. We were a well-trained and highly proficient crew, flying a good jet on a routine mission, on a clear day. It was the proverbial "milk run." In retrospect, I guess it's easy to understand how we got our butts in the wringer. Perhaps due to our high state of proficiency and the relative simplicity of the mission (we had just returned from a 45-day temporary duty in England), we had grown complacent. Once the initial error was made, we seized upon the first possible cause without an analysis of the possible implications of our actions. If the Boom Operator had not come forward when he did, I honestly can't say what would have happened. I can just see all of our friends and buddies at the mishap briefing shaking their heads in disbelief, uttering things like "Man those dudes were really stupid. How many mistakes can you make on a five minute flight?" And they would have been right.

Bottom line: Fly the jet, use the checklist, fight complacency, watch your buddy, communicate problems early, and debrief your lessons so others don't have to make the same mistakes.

The crew on this aircraft was clearly fortunate. Although the ill-defined problem was self-induced, their process of assessing and solving the problem illustrates several important techniques, as well as hazards, involved with all types of airborne problem solving. The crew's initial solution aggravated the controllability problem by slowing the aircraft down. The problem was not diagnosed correctly until the boom operator came forward and identified the open drain valves on the right wing main tanks. The lesson learned is that it is extremely important to get as many people as possible working on a problem as quickly as you can. In this case, the pilots had two problems not shared by the other crewmembers. First, they were complacent and overconfident, which may have led them to assume that they could not have made such a simple mistake as opening the wrong valves. Second, they had their hands full of aircraft and had to devote a great deal of their attention to maintaining aircraft control and coordinating with air traffic control about their problem.

There are cautions involved with increasing crew involvement. If the team does not function well, it can make matters worse instead of better. Communication and assertiveness are key elements to effective crew coordination under conditions of high stress.

Decision-making hazards

The previous case study dramatically illustrates the importance of situation assessment and a team approach to finding solutions to ill-defined problems. The pilots failed to determine the actual cause of the problem prior to initiating "corrective" action. Because the crew wanted so desperately to find a solution to the ever-worsening problem, the strength of the first idea swayed their judgment and actions. These hazards to good judgment and effective decision making deserve a closer look.

Strength of an idea

The human mind naturally seeks to resolve dissonance or conflict. In flight, this tendency can lead to a dangerous situation in which an aviator's need to resolve internal conflict overrides his or her good judgment and decision-making skills. There are two primary ways in which this phenomenon can occur. The first is in a time-critical situation, when a solution is needed right away. The case study of the KC-135 crew illustrates the tendency to grasp the first seemingly acceptable solution that is offered when confronted with a rapidly deteriorating situation. This phenomenon can be made even more tempting if the individual who proposes the solution is of higher authority or a recognized "expert." This tendency to defer to greater rank or perceived knowledge is also referred to as "excessive professional courtesy" or the "halo effect." (For a more complete discussion of hazardous attitudes, see Chapters 4 and 6.) Individuals on crewed aircraft must be especially concerned with this problem, but the tendency to defer to premature solutions proposed by others can also come from outside of the aircraft, such as air traffic controllers or maintenance systems experts on the ground, who are trying to help with your airborne situation. A similar decision-making hazard occurs when the need for consensus within the group leads to poor judgment.

Groupthink

There are many circumstances in which the group does not perform as well as the individual, and it can be especially critical in airborne operations. Recall from Chapter 9 that the situational awareness of the group is limited to that of the individual controlling the aircraft at the moment of truth. Similarly, in decision making, a single individual may hold the vital piece of information needed to resolve a particular problem. If this person is nonassertive and sees the group consensus moving in another direction, it is likely that they will withhold information that might tend to cause conflict within the group. This problem is compounded in the commercial and military settings, where challenging rank can have profound career implications. Even in general aviation, the tendency to go along with the most experienced flight crewmember—especially if he or she is an instructor—can inhibit good judgment and decision making.

One technique to help resolve this problem is for the pilot in command to make assertiveness and contribution to the team decision-making process a major point in each and every preflight briefing. Additionally, the pilot in command should actively seek individual inputs from all sources of information prior to making a decision if time permits.

Scheduling hazards: Delays and premature actions

Situation assessment and decision making must be followed by action, but many good decisions are delayed beyond the point of usefulness or jumped at prematurely. The adrenaline release associated with inflight emergencies can do funny things to a flyer's mind. This phenomenon, also known as temporal distortion, can make seconds seem like hours. On occasion, it can have the reverse effect. Crewmembers who have survived ejections from combat aircraft will often take several minutes describing every detail of an event that only lasted a few seconds, with each and every detail seemingly stretched out in time. However, temporal distortion seldom affects two crewmembers in the same way, nor does it affect the inner movements of a clock or watch. Therein lies the aviator's solution to dealing with the scheduling function of decision making.

On making an accurate assessment of the situation and analyzing the immediate and potential risks, the pilot in command should establish a schedule for action and execution of the plan. Realizing the tendency for the mind to play temporal tricks, this schedule must include actual clock times and not just "in five minutes we will . . ." type statements. On crewed aircraft, if time and work load permit, one crewmember should be designated as the timekeeper for the plan to ensure that actions are implemented on schedule. If flying in formation, delegate this responsibility to a wingman. If flying alone, take a moment to write down a written schedule for action if you can do so without sacrificing aircraft control. The bottom line here is to expect temporal distortion to occur, so create a backup for your malfunctioning internal clock.

Seeking a perfect solution

Many type "A" personalities are prone to chasing the airborne equivalent of the Holy Grail—a perfect solution to an inflight challenge. But aviators should not always be interested in finding the *best possible* answer to a given situation but rather one that works in a timely manner. Some challenges present us with situations that deny multiple options. In these cases, we have to scramble to find any workable or safe solution. The dynamic interchange between human, machine, and environment makes the "perfect" decision almost indefinable. In fact, the chase for an elusive decision that will maximize safety, effectiveness, efficiency, and save gas at the same time is likely to lead to distraction that can worsen the original problem beyond the scope of solving it at all. Many accidents have occurred while aviators were too busy chasing the mice to see the elephants bearing down on them.

A tactic for dealing with this compulsive tendency for perfection is to *first* seek a safe and workable solution, and then improve upon it after the situation has stabilized and the initial corrective action is underway. Often the need for the perfect fix disappears after a workable solution has been implemented, eliminating the potential for distraction or delay.

Hidden agendas

There is usually an underlying cause for poor judgment, and it is often a drive to accomplish a personal objective or hidden agenda. Aviators must be consciously aware of these tendencies and take a mature approach when competing personal interests conflict. The classic example is the case of "get there-itis," in which the need to arrive at a location interferes with good risk assessment, often causing a pilot to take foolish and unnecessary actions to get somewhere on time. (For a more detailed description and antidote, see the discussion on hazardous attitudes in Chapter 4.) The traditional antidote for this judgment disease is to tell yourself that it's "better late than never" and to realize that these words could have saved the lives of hundreds of flyers who tried to press too far to get somewhere.

Hazards associated with knowledge and experience

Knowledge, procedures, and experience can be double-edged swords, causing decision-making problems if the aviator is not aware of the potential pitfalls associated with each. Our decision-making process tends to gravitate to areas of knowledge that we are comfortable with. For example, an expert on an aircraft's electrical system will look for ac/dc issues in every problem, whereas a golden-hands type will try to fly out of his or her problems. Unfortunately, the problems that arise in flight are certainly not constrained in this manner. In fact, it often seems as if inflight problems have an eerie way of finding exactly what it is that we don't know and then exploiting it.

Experience, while usually viewed as a positive, can create its own set of unique problems. With experience, aviators develop internal models, or *heuristics*, with which to make routine decisions more automatic. The explanation of heuristic decision making is a rather complex aspect of aviation psychology and goes beyond the scope of this text. Consider it simply a cognitive shortcut. While this shortcut can be valuable, it also creates a potential for danger. Under certain conditions, experience can create a mindset of expectations in which the pilot sees what he or she believes instead of believing what he or she sees. The military literature on "friendly fire" fratricide incidents is full of accounts where pilots fired on friendly forces because they expected to see the enemy at a given time or location. Commercial and general aviation are not immune to this problem, and many highly experienced flyers have fallen into the trap of not giving appropriate attention to a condition that they were accustomed to seeing or a task that they had accomplished successfully in the past. This phenomenon, when compounded by the complacency that often accompanies high experience levels, shows experience to be truly a double-edged sword.

Summary of effective judgment and decision making

The foundations and pillars of airmanship form the basis for good judgment and decision making. There is no room for imperfect flight discipline, and there are no shortcuts to adequate preparation in the skill, proficiency, and knowledge arenas. Once these prerequisites are in place, an aviator can successfully and systematically improve judgment and decision making by understanding and practicing proven concepts and by avoiding common pitfalls.

All good judgment and decision making is grounded in accurate situation assessment. For a pilot or other crewmember to take appropriate actions in flight, this assessment must answer three crucial questions. The first question is "What is the precise nature of my problem?" The pilot of the ill-fated KC-135 who lost water injection on takeoff was focused on a performance rather than a control problem, a distraction that contributed, if not caused, the disaster. The second question a flyer should ask is "What are the risks associated with this situation?" The new F-16 pilot that continued flight into IMC in an area of known high terrain clearly did not effectively analyze the inherent risks associated with his decision. The final question in situation assessment is "How long do I have to work the problem?" The importance of this scheduling function was clearly illustrated by the captain and crew of United Flight 173, an oversight that ended with fuel exhaustion and disaster in a Portland, Oregon, suburb (see Chapter 9).

Understanding the types of decisions made in flight allows an aviator to focus on key ingredients of success for each decision type. Aviation decisions generally fall into three basic types. Rule-based decisions are seen as the most straightforward, but they require at least two airmanship prerequisites for effective usage. First, the aviator must possess knowledge of the applicable rules and directives. Second, he or she must have the personal integrity and discipline to apply them when appropriate and not just when convenient. The failure of the Evergreen International crew to apply well-known regulatory guidance demonstrates the potential for disaster from even a single intentional oversight during a rule-based decision.

Knowledge-based decisions fall into the well-defined and ill-defined categories. Well-defined problems require checklist and procedural knowledge and discipline. Ill-defined problems require more complex problem-solving skills. The KC-135 crew that inadvertently drained fuel into the wrong tanks failed to practice good checklist discipline, which created an ill-defined problem with a tight timetable for resolution. The value of a team approach to problem resolving was also seen in that example, with the input of the enlisted boom operator seen as the key to breaking the error chain and avoiding disaster.

Several pitfalls should be avoided if one is to consistently practice sound judgment. A pilot must take care not to grasp at the first available solution or one that is offered by a higher-ranking crewmember or so-called "expert." The strength of an idea should be based on its validity for the given situation and not its origin or timeliness. Care should also be taken to avoid groupthink, a phenomenon in which a lack of assertiveness prevents key information from being presented in an effort to promote group harmony.

The scheduling function is critical to problem solving in flight. Temporal distortion should be expected and accounted for. Use the clock to help, and, if possible, jot down a written schedule for action on your knee board, or select a wingman or crewmember to perform the role of timekeeper until the problem is solved. Expect your internal clock to malfunction, so don't be phased when it does.

The tendency for pilots in command to seek a perfect solution often results in delaying necessary actions. A recommended tactic is to find a workable solution and then set about improving it if time and conditions permit. All aviators must also be aware of hidden agendas that may drive inflight decisions and actions out of the realm of good judgment. Finally, caution must be exercised by those with expert knowledge and experience to prevent expectations from weighing in too heavily on actual circumstances or creating an atmosphere of complacency.

A final perspective

Judgment is inexorably tied to the individual, and each decision we make is affected by our personal preparation, experience, and personality. Therefore, each of us must make a searching self-assessment to determine our current state of readiness to make quality decisions. There is no magic in this process, only a thorough and self-disciplined approach to improvement across all areas of airmanship. Because of the complexity of aviation decision making, judgment is best viewed as a combination of art and science. The science can be learned and, in fact, was a stated goal of this chapter, but the art must be practiced and refined to approach your own personal judgment potential. This improvement can be facilitated by actively critiquing your own inflight decisions. After each flight in which an important decision had to be made, ask yourself what factors were present, what could have happened if events had unfolded differently, and how your course of action could have been improved. These questions should be asked following every flight, but improvement and understanding can also be accomplished vicariously, through case study analysis, hangar flying, and just listening to others tell their stories.

Judgment is not a magical quality; it can be taught and learned. The improvement process demands disciplined attention and constant self-critique, but if one is willing to seek serious improvement, judgment can be developed just like any other inflight skill. In the final analysis, our judgment is as good as we want to make it.

References

Federal Aviation Administration (FAA). 1988. *Introduction to Pilot Judgment.* FAA Accident Prevention Branch. FAA Pamphlet P-8740-53.

Hughes Training Inc. 1995. *Aircrew Coordination Workbook: Military Mishaps.* Abilene, Tx.: Hughes Training Inc.

Jensen, Richard, and R. S. Benel. 1977. Judgment evaluation and instruction in civil pilot training. *Aeronautical Decision Making for Helicopter Pilots.* Eds.: Richard Adams and Jack Thompson. Federal Aviation Administration: DOT/FAA PM-86-45, p. 3.

Klein, G. A. 1993. A recognition-primed decision (RPD) model of rapid decision making. *Decision Making in Action: Models and Methods.* Eds. G. Klein, J. Orasanu, R. Caulderwood, and C. Zsambok. Norwood, New Jersey: Ablex.

Magnuson, Kent. 1995. Human factors associated with mishaps database. Kirtland AFB, N.M.: Air Force Safety Center.

National Transportation Safety Board. 1989 Report brief. Accident ID #FTW89MA047, File #944.

Orasanu, Judith. 1993. Decision making in the cockpit. *Cockpit Resource Management.* Eds.: E. Wiener, B. Kanki, and R. Helmreich. San Diego: Academic Press.

————. 1995. Situation awareness: Its role in flight crew decision making. NASA Ames research paper. In nasa.gov/publications/OSU_Orasanu. Internet.

Rasmussen, J. 1993. Deciding and doing: Decision making in natural context. *Decision Making in Action: Models and methods.* Eds.: G. Klein, Judith Orasanu, R. Caulderwood, and C. Zsambok. Norwood, N.J.: Ablex.

Webster's Ninth New Collegiate Dictionary. 1990. Springfield, Mass.: Merriam-Webster, Inc.

11

Airmanship on the edge: The total package in action

The man who is prepared has his battle half fought.
Cervantes, *Don Quixote*

A personal view from Jim Simon, Major, USAF

In aviation, you never know when or where your next trouble is coming from, a fact I learned vividly on a cool moonless night in the mountainous terrain of southeast Turkey. The war had been on for several weeks, and our mission was one we had practiced and briefed many times. The helicopter and crew was prepped to infiltrate a special forces (SF) team into Turkey and provide close air support should the team get in trouble. The cloudless sky was giving way to a thick valley fog that made the naked eye blind below 200 feet—creating conditions in which our night-vision goggles, our MH-53J Pave Low III, and our training and teamwork would be in high demand.

As we descended into the mist toward our landing zone (LZ), our gunners and flight engineers in the cabin began their chant as they talked us down toward the ground. "Two hundred feet, 60 knots—clear down right—left—tail. One hundred feet—30 knots . . . ." All the while our focus stayed on the mission to get the SF team to the right place at the right time—always the team. So far it had been a routine mission, but that was all about to change.

All of our training paid off when we suddenly started receiving indirect machine-gun fire out of the fog from our three o'clock position. The aircraft commander, Captain Dennis "Jonesy" Jones, made a split-second decision only years of training could have prepared him for. The crew reacted to his commands with precision and symmetry. He quickly ordered the crew to lower the ramp and directed the SF team to rapidly disembark. He knew that the elite Special Forces troops were safer on the ground, where they knew how to fight the best, and that we, with our lights-out, comm-out configuration, had a few more seconds before the shooter could home-in on the roar of our Pave Low in the thick fog. We shot out of the LZ at low altitude and made radio contact with the SF team in anticipation of having to engage our unknown attackers. But the firing had stopped, and our 50-caliber services were not required.

We had been caught slightly off-guard, because we did not expect to be fired on that night. This was not because we were cocky or unprepared for combat, but because it was a training mission in friendly territory! We were never told whether the bullets that were meant for us were Kurdish rebels fighting the Turks or Turkish soldiers who might have been engaging what they thought to be an Iraqi helicopter. The lesson is, you never know where your next challenge will come from. Needless to say, the debrief that night centered on expecting the unexpected, flying like you're going to fight, and keeping a round in the chamber.

Major Jim Simon is a special operations pilot in the United States Air Force.

Aviation is ideally suited for heroes. Most flyers occasionally daydream about making a miracle recovery of a disabled aircraft or flying a perfect military mission. The would-be heroes see themselves performing coolly under intense pressure—their trained mind and hands deftly manipulating the stricken craft to save the day. Sometimes these stories are not just daydreams. Sometimes they are all too real.

The following case studies illustrate ultimate airmanship under extreme conditions. In both cases, the situation is grim and lives are on the line. They illustrate circumstances so severe that anything less than the total airmanship package would have been insufficient. But these airmen, on these days, had what it took. The chapter concludes with an analysis of the airmanship characteristics exhibited by the heroes of these scenarios and poses the question: If you were placed in a similar extreme situation, would you have what it takes?

We begin with an example from commercial aviation. The recovery of United Airlines Flight 232 has been called the "Iowa Miracle," but to Captain Al Haynes, it was no miracle, but rather the end result of years of preparation—and a little luck.

Case study: Coping with a "one in a billion" loss of all flight controls

by Capt. Alfred C. Haynes, United Airlines

At 1516 hours on July 19, 1989, the author of this case study was captain of United Flight 232, a McDonnell Douglas DC-10-10. While cruising at 37,000 feet, the aircraft suffered a catastrophic engine failure. The uncontained disintegration of the number-two engine's fan rotor caused the loss of all three of the aircraft's redundant hydraulic flight control systems and made the aircraft almost uncontrollable. Captain Haynes and his crew, augmented by a DC-10 instructor pilot who was aboard as a passenger, were able to navigate to the municipal airport at Sioux City, Iowa, where the aircraft was crash-landed approximately 45 minutes after the hydraulic failure. Of the 285 passengers and 11 crewmembers aboard, 174 passengers and 10 crewmembers survived.

That the aircraft was controlled at all and that anyone survived in this unusual circumstance was recognized by the industry as extraordinary airmanship by the crew. Among many other accolades, Captain Haynes and his crew were awarded the Flight Safety Foundation President's Special Commendation for Extraordinary Professionalism and Valor during the foundation's 42nd International Air Safety Seminar in Athens, Greece, during November 1989, the first formal international recognition of their accomplishment. Captain Haynes subsequently assembled his reflections on what happened during the ordeal along with input from a paper on the Sioux City area's response to the disaster by Michael T. Charles, Ph.D. The following is what he considers five primary factors involved in making it possible to cope with a major inflight emergency such as the one-in-a-billion loss of all flight controls. It is adapted and reprinted by permission of the Flight Safety Foundation and the author.

Factor number one is called luck

There are five very important factors that contributed to the degree of success that occurred during the events that led to our landing at Sioux Gateway Airport: luck, communications, preparation, execution and cooperation.

The occurrence of good fortune may have differing connotations according to individual personal beliefs, but for this discussion we will call it luck. You could say we had bad luck in the form of an occurrence that, according to all the odds, could not be expected ever to happen. But we also had good luck, in that we were left with a chance to survive.

For instance, how did we get the aircraft in Sioux City to begin with, after the six-foot diameter fan on the number two engine (the center engine located on the vertical tail) failed? In less than one revolution, it instantaneously sprayed fan blades and pieces out through the right side of the engine housing, cutting a large gash in the forward edge of the horizontal stabilizer, separating a 12- by 10-foot cone section of the tail and peppering the horizontal stabilizer with more than 70 pieces of shrapnel.

The hydraulic lines of one of the three systems were torn out with the accessory section of the failed engine and the shrapnel from the disintegrated fan instantly severed the hydraulic lines of the remaining systems. The staggering import of that instantaneous and absolute occurrence can be better realized by an understanding of the workings of the flight control system on the DC-10.

In company with the Lockheed L-1011, Boeing 747, some military aircraft and others that are coming along, the DC-10 does not have cables running between the pilot's control column and the control surfaces that, in jumbo jets, can involve extreme distances. Long cables can develop slack, can bind and require considerable attention to lubrication and maintenance. This older technology is being replaced by electrical/electronic and hydraulic connections between the pilots' controls and the flight surface actuators.

The cables from the control columns in the DC-10 lead only a short distance to the hydraulic controls that actuate the control surfaces through hydraulic piping that leads to actuators at each control in the wings and tail. There is no provision for manual operation of the flight controls. This is because of the triple redundancy afforded by the three independent and continuously operating hydraulic systems. These systems are intended to provide power for full operation and control of the aircraft in case one, or even two, of the hydraulic systems fail.

A catastrophic failure of all three systems was considered by designers to be almost infinitely remote; we had enough redundancy in the system to feel that this was not going to happen. I was told one time that if I have odds of ten to the ninth power against something happening, that is a one in a billion chance. And that is what the designers said was the case here. In the DC-10, the three hydraulic systems are not connected, fluidwise, so if the fluid is lost in one system, it will not lose the fluid in the other two. The systems are connected mechanically through an electric pump so that if the pressure in one system is lost, the pressure in another system will immediately start a motor that will drive a pump in the failed system and the three systems will be restored to service.

All of the separate hydraulic lines are routed to different control surfaces, which themselves are split for actuation by different hydraulic systems for even further redundancy. Hydraulic system number one operates some sections of the major controls such as ailerons, rudder and elevator and flaps; system two runs other sections of the same control surfaces; and system three adds further backup. So the three systems can actually be reduced to one hydraulic control system and the pilot can still control the airplane. And if all else fails, there is an air-driven generator that is extended from the underside of the airplane to drive one of the emergency hydraulic pump motors in the tail.

So, everyone was confident that the "impossible" complete loss of all flight controls was not going to happen.

But on July 19, 1989, Murphy's Law caught up with the airline industry and our aircraft lost all three hydraulic systems. That left us at 37,000 feet with no ailerons to control roll, no rudders to coordinate a turn, no elevators to control pitch, no leading edge devices to help us slow down for landing, no spoilers on the wings to slow us down in flight or to help braking on the ground, no nosewheel steering and no brakes.

That did not leave us a great deal to work with.

What we did have was control of the number one throttle and number three throttle, so by adding thrust on one side and reducing thrust on another we could force the airplane in a skid to turn one way or another. Our biggest problem was pitch control. With no pitch control, and just the slight amount of substitute steering capability we had, it is a wonder to me that we ever got the airplane on the ground, and I attribute that to a great deal of luck. The things that we happened to try that day (not having any idea of what would result because this situation had not been expected or practiced for) happened to be the right things, and they happened to work. So luck played a very big part in even getting the airplane to respond.

The second lucky thing was the location. We could have been half-way to Honolulu, or over the middle of the Rocky Mountains, or we could have been taking off right over a city. But as it was, we were over the reasonably flat lands of Iowa, which gave us a little bit of confidence in our minds about survival. I had serious doubts about making the airport at times, but the four of us in the cockpit did have some feeling that if we could just get the airplane on the ground, because of the flat farmland below, we could expect survivors. That helped in the back of our minds as we were trying to fight this problem.

Luck with the weather was another important factor. If you have ever flown over the U.S. Midwest in July, you know there is usually a line of thunderstorms that run from the Canadian border down to Texas, and it would have been absolutely impossible with marginal control of the aircraft to get through a thunderstorm safely. As a matter of fact, one year to the day later, when we were in Sioux City for a memorial service for those on Flight 232 who did not survive, there was a huge thunderstorm directly overhead; had that storm been there the day of the accident, there was no way we could have gotten the airplane in to the airport at all. So the favorable weather was very important.

The time of day that our engine failure occurred was another very lucky circumstance. Almost four o'clock in the afternoon, it was approaching shift change at Marion Health Center and St. Luke's Hospital and all the other emergency services around the Siouxland area (Sioux City and surrounding communities). By the time we did arrive in Sioux City, our plight having been reported, the morning shifts were just going off duty, so both hospitals were double staffed. Further, there were so many volunteers from the various emergency units and health clinics around the area that the hospitals had to turn some of them away.

And, as a final piece of luck, it was the only day of the month when the 185th Iowa Air National Guard was on duty, and there were 285 trained national guard personnel standing by waiting for us when we got to Sioux City. So, put all of those things together and it is just an unbelievable amount of luck that helped us get the airplane there, to receive the level of help we did and to experience the survival rate that we had.

Factor number two is communications

A second big factor in our favor that day was communications, which played another very important part in our having the survival rate that we had. It started in the cockpit with communications to air traffic control in Minneapolis Center and then with Sioux City approach control and control tower.

Communications within the cockpit intensified as soon as we realized that we were in very serious trouble. I turned to our second officer Dudley Dvorak, and said, "Dudley, get in touch with San Francisco aero maintenance (SAM) and see if there is any way they can help us, in case they know something we don't." We had run out of ideas in about 20 seconds, and we needed some help. Dudley got on the radio and spent that entire time communicating with our SAM facility that had a group of experts who were immediately brought in. They got on the computers and checked through the log books to see if there was any information they could find that could help us. As it turned out, of course, there was nothing they could do to help us.

However, the communications that were established with SAM had the secondary benefit of allowing our dispatch center in Chicago find out where we were and what we were going to do. We certainly did not have time to call them separately, as well as SAM and ATC (air traffic control) to tell them what was going on, so they monitored what Dudley was saying to SAM. Therefore, the personnel in our Chicago flight center were so prepared for us to go to Sioux City that they pulled an airplane out of a hanger in Chicago, loaded it with emergency supplies and people, flew it to Sioux City, and some of our staff were in the hospital before I was admitted to my room. That is how quickly they responded. And it was done through communications, good positive communications on Dudley's part.

Kevin Bauchman, who happened to be on the radar console at Sioux City Approach Control at the time, became our primary contact with ATC services after we were handed off from the center. He was backed up by a team of five controllers in the Sioux Gateway Airport control tower, co-located with the approach control, who worked together to coordinate the many aspects involved in preparing for the arrival of our aircraft.

If you have a serious problem like we did, and you need the kind of help that does not add to the tension level, a voice like Bauchman's, as calm and as steady as he was, certainly was an influence on us and helped us remain composed. The only time that Kevin's voice ever cracked was when he found out that we were in position to land on runway 22 instead of runway 31 and he had emergency equipment sitting in the middle of our intended landing area—and only had two minutes to get it out of the way before we arrived. He raised his voice just slightly and then fell right back into his calm, soothing, here-it-is voice. When I had the opportunity later to compliment Bauchman on his coolness throughout the tense situation, he told me, to my surprise, that he had transferred to Sioux City because he found his previous duty station too stressful.

Meanwhile, extensive communications were going on between the emergency response dispatcher and the individual ground units. They worked extremely well together. We heard controller Bauchman tell us, "We have equipment standing by at the airport and we also have them out in your direction." When we reported that we may not make it to the airport, emergency dispatchers actually called local communities in the area we were flying over and had them dispatch emergency vehicles out to the highways to find us and follow us, and if we landed out there somewhere to be as close as they could possibly be. Good, very good communications.

The communications between the cockpit and the cabin crew were good, as good as they could be, considering the intense work load in the cockpit that precluded extensive contact with the cabin crew. And the communications between our flight attendants and their passengers were good all the way around—a very important factor.

Preparation is the third factor

Preparation also was an important factor for the members of the cabin crew of Flight 232. Their preparation was done through recurrent training every year, where they are taught to inform passengers how to prepare for an emergency landing. To be thrust into an actual emergency situation with disastrous implications was a great shock to all eight of the flight attendants (nine, actually, because one who was deadheading helped out). They had been practicing this procedure during their entire time in service from one month for our most junior flight attendant to somewhere between 15 to 20 years for the most senior one, but they never dreamed they would ever have to do it. Through proper preparation by our training center, which I am very confident is also practiced by other carriers, they were able to do what they had to do because they were prepared.

And then you come to us — how does the cockpit crew prepare for a situation like this? Well, there is no training on earth that can prepare you to do this. This is something that conventional wisdom said could not happen, would never happen, and therefore there was no procedure in the book for doing it: so preparation in that respect was not possible. But through the constant recurrent training that we accomplish every year, practicing in the simulator types of emergencies that we can expect, we were supplied with some clue as to how to start. I am firmly convinced that the best preparation we had is a program that United Airlines started in 1980 called Command Leadership Resource Management (CLRM) training. It is now referred to as Cockpit Resource Management (CRM).

The CRM training program was instituted following a couple of accidents that debunked the old axiom which states that what the captain says is law and implies that he is omnipotent. When the captain does not know how to resolve a situation, why should he be the only one involved in the problem-solving effort? During the development of this training program, we found that there is a lot of experience sitting in the other two seats, so why not use it? Why not train the other flight crew members to respond in such a way that the captain will consider their advice and utilize their knowledge?

This program was started as a kind of a game. Three chairs were arranged in a room, one each for the captain, the copilot and the second officer. A scenario was given to the crew members for them to follow. In a training session I participated in, I played the part of the copilot and I was to be the only one with any sense. The captain was what we call a demigod, he played the part of the autocrat who said, "If I say we're going, we're going," and his intention was to take off into a thunderstorm. My job, according to the script, was to keep him from doing it any way I could. The second officer was to act like a wimp. He did not care what the captain did; he would go along with it.

We all played our parts well. The captain was a complete tyrant. We did not like him, and I really almost did hate him by the time we completed the exercise. But what we did, was work together to try and find a way to stop the captain from putting the aircraft and its occupants into an unsafe situation. It reached the point where the only way I could stop him from taking off was to just stand on the brakes. Now, it is very difficult to make an airplane take off while the copilot is sitting on the brakes, and the goal of the exercise was attained. Thus, we focused on the main point of the CRM training program, which is that it teaches the crew to work together. It showed that we could have resolved this problem much more amicably and efficiently among the three of us if we had all applied the principles of CRM.

I am firmly convinced that CRM played a very important part in our landing at Sioux City with any chance of survival. I also believe that its principles apply no matter how many crew members are in the cockpit. Those who fly single-pilot aircraft sometimes ask, "How does CRM affect me if I fly by myself?" Well, CRM does not just imply the use of other sources only in the cockpit—it is an "everybody resource." To these pilots I say there are all sorts of resources available to them. Ask an astronaut if he thinks he got to the moon by himself. I don't think so—he had a great deal of help.

All pilots have a lot of help; all you have to do is ask for it and use it when you get it. What would we on Flight 232 have done without Kevin Bauchman, the ATC controller who did so much to get us to the Sioux City airport? The DME (distance measuring equipment) at the Sioux City VOR (visual omni range) navigational transmitter was not working that day, and we had no idea how far we were from the station. By informing him of our situation and asking him for help, he kept us constantly informed as to where we were, and adding the height information from our altimeters, we were able to get to Sioux City. The bottom line for pilots is that you have resources available to you. Use them as team members—you are not alone up there. If you do have a copilot, listen to her or him. They are sure to have some advice for you. There were 103 years of flying experience in that cockpit when we faced our nemesis and it came through to help—but not one minute of that 103 years had been spent operating an airplane the way we were trying to fly it. If we had not worked together, with everybody coming up with ideas and discussing what we should do next and how we were going to do it, I do not think we would have made it to Sioux City.

Execution comes next

Now for the execution. How did everyone accomplish what each was trying to do during the emergency over Iowa? We will begin with the cockpit crew. When the engine blew, William R. Records, the first officer, was flying the aircraft in the copilot's position. It was his leg, and the aircraft was on autopilot. Bill has about 26 years of flying experience with National, Pan Am and United.

The rest of the flight crew members were sitting there, on this beautiful day after lunch, having a cup of coffee, watching the world go by, when without any warning whatsoever there was a very loud explosion. At first, I thought it was a decompression. It was that loud and that sudden. But there was no rush of air, no change of pressure and no condensation of the air in the aircraft. So I had to figure it was something else.

I saw Bill immediately grab the control yoke and the red warning lights illuminate for the autopilot. He had cut the autopilot off, I thought, and I assumed that he was taking over manual control of the airplane. Now, I thought, we have taken care of step one in any emergency and that is that someone flies the airplane. We have had a number of accidents in commercial aviation because everybody was working on the problem, which sometimes is not a big problem in the first place, and no one is flying the airplane. So step one, in any training center, is that somebody flies the airplane. That is a little difficult if you are going to be by yourself. But that is still the first thing you have to do: fly the airplane.

I thought next, now that Bill is flying the airplane, I can divert my attention to Dudley—second officer Dudley J. Dvorak—and we can shut the engine down which is our job. So he and I determined that the number two engine had failed, and at the time we thought that was all that was wrong. I called for the checklist and Dudley got out his book, laid it on the console, and read the first item of the engine shutdown procedure. He said, "Close the throttle."

And the throttle would not close.

Now, I have never shut a jet engine down before in flight, because they have become too reliable. This was my first experience of losing an engine in flight on a jet aircraft. In a simulator, you pull the throttle back and it goes back. This throttle would not go back. That was the first indication that we had something more than a simple engine failure. Number two item on the checklist was to close off the fuel supply to the engine.

The fuel lever would not move — it was binding.

About this time, Dudley said to actuate the firewall shutoff valve. I did that, and the fuel supply to the number two engine was finally shut off.

By then, we were about 14 seconds into the episode, and Bill said to me, "Al, I can't control the airplane."

My focus quickly changed from the engine controls to the copilot. The first thing I noticed as I swung around was that Bill had applied full left aileron, something that you would never see in the air, much less at 35,000 feet. Further, he had the control column completely back in his lap, calling for full up-elevator. That is something else you would never expect to see in flight. But what really caught my eye was that with the control yoke in this condition, the airplane was in a descending right turn and at increasing angle.

With all those pilots on board, I then said the dumbest thing I ever said in my life: "I've got it." Well, I took control of the aircraft, but I surely did not know what I was going to do with it. Bill was absolutely right—the airplane was not responding to the control inputs. As the airplane reached about 38 degrees of bank on its way toward rolling over on its back, we slammed the number one (left) throttle closed and firewalled the number three throttle—and the right wing slowly came back up. I have been asked how we thought to do that; I do not have the foggiest idea. There was *nothing left to do, I guess, but it worked. There is another instance where I talk about luck; we tried something that we did not know what to expect from and we discovered that it worked.*

For the next few minutes we were trying to fly the airplane with the yoke and it took both pilots to do it. One person could not handle the yoke by himself because the pressures on it were just too great. We both had to do it. At the same time and trying not to let go of the yoke, we had to work our hands around the frozen number two throttle and make a number of quick adjustments to the number one and number three throttles to help control the aircraft. We had to close one, open one; open one, close one. We did this for a while and, in the meantime, Dudley was on the radio trying to get us some help.

After about 15 minutes of this, and talking to the ATC center and getting directions to Sioux City, we were advised that Capt. Dennis E. Fitch—an instructor pilot for the DC-10—was a passenger in our aircraft. Considering the aura that surrounds flight instructors, we, naturally, invited him to the flight deck. Maybe he knew more about the systems than we did and could help us out of our dilemma. He arrived, took one look at the instrument panel and that was it—that was the end of his knowledge, too. He also had not faced this situation before.

I asked Fitch if he would go back into the cabin and look at the controls. He came back and said, "The controls aren't moving, how can I help now?" We were still struggling with the yoke and the throttles at the same time, and once again, out of the blue came a decision, "Take the throttles and operate them in response to our commands. Take one throttle lever in each hand—you can do it much smoother than we can—and see if we can't smooth this thing out and get a little better control of the airplane." For the next 30 minutes, that is how we flew the airplane to Sioux City's Gateway Airport.

In the meantime, we began to realize that we had to work up some kind of system so that the airplane and the Sioux City airport would be at the same place at the same time. We took the formula normally used during descents in a DC-10: for every thousand feet you descend, you will travel three miles. Using that formula as a guide, because we certainly were unable to maintain a consistent rate of descent, we began a series of right turns because the airplane wanted to turn right all the time. That was one of the other little problems we had, the fact that we could not keep power the same on both good engines. If we ever allowed the throttles to remain at equal power settings, the aircraft would roll over, so we could not close or open them both at the same time.

To add to our problems, as if we did not have enough of them, there were the phugoids. These are longitudinal oscillations that are induced when an aircraft is displaced longitudinally from a stable and level flight condition. Generally they are supposed to dampen themselves out after a few nose-up and nose-down cycles if the airplane is trimmed and power setting is constant. However, in our case with continuing wide variances in the power settings of the two remaining engines that are mounted below the center of gravity and with no power from the engine mounted above it, the phugoids, induced by the loss of number two engine power and initial dive-roll of the aircraft, never became completely controllable. We did our best to minimize them while trying to keep the airplane right-side-up and trying to navigate to the airport.

Attempts to control the phugoids became a delicate balancing act. In an aircraft that is trimmed to fly in level flight, if you should push the nose down a little bit, it picks up airspeed, but it wants to fly at the speed it was trimmed for, so it will seek its trim speed all by itself. When you release the nose-down pressure the airplane will come back up, overshoot the airspeed, slow down and the nose will lower again. After a few repetitions, it will level off and fly right where it was before being disturbed.

Our aircraft was trimmed for 270-knot cruise flight before the engine failed, and this is the trim speed it sought to return to for level flight, regardless of the fact that the top-mounted engine was producing no thrust (which would have intended to add a slight nose-down influence) and we were often adding large amounts of thrust in the two engines that, mounted beneath the wings, tended to raise the nose. Each phugoid oscillation took between 40 and 60 seconds, so the airplane was not really behaving erratically, it was fairly steady as far as the passengers were concerned.

However, the technique to dampen out the phugoids is to react just the opposite of what you think is normal. When the nose starts down and the airspeed starts to build up, you have to add power, because you want the pitch-up tendency created by the two, underwing-mounted engines to bring up the nose; but the hardest part is when the nose starts up and the airspeed starts to fall, you have to close the throttles, and that's not very easy to accept doing.

Adding to the problem, when we needed to add power or to close the throttles for phugoid control, it was necessary also to add or take away power on either side to keep the airplane from going over on its back. As a result, we could never eliminate the phugoids, and they went on for 41 minutes.

At one point, Fitch traded seats with Dvorak about the time we were ready to land. I said, "Let's get ready for landing. Denny [Fitch], you sit here and strap yourself in." Then, Dvorak swung around into a position where he could reach the throttles, and began to manipulate them. Fitch had been handling the throttles for about 20 minutes and he had a feel for what it took to give us what we needed. Dvorak, of course, did not have the benefit of that practice. He responded according to our calls such as "we need a wing up," "need a little more," "that's not enough," or "we need a little less turn," "you need a little more turn."

It soon became obvious that, although Dvorak was the regular crew member of the two, Fitch had developed a level of expertise at this entirely new skill. In another instance that illustrated the benefits of CRM, at Dvorak's suggestion, we decided it was better that Fitch sit by the throttle controls because he had been doing it. Dvorak got up and gave Fitch his seat and Dvorak took the seat behind me, the jump seat. That is why the deadheading captain was sitting in the second officer's seat when we landed.

Bauchman was vectoring us for runway 31 and when we got down to about 3,500 feet and saw a runway straight in front of us, we could not believe it. We were shocked. There was a runway and that was what we were going to land on. That is when the controller lost his cool a little bit, because it was not the runway he expected us to line up with. He had about three fire trucks sitting on runway 22 and we were lined up to land there.

Unfortunately, just as the airplane came over the trees as we approached the airport, the airplane began one of its down phugoids. We were about 300 feet in the air and the DC-10 decided that it was going to start down—the nose went down, the rate of descent increased, the airspeed increased, and we hit the ground.

We first touched down on the right main gear, the right wing tip and the number three engine. The nose wheel made contact just about simultaneously, and then the left main gear was slammed into the ground.

As we hit the ground, the tail broke off. The right wing tip also broke off spilling fuel on the ground causing a fire as we slid along on the runway. At this point, the left wing began to fly again, I think, and it came up. With no weight in the tail, the tail came up and the airplane bounced on its nose three times. We became airborne again, came back down, and fortunately for the four of us flight crew members, the cockpit broke off from the fuselage. It was unfortunate for most of the first class passengers, because their then-exposed section of the fuselage bore the brunt of the damage. The aircraft came to rest in a field to the right of the runway.

Where we had first touched down, the right main landing gear gouged an 18-inch hole through the 12-inch-thick-concrete of the runway. The reason for such force is that the airspeed for a normal touchdown in a DC-10 is 140 knots— because of our phugoids we were doing 215 knots. Also, the normal rate of descent at touchdown is about 300 feet a minute. Our rate of descent at touchdown was 1,854 feet a minute.

Further, in any kind of airplane, it is expected that the aircraft will travel straight along the runway after landing. Without flight controls, we had no way to crab or to otherwise correct a drift, plus we had a quartering tailwind that gave us an additional 10 knots of speed.

The final step is cooperation

The fifth and final item is cooperation. The excellent cooperation in the cockpit has already been mentioned, but we also benefited from tremendous cooperation between the cabin and the cockpit crew, especially considering that we did not have a lot of time to talk to them. Fortunately, for us, the senior flight attendant, Janice T. Brown, was very experienced and rose to the occasion. She mentioned later that when I had called her to the flight deck for the first time, she recognized immediately that we did not have an emergency, that we had a crisis as soon as she opened the cockpit door. Another flight attendant, Virginia A. Murray, who came up a little later, also took one look into the cockpit and knew we were in very serious trouble. A lot of communication was not necessary to accomplish the level of cooperation we needed; it happened, to a large extent, spontaneously. Spontaneous action was required because of the complication of the emergency—hint at prior CRM training.

For something that you have practiced for but never are called upon to do, I was very pleased, rather than surprised, that the whole effort worked as well as it did. We crew members talked later about what ATC did, what the National Guard did and about the cooperation from Marion and St. Luke. We realized the benefit of advance notice to emergency services. We had announced the nature of our emergency early, and once we made the decision to head for Sioux City, the leaders of the response group were given about a 25-minute notice that we were going to crash.

The passengers ended up upside down because the fuselage was on its back, with smoke, fire and debris all around. When they finally got out of the airplane, they found themselves standing in a corn field, surrounded by corn eight feet high. I cannot imagine what they must have felt like. But they stayed calm and they helped each other. One of the survivors started climbing out of the airplane and heard a baby crying; he went back inside, found the baby in an overhead bin where she had been tossed, took her out of the airplane and brought her to her family that had been driven out by the thick smoke. This type of thing occurred in a number of instances—passengers were helping each other and the flight attendants were continuing to carry out their duties even though they were victims as well.

Rescuers initially ignored the separated cockpit because it had been compressed to a waist-high section of the wreckage and looked like an uninhabitable piece of junk. When we were discovered inside and alive after 35 minutes by 185th guardsmen, however, they pried us out—very carefully because the four of us were confined in a small area and had a wide range of injuries.

Records broke both hips, eight ribs and a toe, and suffered numerous bruises and contusions. Dvorak suffered a shattered right ankle, multiple bruises and contusions; he has three pins in his right leg, his ankle is aimed a little bit off to the right, he has a permanent limp and will eventually have to have the ankle fused. Fitch suffered multiple bruises, contusions, a broken rib, internal injuries, a severed nerve in his right hand, a broken right arm and a dislocated left shoulder. I was relatively uninjured, suffering a slight cut on my right ankle, bruises, contusions and a black eye; I had no broken bones but needed 92 stitches to close lacerations on my head. Records, Dvorak, and I were back at work approximately three months later; Fitch with the most serious injuries of the flight crew, was back approximately 11 months later.

Of the flight attendants, one was killed. The rest suffered varying degrees of injuries, but they eventually returned to work. Right after the crash, they continued to do their job. Fortunately, the emergency services at Sioux City recognized that they were victims also and quickly relieved them.

Luck, communications, preparation, execution and cooperation— these five factors will not guarantee survival during a serious inflight emergency. It is regrettable that

*111 passengers and one flight attendant did not survive the
crash landing at Sioux City, and my deepest sympathies go
to their families and friends. However, when the five factors
involved in training for emergencies can act in concert as
they did on our case, they can make the difference between
a complete catastrophe and a survivable accident.*

About the author

Captain Alfred C. Haynes has been a line pilot for United Airlines
since he began work with the company in February 1956. He has
logged 29,967 hours of flight time, of which 7190 is in the McDonnell Douglas DC-10. He holds an Airline Transport Pilot certificate
with type ratings in the DC-10 and the Boeing 727.

Haynes was initially trained in the DC-10 as a first officer and completed this training early in 1976. He became type rated in the aircraft in 1983. He served as a Boeing 727 captain between 1985 and
1987, after which he was requalified as a DC-10 captain and has
flown that aircraft until retirement.

A final perspective on Flight 232

Captain Haynes's excellent analysis leaves little to add. It is clear
from his explanation that his internal model of airmanship closely
parallels many aspects we have discussed in the preceding chapters.
One would expect to find high levels of diverse skills in professional
airline pilots, and they were demonstrated dramatically in the recovery of Flight 232. The preparation, communications, cooperation,
and execution discussed by Captain Haynes illustrates high levels of
skill and proficiency. The extended crew, which included the flight
attendants and outstanding Air Traffic Control (ATC) support, used a
deep knowledge base to bring to bear all available assets in an attempt to find a workable solution to their unique problem. Everyone
involved maintained outstanding situational awareness throughout
the 45-minute emergency. Finally, the crew exercised outstanding
judgment throughout. Although Captain Haynes credits much of
their success to luck, this was certainly overstated, perhaps based on
the modesty of the senior captain. A "one-in-a-billion" loss of all
flight controls does not meet most flyers' definition of good luck.

Nineteen months after the crew of Flight 232 completed their mirac-
ulous recovery, airmanship challenges of a different type were being

faced on the other side of the world. The next case study illustrates another example of exemplary performance under stress. In this instance however, the enemy was not an aircraft malfunction, but a thinking and aggressive Iraqi military.

Case study: Military airmanship on the edge

On February 24, 1991, the coalition ground war had just begun, and a situation arose that required immediate attention. A U.S. Army special forces "A-team," consisting of eight Green Berets, had been compromised deep behind enemy lines and were requesting immediate exfiltration. This team, under the command of Chief Warrant Officer 2C Chad Balwantz of the 5th Special Forces Group, had been covertly inserted the day before to monitor vehicle flow on Highway 8, south out of Baghdad. A little Arab girl accidentally discovered the team while playing with a soccer ball. Athough the A-team could have easily maintained their cover by eliminating or capturing the girl, Balwantz gave the order to let her go, stating "We're not going to kill any civilians, fellas. That's not going to happen" (Balwantz 1994). As a result, the team was soon confronted by armed civilians and Iraqi regulars, sometimes outnumbered by a 50-to-1 margin. Even the outstanding training of a crack special forces unit would not be able to hold out for long against these odds. If they were going to avoid capture or death, they needed help—and they needed it quickly.

Several hundred miles to the south, Lt. Col. Billy Diehl was just completing an aerial refueling with his four-ship of F-16C Fighting Falcons. Diehl and his wingmen—a full colonel, a lieutenant colonel, and a first lieutenant, were a part of the 363rd Fighter Wing based out of the United Arab Emirates. Armed with CBU-87 cluster munitions, their mission was to attack an Iraqi Republican Guard armor position just north of the coalition ground offensive. After coordinating with "Killer Scout" forward air control aircraft, which are specially equipped F-16s that are used to identify and mark targets for incoming attack aircraft, Diehl became aware that there was perhaps a higher-priority mission that was in the making. "We had just been cleared on an array of Iraqi armor, when we heard some guard transmissions ["Guard" is a UHF emergency frequency (243.0)]. It sounded as if there was a rescue effort being kicked off" (Diehl 1994). Diehl talked to AWACS and relayed that "we were full up on fuel, and had all 16 canisters of CBU-87 left, so we were definitely a player for any sort of SAR [search and rescue] mission that was going on." The AWACS immediately passed Diehl's flight inertial navigation system coordinates [INS] and diverted them off of their mission and onto the SAR effort. "As I typed in the coordinates, I recognized that we were headed to Baghdad. My first thoughts were that we must have lost an aircraft up north. I realized shortly, however, that these were ground troops we were going in to help. I couldn't help but wonder what they were doing way up there."

The four-ship was led north by two Killer Scout aircraft, who were already talking to the A-team on the ground. It was apparent that the Green Berets were already engaging in heavy combat, and the situation was heating up. "When we arrived on station, the Killer Scouts were already doing some outstanding work. They were communicating with the team on the ground, who were using their PRC-90 survival radios to coordinate for the close air support [CAS]."

As Diehl approached the area where the A-team was engaged, the communication between all members of the effort became very complex. At the top of the communication chain was the AWACS and Airborne Command and Control Center (ABCCC), both of whom were operating on VHF frequencies. The AWACS and ABCCC were coordinating the entire effort on a macro scale. Their mission was to ensure continuous close air support was available until such time as the rescue helicopters could arrive from several hundred miles to the south. This included identifying follow-on Killer Scout and close air support sorties, coordinating air refuelings, and maintaining contact with the rescue part of the mission. Next in the communication chain came the Killer Scout aircraft, who operated on VHF with the AWACS, UHF with the CAS aircraft, and finally, the UHF guard frequency with the troops on the ground. Diehl immediately surmised that, as the lead shooter, he could not risk the confusing chatter of multiple frequencies interfering with the command and control of his four-ship. He decided to send his wingmen to a "discreet" UHF channel to minimize the confusion. This proved to be an important decision as intraflight communication became essential to the eventual success of the mission.

To further complicate the situation, the F-16s began to take antiaircraft artillery fire (AAA) from the ground as soon as they arrived in the area. This forced Diehl to reevaluate the risk factors associated with the attack and mandated a high-altitude attack to keep the

F-16s above the lethal altitude of the AAA. Diehl, who was a combat veteran with 172 missions in Vietnam (complete with a MiG kill) had been hit by AAA before and knew the capabilities and dangers posed by the enemy systems. "They were harassing us with 37-mm AAA, and also tracking us with a 57-mm system. It would have been extremely hazardous to spend much time below 10,000 feet." This was especially true in a tactical situation like this, where the fighters were required to make numerous orbits over the enemy to locate the friendlies' position. On literally every pass, Iraqi 37-mm AAA would fire upon the American fighters. The altitude also played a factor in a more significant problem—distinguishing between the good guys and the Iraqis.

On arrival in the area of operations, the first priority was for the Killer Scout, call sign "Pointer 73," to help the incoming F-16s locate the Army team on the ground. This was not an easy task. Although both the Killer Scout and the Green Berets were equiped with Global Positioning System (GPS) receivers, which can pinpoint any spot on the earth to within 50 feet, Diehl's aircraft were not. This made Pointer 73's mission to "talk" Diehl's eyes onto the target. Diehl relates how difficult the task was. "He (Pointer 73) kept referring to a football-shaped field and a drainage ditch, but the terrain was basically featureless and I couldn't make out what he was referring to. My biggest fear was accidentally putting ordnance on the friendlies, killing the good guys." I made three laps around the area and still wasn't certain where the friendlies were. I knew the situation was getting desperate down there, and then we got a break."

The break Diehl referred to didn't seem like one to Chief Warrant Officer Balwantz and the rest of his team. Already engaged in a heavy firefight, they had just spotted a convoy of what appeared to be Iraqi regular infantry coming off the road and barreling across the desert floor directly at them. The A-team radioman, Sergeant First Class Robert "Buzz Saw" Degroff, radioed to Pointer 73 that the situation was desperate and they needed fire on the incoming convoy. Although Diehl had been unable to locate the ground team by a verbal description of the terrain features, he could clearly see the dust trail being kicked up by the line of trucks moving across the desert floor below. Had the truck drivers known what was watching from above, they might not have been in such a hurry. The following transmissions were taken from the gun film recorder aboard an F-16 in Diehl's flight.

Pointer 73 (to ground team): "You've got a four-ship with CBU-87s on those vehicles."

Sgt. Degroff: "Roger, we certainly appreciate it."

Pointer 73 (to ground team): "What I need you to do is give me directions where the bombs fall, if possible."

Sgt. Degroff: "We'll do the best we can. Obviously we have our heads down pretty low at this time."

After ensuring that Diehl had a visual identification on the moving convoy, the Killer Scout cleared him in "hot."

Pointer 73 (to Diehl): "Stop the vehicles on that road."

Lt. Col. Diehl: "I'm in."

Accomplishing what he would later describe as the "best pass of my career," quite a statement for a pilot who already possessed a Silver Star and two Distinguished Flying Crosses, Diehl obliterated the convoy with two cluster bombs. At the completion of the pass, he coolly declined the need for bomb damage assessment (BDA) from the ground, stating "Those vehicles should be down. What's next?"

Chief Balwantz, the A-team leader, recalled the significance of the successful strike on the convoy. "Things weren't looking too good for us until then. It was like being in a street fight and your big brother shows up."

The fight continued on the ground, and it appeared as if Balwantz and his team were in imminent danger of being outflanked by enemy riflemen. Degroff radioed for more help.

Degroff (to Pointer 73): ". . . from our position, could you please drop cluster munitions 500 meters due north of our position."

Diehl, hearing the close proximity of the requested strike, was immediately concerned. The CBU-87 cluster munition uses small but extremely lethal antipersonnel bomblets to cover an area of *5000* square meters. He wanted no mistakes. "Make sure he understands that these are cluster munitions," he radioed to Pointer 73. "This is danger close." The Killer Scout confirmed the request and cleared him in for the attack. To further complicate the equation, an Iraqi 57-mm tracking system began tracking the flight, and

Diehl was required to begin defensive countermeasures. He ordered the flight to expend chaff in hopes of decoying the enemy AAA system. In spite of these interruptions, the second pass was as effective as the first.

Although Diehl now knew where the "football-shaped field" and the A-team were located, his wingmen, who had remained in orbit at altitude, were still unsure. Demonstrating maturity and assertiveness, it was Lieutenant Brian Turner, the Number-4 man in the formation, who spoke up first. He stated that he thought he could identify the terrain features from the discussion and may have had a visual on Diehl's earlier CBU attack. Diehl had him describe the features to be certain that the young flyer was looking at the right spot, and once assured that his wingmen knew the locations of the friendlies, he cleared him for the next attack. "Brian really came through for us," Diehl stated. But the fight on the ground was still far from over, and the Number-2 wingman, Lt. Col. "Rosy" Rosenthal, was having difficulty identifying the friendlies.

Rosenthal had been orbiting 20,000 feet above the desert floor and had been unable to see the CBUs impact to help him identify the area. However, the situation on the ground was once again becoming critical. Diehl and Rosenthal discussed the terrain features in detail, until Rosenthal was "fairly certain" that he had pinpointed the friendlies, location. To be sure, Diehl maneuvered into a close chase position off his Number 2 and "followed him down the chute." This was to be another "danger close" pass, and both pilots wanted to be as certain as humanly possible of the location to avoid fratricide. It appeared to the flight leader that Rosenthal was lined up on the correct target, and Rosy released his weapons. Diehl's heart initially sank as he saw the weapons' impact point. He feared the worse. "My God, that was awful close!" he radioed to Rosenthal. The lack of the familiar radio transmission from the ground team spoke volumes.

As the two pilots pulled off the attack run, uncertain of the fate of those below, an exuberant voice crackled over the Guard frequency. "On target! On target!" Degroff exclaimed. Rosenthal had put the weapons right on the enemy location, a mere 200 meters from the Green Berets. As Diehl cleared his Number-3 wingman, Colonel Robert Van Sice, for the next attack run, he added an additional piece of guidance: "No closer."

As the sun began to set, Diehl's flight was out of munitions and nearly out of gas. Three subsequent four-ship formations would follow and provide close air support for the stranded team until finally two MH-60 Blackhawk helicopters picked the team up—all alive and unwounded.

This mission represented one of the U.S. military's finest hours. Co-operation, innovation, and valor were all displayed in abundance. Lt. Col. Billy Diehl received the Distinguished Flying Cross—his third—for his efforts on this mission, as did the pilot of Pointer 73. Chief Warrant Officer 2C Balwantz received the Silver Star. The remainder of the ground team received Bronze Stars with "V-devices" for valor. They all—flyers and ground troops alike—got to go home in one piece.

An airmanship analysis of a near-perfect mission

What does it take to pull off a mission like this one? Clearly, Lt. Col. Diehl made the right call at nearly every opportunity in this scenario. He consistently exercised good judgment, the capstone of the airmanship model, but what allowed him to reach this level of expertise? Using the airmanship model, let's use this near-perfect mission to take a closer look at the subcomponents of airmanship.

Judgment and situational awareness

We see the first example of superior airmanship when Lt. Col. Diehl pulls off the tanker following air refueling. Although he already had a scheduled mission to accomplish, he determined that a higher-priority effort was taking place and that his flight might be able to contribute. This step in and of itself demonstrates a level of situational awareness and confidence that many aviators do not have. Changing a combat mission in flight is complex and full of risks, but Diehl was confident in himself and his team, so he made the call to offer help. It was not the brash overconfidence of an airman out to make a name for himself, but rather a mature decison based on a quick analysis of the situation. From what well does this type of confidence spring?

Pillars of knowledge

The flight leader's decision was based on several known factors. First and foremost, Billy Diehl knew his own capabilities. As a combat veteran, he knew that war was fluid and often required aircrews to change plans to take advantage of the tactical flexibility inherent in airpower. Having been there before, he knew he could hack it and was prepared to respond to changing situations and scenarios.

He also knew the capabilities and limitations of his aircraft and weapons. As a career fighter pilot, he knew that an F-16 armed with cluster munitions is about as good a multirole aircraft as has ever graced the skies. Having just topped off his fuel tanks, he understood that he may have been in the best position to aid in the search-and-rescue effort. As the situation progressed, it became apparent that he also understood the limitations of his jet and weapons. Without the Global Positioning System or precision-guided munitions (PGM), Lt. Col. Diehl intuitively realized that the pilots, and not high technology, would be at issue in this effort, which leads to a discussion of the most significant element in this success story, the team.

Lt. Col. Diehl knew and effectively utilized the whole team. This included the AWACS, the Killer Scouts, the team on the ground, and his wingmen. Without the GPS or PGMs, he was compelled to continuously communicate with all the elements of the team to keep abreast of the ground situation, as well as to maintain a picture of what *the other flyers were seeing*. Time and again, he took the extra step to ensure that his wingmen had the proper sight picture. He verified and revarified ground references with Lt. Turner prior to clearing him on target and took the innovative measure of following Lt. Col. Rosenthal "down the chute" to take every measure to ensure mission success. Lt. Col. Diehl understood what every component on this team could do for the mission, as well as what he could do for them. In this manner he led and facilitated an extremely complex team effort to a successful completion.

Knowledge of the enemy was also a factor. Although at this point in the war there was no longer an airborne threat from the Iraqis, Diehl was acutely aware of the capabilities of the enemy AAA systems. He kept his aircraft above the lethal altitude of the AAA whenever possible and utilized effective countermeasures when it appeared that the larger 57-mm system was tracking his flight. More than one overzealous pilot did not give the Iraqi AAA threat enough respect in other Desert Storm scenarios, and these mistakes resulted in the loss of several coalition aircraft. Even in the heat of battle, Diehl never lost track of the threat.

The foundations: Discipline, skill, and proficiency

Throughout this mission, the flight leader maintained strict command and control of his flight, providing constant reinforcement of the need for positive identification of the American troops on the ground during the "shooting phase" of the operation. This strict discipline was necessary in an environment where the calls for help from the ground were becoming increasingly desperate. Throughout the intense battle, Diehl resisted the temptation to turn his wingmen loose without being certain that they understood his requirements for positive identification. "My biggest fear was that one of us would put one (CBU-87) in too close and hurt those guys."

In this scenario, skill and proficiency were essential prerequisites for success. The combination of the close proximity of friendly troops, high altitude, and nonprecision weapons made it an extremely challenging bombing effort. That the enemy was kept at bay and that all the Americans came home unscathed is a testimony to the skill and proficiency of the four F-16 pilots. All of the flyers were highly qualified and had recent combat experience in the air phase of the war. The importance of pure applied skill to the success of this mission cannot be overemphasized.

An airmanship perspective: What it takes to be a hero

We began this chapter by stating that aviation is ideally suited for heroes. A closer look at the two case studies in this chapter illustrate two prerequisites for a hero to step forward—readiness and opportunity. Both heroes of these stories share similar characteristics with regard to these two elements. Both men had spent a lifetime preparing for a situation that they could not foresee. Both were highly disciplined, skilled, and proficient. They were veterans in their respective aircraft and were intimately familiar with the aircraft systems, capabilities, and limitations. Both Captain Haynes and Lt. Col. Diehl credit the team for their successes, and both are correct in their statements, but the key ingredient to the success of their teams was the pilots' familiarilty and efficient utilization of all aspects of the team. The final aspect of this equation is opportunity. When the chance arose, these men were ready. Although few flyers look for opportunities to become heroes, we must be ready. In aviation, one does not choose the opportunity to excel. It chooses you.

References

Balwantz, CWO Chad. 1994. Personal interview. May.

Diehl, Lt. Col. Billy. 1994. Taken from videotaped interview. May.

12

Inhibitors and obstacles to achieving airmanship

with Glenn Hover

Do you see difficulties in every opportunity or opportunities in every difficulty?
Anonymous

A personal view from John Lauber, Ph.D.

The recognition that we needed to improve human performance in flight operations came about as a result of three primary inputs. First, of course, were highly publicized accidents, like the DC-8 at Portland. Beyond that, we were seeing the same kinds of human-factors problems reported through the confidential aircrew safety reporting system (ASRS). Leadership problems, followership problems, communication, and assertiveness issues were all bubbling to the surface. The final key was the Ruffel-Smith study, which used 747 aircrews to identify key characteristics and behaviors associated with good and bad crew performance. As we looked at these characteristics, the performance of the crew as a whole seemed to be dependent on something so tangible you could almost reach out and touch it. Effective leadership seemed to be the key. It set the stage for crew success. This study was the direct source of much of the early thinking that eventually resulted in the early cockpit resource management (CRM) curricula.

The big issues presently are to extend these concepts beyond the cockpit so that cooperation, appropriate assertiveness, leadership styles, etc., become part and parcel of the daily activities of both the individual and the organization. The principles of CRM should become embedded in the way of doing business for the entire enterprise, and it doesn't matter whether we are talking about a major airline, a military squadron, or a local general aviation organization. We are doing that at Delta right now, developing CRM programs with our airport and customer service representatives and our maintenance people. Of course, we've already done it long ago for cockpit and cabin crews. Habit patterns that are reinforced in day-to-day operations are far more likely to be second nature in flight. That's the goal — safe and effective operations at all levels.

The concepts of effective teamwork, leadership, followership, and the other principles of CRM need to be taught from day one — if not sooner. We may even want to begin to select our people based on these characteristics and qualities. It's that important to success.

The founder of cockpit/crew resource management, John Lauber is a former member of the National Transportation Safety Board and vice president of corporate safety and compliance for Delta Airlines.

If perfect airmanship were easy, everyone would have it. Unfortunately, there are many obstacles and inhibitors to effective airmanship, and they must be identified, understood, and overcome or avoided if we are to reach our potential as airmen. There are really two levels of inhibitors. First are inhibitors that affect the whole idea of airmanship, such as motivation, personality traits, and attitudes. Next are inhibitors that affect one or more of the elements of airmanship, which necessarily affect the whole. For each foundation, pillar, and capstone in the airmanship model, there are forces that deny their development or erode their effectiveness over time. This chapter identifies some of the common reasons we don't always perform to our maximum capabilities, by looking first at airmanship as a whole and then addressing each aspect of airmanship and illustrating common inhibitors and obstacles to overcome. The goal is obvious — to identify and conquer obstacles in our path to personal improvement and achievement.

Obstacles and inhibitors to "big picture" airmanship

Some factors affect the whole of airmanship, as opposed to individual parts of it; for example, improper motivation. Aviators choose to fly for a variety of reasons, and although there is no single "right" reason for wanting to fly, motivational issues are involved with good airmanship.

The adrenaline junkie or daredevil

The adrenaline junkie flies for the "rush." He or she loves the thrill of it, and approaches each flight as another opportunity to cheat death. I suspect that there is a little bit of this type of motivation in all of us who fly, or else why would we have chosen something as inherently dangerous as flying as a hobby or profession in the first place? However, there is a difference between a healthy approach to mastering a challenging environment—the approach that most of us take—and intentionally placing yourself in dangerous conditions to provide the adrenaline fix that seems to "hook" these aviators. The problem with this motivation is that it runs contrary to a logical and professional approach to airmanship. True airmen strive for control, predictability, and precision. Thrill-seekers intentionally flirt with out-of-control conditions, courting disaster and always trying to get closer to "the edge." Thrill-seeking is not conducive to the airmanship approach advocated by this book.

A Navy Blue Angel pilot, who had what many would consider to be the ultimate job for the thrill-seeker, once explained to me that precision doesn't leave much room for poor airmanship or thrill-seeking. When I asked him what it felt like to fly formation aerobatics so close to each other and the ground, his reply was surprising and gives us a unique insight on the approach taken by some of the finest pilots in the world. He told me, "I can't really say what it feels like, because I'm always too busy to notice."[1] He explained further that even on reflection after the flight, his mind naturally gravitated to pictures of maintaining position, adjusting power, and preparing for the next maneuver. It is a far cry from the "hair on fire, devil-may-care" approach taken by the adrenaline junkie.

---

1. The author hosted the Blue Angels at the McConnell AFB airshow and open house in 1990. The conversation took place following a practice session.

The control freak

A second motivational problem is seen in those who use the flight environment as their personal kingdom, an approach that does not leave much room for participative decision making and airmanship. These individuals are often seeking to bolster low self-esteem or making up for a lack of control in other areas of their lives. An obsessive need for control can prevent airmanship growth in several ways. First, it prevents timely and accurate feedback from others, who often view the controller as unapproachable. Second, it inhibits tactical execution based on all available resources, since the controller decides which resources are relevant. Finally, controllers are often less interested in improvement than they are self-gratification through the control of others.

I am what I fly

Some aviators see flying as an identity. Their goal is simply to be an aviator or perhaps the aviator of a *specific* type of aircraft. When asked what they do or who they are, the response is typically something like, "I'm a Navy fighter pilot" or "I'm an Eagle driver." Identity motivation is not always contrary to good airmanship, however, because many make this work towards improvement. For example, someone who sees his or her identity as "fighter pilot" might well decide that a better identity would be "great fighter pilot" or even "the best fighter pilot." The problem arises in those who see the label as an end unto itself. After they assume the identity of their aircraft, they have no incentive to continue improvement. This lack of motivation cuts airmanship off at the knees.

Inhibitors to discipline

Beyond big-picture inhibitors are issues that affect each aspect of the airmanship model, the foundation of which is discipline.

There can be no true airmanship without discipline, and factors that inhibit personal self-discipline should be at the top of our list of targets for removal. Overcompetitiveness, fear of looking bad, peer pressure, a macho attitude, the airshow syndrome, and "finis-flight-itis" can all affect discipline in a very negative way.

Excessive competition

Although healthy competition is good for development, obsessive overcompetitiveness can lead to a flyer who is willing to "cheat" to win. These flyers are willing to cut corners on safety and regulatory compliance to achieve a competitive edge over their contemporaries. A classic example occurred during a military navigation competition, when a bomber and a tanker decided to fly an illegal formation in an attempt to use the bomber's better navigation system to get the tanker closer to the scored "end nav point" at the conclusion of the exercise. In their zeal to win the competition, the crews did an incomplete formation briefing, which resulted in *both* aircraft thinking that they were the flight lead and neither attempting to fly off of the other. The result was predictable — a severe midair collision that all were lucky to survive.

Peer pressure

Oversensitivity to peer pressure can also be an obstacle to good air discipline. This tactic was used by the president of Downeast Airlines (see Chapter 7) to coerce his pilots to bust weather minimums and turn his company into a money-maker at the expense of passenger and crew safety. He established a rivalry between pilots so that they would pressure each other to "get the job done." Similar occurrences happen throughout the aviation arena, where pilots kid each other about trying something new or different. Once again, peer pressure can cut both ways, depending on how it is applied or interpreted. Aviators can exert influence for compliance, or they can inhibit good discipline by pressuring others to attempt questionable or even prohibited maneuvers. Stories of flying clubs based on daredevil maneuvers abound, such as flying under a bridge, doing a touch-and-go off a mesa, or flying through a particular canyon. These antics are sometimes seen as rites of passage or something that must be done if you want to be "one of the boys." Don't fall victim to these traps. Brigadier General Chuck Yeager recounts a lesson that he learned early in his flying career:

> *I would double or cut in half whatever it was that they (the peers) said they had done, taking the conservative approach. If they said that they did a Split-S from 3,000 feet, I would start mine from 6,000. Even then sometimes I'd barely make*

it. The point is that flyers have a tendency to embellish what they really did when they get back on the ground. I knew many who have fallen victim to that trap. Keep in mind, you can't do anything that hasn't been done before, and that includes making a smokin' hole (Yeager 1996).

Airshow syndrome

It seems that everybody wants to be a Thunderbird or a Blue Angel. The desire to impress friends or acquaintances on the ground is probably responsible for more sudden losses of judgment than any other cause. For these flyers, it is not enough to be a pilot, they feel that they must "strut their stuff" in front of others to make bystanders fully appreciate how impressive they really are. Even more tragically, many of these accidents take place in front of family members. The typical setup is a "departure show" from a field near home or a prearranged spot on the ground where the pilot has positioned observers so they can get a good look at what he or she can really do.

Demonstration flying takes special training, training most of us never receive. During most military demonstrations by the real pros, for example, the Thunderbirds or Blue Angels, the narrator usually makes an announcement during the performance that goes something like this: "Ladies and gentlemen, the maneuvers you see being accomplished here today are standard training items performed by all of our pilots in uniform." There is a grain of truth to this statement. Military pilots are trained to perform aerobatics and formation flight, but saying that all pilots are trained like the Thunderbirds is like saying a Boy Scout troop is trained like a Navy Seal. Yet many pilots still believe that they are qualified to perform low-altitude aerobatics, and the tragedies continue.

Closely related to the airshow syndrome is the unnatural fear of looking bad in front of others. An old pilot cliché says, "I'd rather die than look bad." Unfortunately, that has become true for many erring aviators. A combination of the airshow syndrome and the unnatural fear of looking bad occurred at a military airshow in the Midwest in the late 1980s. Two B-1B bombers were conducting a demonstration on parallel runways, performing synchronized maneuvers at the base open house and airshow. After initial low-speed passes, the crews were going to fly a short box pattern in which they would accelerate and sweep back the wings for a simultaneous high-speed pass. This would take a few minutes, because the

B-1 has an automatic center of gravity (CG) system that must pump fuel forward or aft to keep the CG within operational limits. One of the pilots was known to be very competitive and aggressive and made his turn a little tighter than he had in practice sessions. It would mean that the two-ship flyby might not be exactly simultaneous, but it would certainly still be impressive. To make a safe approach to the field, he would either have to extend the base leg turn or modify his box pattern to provide time and room for a safe turn. Neither was acceptable to this pilot, who attempted to make a tight turn to the final heading and preserve the integrity of the performance. During the turn, the pilot increased the bank and back pressure on the stick and tried to "pull" the big bomber around. A severe sink rate developed, and the aircraft dropped to less than 100 feet agl, according to one of the navigators. After several "watch the altitude" calls from the rest of the crew, the copilot took control of the aircraft, rolled out, relaxed back pressure, and used afterburners to keep the $200-million aircraft from hitting the ground. The crowd never knew the difference.

The airshow syndrome seems to affect the military and general aviation sectors far more than commercial operations. Keep in mind, however, that many commercial aviators also fly with a National Guard or Reserve unit, and many more are also private pilots, so none of us are immune.

Finis flights or "getting short"

Another common cause for sudden losses of air discipline occurs routinely on what has come to be called a "finis" or last flight in a given aircraft. For some reason, the secret undisciplined desire to do something stupid leaps from the depths of subconscious directly to our hands and feet during the last hours in a given aircraft. It does not necessarily have to occur on the last flight, but the risk increases as aviators grow "short" in time remaining.

During her final flight in the T-38 Talon, a young Air Force captain decided to do an aileron roll on the departure leg from a West Coast airfield. This maneuver impressed her commanders so much that they delayed her upgrade to aircraft commander in her primary aircraft for some time, and her airmanship reputation was tarnished forever. In another, almost humorous example, a military pilot who was on his last flight before going to work for the airlines did an unbriefed afterburner climbout of a low-level bombing range and attempted to accomplish a 180-degree roll to level off at the assigned altitude. He botched the maneuver, resulting in a negative-G condition, which caused his checklist to fly off of his leg and land on top of the interphone panel, disconnecting him from communication with the rest of the crew. The weapons systems officer, who was caught off guard by the maneuver, asked repeatedly what was wrong and, thinking that the pilot was incapacitated, grabbed the yellow handles to initiate ejection. The pilot came back up on interphone at the last moment to prevent what would have been an extremely embarrassing incident.

Your final flight in any aircraft should not be viewed as an opportunity to try something crazy but rather as the culmination of a professional airmanship relationship between human and machine. Try to fly a perfect mission, not an undisciplined one.

Lack of oversight

For some pilots, a lack of adult supervision is just too much temptation for some to resist. These aviators use the opportunity not to show their professionalism and maturity, but to attempt undisciplined and often illegal maneuvers. It often occurs when a flyer is operating away from his or her home airdrome and is frequently done to show off the pilot's capabilities and those of the aircraft to new acquaintances. This problem shows once again that real airmanship comes from within. Organizational oversight should not be necessary to counteract poor discipline once airmanship is instilled at an individual level. Individual airmanship won't stop the temptations, but it will give you the power to resist them!

All-instructor crews

There may be nothing quite as dangerous as two instructor pilots (IP) flying together. It would seem to be a contradiction, but a couple of factors make this apparent contradiction more understandable. The first is a heightened sense of competition. Most instructors have a healthy sense of pride in their own capabilities and love to demonstrate their superiority in the presence of others of the same cut of cloth. This can and often does lead to the invention of new and creative ways to prove their expertise. The second reason for the increased risk is the complacency that exists when two pilots both think that the other guy will watch out for them, and, at the same time, neither is comfortable questioning the other's actions! It can lead to an unwarranted feeling of invulnerability and, in turn, deadly complacency.

On one of my first dual-instructor flights, an "old head" IP attempted to show me his new instructional innovation, a "high sink rate touch-and-go demonstration." His idea was to intentionally induce a high sink rate on short final approach to demonstrate to the student what it looked and felt like. He would then add maximum power to break the sink rate at the last moment and grease the aircraft onto the runway. It sounds stupid, doesn't it? But it didn't seem like it that night. (Night? Yikes! How stupid were we?) To make a long story short, we lowered the field elevation by a couple of feet with the first attempt and had to write up the aircraft for an excessively hard landing. Luckily, the maintenance inspection revealed that nothing was hurt but our egos.

Case study: Halo effect on a dual instructor mission (Hughes 1995)

A much more tragic ending occurred when two instructor flight examiners teamed up on a special-operations low-level (SOLL) upgrade flight in a military C-130. The mission was scheduled for a five-hour, day-into-night sortie and involved many advanced maneuvers, including several 60-degree bank turns during the low-level route, a 60-degree bank course reversal turn, a maximum-performance climbing turn, and a four-engine obstacle-clearance climb that resulted in "a little burble" (approach to stall). Three pilots were on board: the acting instructor pilot/aircraft commander (IP/AC), a fully qualified copilot, and the IP receiving the SOLL training, who was at the controls.

Following the low-level obstacle-clearance climb, the IP/AC commented briefly on the minimum radius turn and the obstacle clearance climb, but beyond that gave little or no instruction during the maximum-performance maneuvers by the IP in training. The IP in training then briefed and set up a simulated three-engine obstacle-clearance climb—a maneuver he had never before attempted in flight—beginning at approximately *100 feet* above the surface of a lake. The pilot retarded the number-one throttle to flight idle and established five degrees of bank while pulling up the nose of the C-130 Hercules. At about 30 degrees nose high, the IP requested 50-percent flaps and began to describe the actions he was taking in an instructional manner. "We get her down to obstacle clearance speed. You can see that I don't center by ball (on the turn and slip indicator) right away." Immediately thereafter, the copilot made four separate "speed" or "airspeed" calls, indicating that the aircraft was getting slow while the upgrade IP was giving his instructional litany. Finally, the acting instructor pilot verbally intervened, stating "We're starting to turn, we're below VMCA (inflight minimum control speed), you can see that." Seven seconds later, the IP/AC became more assertive and directed "Let's knock it off." He was too late, and the aircraft turned left momentarily and then entered a right spiraling descent, impacting in 12 to 17 feet of water. All nine crewmembers were fatally injured.

The IP/AC was an extremely knowledgeable pilot and was obviously confident in his own abilities and those of the upgrading pilot. Both were flight examiner qualified, a fact that might have led to complacency on the part of all aboard the aircraft. But even the best of crews can get in over their heads, and it appears as if the pilot at the controls fixated momentarily on the turn and slip indicator while performing the high-risk maneuver, allowing his airspeed to decay. The call to "knock it off" was simply too little, too late. Based on their confidence in each other, they had failed to recognize and correct a simple airspeed decay, a situation that all three pilots would have likely pointed out and corrected immediately if flying with a less-experienced pilot team. It is ironic that high skill, experience, and confidence levels can be so effective on an individual basis but result in overconfidence and complacency when combined.

Inhibitors to skill and proficiency

A lack of skill or proficiency can be traced to many causes, not all of which are within the control of the aviator. The aviator is the final control, however, and is ultimately responsible for his or her own training. Let's look at a few reasons why an aviator might not develop skill or proficiency, or why they may atrophy once they are in place.

Poor self-assessment

The words "current" and "qualified" do not mean the same thing as "skilled" and "proficient." Individual skill and proficiency levels are dynamic and require disciplined and constant assessment to maintain safe and effective flight operations. The question of personal readiness is a difficult one and is largely ignored by modern aviators, who rely on organizational measures such as the annual or biannual check ride for their assessment. The airmanship model provides a means for well-rounded self-assessment in multiple areas of airmanship. The discipline to critique yourself after every flight across the airmanship areas will provide valuable feedback and trend information for personal improvement. "Gut-feel" assessment is often incomplete and misleading. You may well be getting better at instrument approaches and crosswind landings at the same time your situational awareness and team skills are deteriorating. Disciplined self-assessment is the key to steady airmanship improvement across the spectrum of your aviation skills.

You would never attempt to fly an aircraft with reference to only a single instrument. Likewise, a multiple cross-check of the operator is a very important and beneficial necessity to maintaining high levels of skill and proficiency.

Poor instruction

Unfortunately, some instructors in the field do not live up to their title. These types are better referred to as "destructors," because they can actually do more damage than good with their lackadaisical or less-than-professional approach to teaching. Although most flight instructors are well-intentioned and dedicated, and poor instructors make up only a very small percentage of active instructors, if he or she happens to be *your* instructor, you could care less about the statistics. So, how can you tell if you have a poor instructor, and what can you do if you end up with one? (For a full discussion on instruction and evaluation, see Chapter 13.)

Your first impression of an instructor is usually a good one. Good instruction should be focused on you, not the instructor. The instructor should present an organized plan for your improvement and lay out clear objectives and time lines. Good instructors know when to talk and when to shut up and let you fly. Good instructors prebrief and debrief well and consistently model good airmanship. Good instructors are not biased and are able to work through personality differences to meet you where you are and then move you towards good airmanship.

You and you alone are ultimately responsible for your learning. If you are unable to learn from a certain instructor, you are wasting both your time and the instructor's by continuing to fly with that person. Nearly all instructors realize that it is often better to make an instructor switch than to continue down an instructional path of diminishing returns. But you as the student must take the first step, because most professional instructors will continue to try and teach as long as the student is willing to show up. Let your old instructor suggest his or her replacement. He or she already has a feel for who you are and what your learning style is and likely knows the other instructors in the local area better than you do. You are not bound by the suggestion, but it is often good guidance.

Regardless of the nature of your problems with an individual instructor, part on good terms. Thank him or her for the time and effort and move on in your training with no hard feelings or regrets.

Lack of focus for improvement

Perhaps the single greatest barrier to improving skill and proficiency is the lack of a specific focus for individual improvement. It is natural to say "Let's go fly" and assume that you are accomplishing quality training, but unless you know what you need to work on, you might as well be lying on the beach. It's fun, but you're not going to get much better because of it. Flying can be fun and provide improvement at the same time, but it requires a focus that only you can provide. That focus requires personal discipline to systematically seek continuous improvement.

Inability to accept criticism

Learning requires the ability to accept and act upon constructive criticism. Those who cannot do so are severely inhibited in their airmanship growth. There are two common results to this all-too-common obstacle to learning. The first is the defense mechanism of argument and excuse. Like the Little League right fielder who always has the sun in his eyes, this would-be pilot has an excuse for every failure. See if any of these lines sound familiar.

- "My last instructor told me not to do it that way."
- "This plane doesn't seem to want to trim up today."
- "I think there is a little windshear on short final."
- "Damn turbulence."
- "You must have seen it wrong; I don't think I did that."

A second, more damaging trait is what I call "emotional jet lag." It is the tendency of a student to mentally stop flying at the geographical coordinates of the last error. The student gets so upset about making an error or being criticized that he or she cannot recover composure enough to continue to learn or sometimes even to fly. It often happens after an instructor has to take control of the aircraft for safety reasons or if the "student" is highly experienced in another aircraft and unused to being criticized.

Neither of these defense mechanisms are helpful, and flying time is too valuable to waste in argument or self-pity. Learning, by its very definition, means that you haven't yet mastered the skill you are attempting. Expect mistakes. Accept mistakes. Learn from mistakes. Give yourself the best opportunity to learn.

Resources

Flying is hard enough, and a lack of appropriate resources can make achieving and maintaining perishable flying skills next to impossible. No airmanship formula can fix this challenge, except to say that (as always) you are responsible for your own training, and when the resources become available and the training time comes, you have to make the call on where you need the work the most.

Inhibitors to knowing yourself

Individual improvement has the prerequisite of personal understanding. To accurately view the airman's world, you need to know what lens you are looking through—what your personal take on the world is. You should be able to objectively take a look at yourself, at least as far as that is humanly possible. It is often difficult for achievers, who tend to see the world as something to be overcome or worked through or around. Many flyers have little time or inclination for self-reflection, but successful airmanship demands that you understand yourself. Your motivations, fears, biases, capabilities, and limitations are all key to unlocking your potential. While feedback from others is an essential part of overall self-assessment, it can often be motivated by feelings of envy, selfishness, or other negative feelings that won't provide a true reflection. Obstacles to self-awareness often come in the form of overzealous critics or, on the flip side, sycophants who wouldn't say a bad word about you if their lives depended on it—which it often does.

Overzealous critics

Criticism is necessary for well-rounded self-assessment, but it can be unreliable. If the critic is motivated by negative feelings towards you, his or her feelings can cloud your self-assessment picture considerably. An old instructor pilot once told me after giving me a particularly brutal check ride critique, "Son, criticism is like chewing up a peanut with the shell still on it. You eat the good parts, and spit the rest out and forget about it." A corollary to this most excellent metaphor is that if you eat the peanut shells (the bad criticism) too, it could be detrimental to your health or your airmanship improvement. A second point is that if you are careful to eat only the nourishing portions of the critique, it's a good idea to keep eating, as it stimulates growth. The bottom line is that criticism should be listened to and analyzed for validity, acted upon if required, and ignored if not.

Sycophants

The flip side of the overzealous critic is the yes-person sycophant, who, for whatever reason, will not give you any negative feedback whatsoever. This problem is very real in both the commercial and military environments, where promotion can be tied to one's ability not to make waves. In commercial aviation, it is especially true during early probationary periods, when a single black mark on a pilot's record can scuttle any chance of an airline career with that company. Although it is not as severe in the military, a similar situation exists when senior-ranking officers fly with more junior crewmembers. The aviation psychologists have even given this phenomena a name: "excessive professional deference."

There are numerous examples of young crewmembers hesitating or refusing to provide real-time feedback to more senior crewmembers. The following case study illustrates a unique example of excessive professional deference.

Case study: The two "lieutenants"[2]

The first lieutenant instructor pilot (1st Lt. IP) had a tough decision to make. The lieutenant general VIP he was responsible for transporting in his military C-21 Learjet wanted to fly a tough nonprecision approach to an unfamiliar base at night, which didn't sound like a good idea. Existing regulations strongly recommended flying a precision approach under these conditions, but eight steps of rank

---

2. This story was related to the author under the condition that the source remain anonymous.

separated the two pilots, and the lieutenant was more than a little nervous about questioning the senior officer's judgment. To further complicate the issue, the general had not demonstrated great proficiency so far on the trip. The conditions were unfavorable at best and possibly dangerous given the wrong sequence of events.

The lieutenant was doing what he had been taught ever since pilot training; he was looking ahead—and he did not like what he saw. As the aircraft commander, he had the responsibility to keep both his passenger and the aircraft safe, but as a mere lieutenant, he was in a delicate situation and wasn't quite sure how to handle it. Although he had suggested the precision approach, the lieutenant general decided he would like to get a little practice on his nonprecision approach procedures. Neither pilot was familiar with this approach, and the IP scrambled to quickly review the approach plate so that they wouldn't miss anything. The IP was starting to see all the elements of this ugly scenario fall into place. The situation went from bad to worse when the general began to get behind the aircraft. As the two pilots worked to simultaneously brief and fly the approach, the general had serious difficulty keeping the Learjet under control. Suddenly a loud voice came from a passenger (also a pilot) who had been watching the scenario unfold from an observer position: "AIRSPEED!" In the attempt to manage the complicated approach, both pilots had dropped the airspeed indicator from their cross-check, and the aircraft had slowed to 15 knots below reference approach speed, developing a severe sink rate on short final. The observer, an instructor pilot himself who had been flying for more than 10 years, felt that the crew had very nearly lost the aircraft.

The funny thing about this mission was that this general was not known as a difficult man to deal with. If approached directly about his decision, he likely would have scrapped the nonprecision approach of his own accord. Yet after his initial suggestion for the precision approach, the first lieutenant failed to even question the decision, in spite of regulatory guidance that indicated it was a poor decision. As the aircraft commander for the mission, he clearly had the authority to do so. All's well that ends well, but sometimes, you just have to shake your head and wonder why pilots will risk their very lives just to avoid conflict and appease others.

No disciplined self-assessment process

Attempting to analyze the man or woman in the mirror can be the most difficult learning task of all. You can't hope to understand yourself if you don't look inside. Take a few moments before and after every mission for self-critique. Before the mission, look for hidden agendas or bias, think about your current state of skill and proficiency, and balance these items against the relative importance and risk of the mission. After the flight, look closely at how well you performed across the spectrum of airmanship, from discipline to judgment. Did you make decisions based on logic, emotions, or both? The whole airmanship picture is viewed through your eyes. So unless you know what is going on inside yourself, your observations could be useless.

Ensure that your confidence, or your lack of confidence, is based on a realistic picture of your performance. Some aviators are too hard on themselves, which inhibits the growth of hard-earned and well-justified confidence. Conversely, others are supremely certain of themselves, in spite of the fact that their skill and proficiency levels do not warrant such confidence. This is why it is important to assess your performance immediately after each mission, before your inner self reinterprets events to fit preconceived expectations.

Inhibitors to knowing your aircraft

Although learning about an aircraft can be a daunting task when you first begin, you must keep in mind that there is a finite amount of information, and that even the most difficult systems can be learned with a disciplined approach over time. There are four primary obstacles to learning about an aircraft. The first is the apparent enormity of the task, which tends to frighten away many pilots from attempting more than a cursory grasp of the knowledge. The second obstacle is improper study techniques, which prevent a systematic assimilation of the information. Third is the challenge of keeping current, as many technical orders and flight manuals change frequently, and staying abreast of changes can be difficult. Finally, there are many unwritten aspects of aircraft knowledge, and these will elude you if you don't know how and where to seek out the information.

Enormity of the challenge

Even the simplest of aircraft are relatively complicated, and the sheer amount of detailed information on any given airplane or jet can be discouraging. Fear not, for even the longest journey begins with a single step, and the best time to start is as soon as you receive the technical orders, flight manuals, and checklist for a new aircraft. In fact, an initial hesitation to study new material may result in a prolonged delay. As your apprehension increases, the pile of books sitting in the corner takes on a menacing and sinister appearance—almost daring you to crack the cover. Although the best time to start studying is at the beginning, it might not be the best *place* to start reading. The dry nature of the systems materials can make it difficult to stick to the task, and it might help to mix up your study habits to include some interesting material along with the sometimes difficult systems sections, which are typically in the early chapters of the tech order. The bottom line is knowledge acquisition, and there are no shortcuts to disciplined study habits and time on task.

Improper study techniques

Everyone learns differently, and there are as many ways to approach the acquisition of aircraft knowledge as there are individuals. Some methods, however, have worked well for many of us in the past. The first order of business is to master three separate sections of the flight manual: normal procedures, emergency procedures, and operating limits and restrictions. There is no way around this basic knowledge. You need this information before you fly, and you need to know it cold. Beyond this core information, several methods can keep your studies moving forward and relevant to your day-to-day flying.

One technique is tie your study to your current motivation and experiences. For example, if you are going to be flying instruments, you may want to review sections of the normal and emergency procedures that deal with your flight instruments and then focus a few dedicated hours to mastering the intricacies of the systems themselves. Make a point to ask a few questions of the mechanics and technicians at the next opportunity. Write down questions that you may have in the margins of your books and seek out answers, calling the manufacturer if you need to. The designers and engineers are always happy to talk to operators. You can also write the manufacturer with any questions that you may have. In my 15 years of this practice, I have never had a question left unanswered by the manufacturer, and once Pratt & Whitney even sent me a hat with its response.

Another part of this approach to learning calls for the study of any system that might have malfunctioned or come to your attention during a recent flight. For example, if you have a maintenance discrepancy with a vent or heater system, it may spur you to study the diagrams of the entire airflow and pneumatic system once again, asking questions until you understand how and why it works like it does. Let your curiosity be your motivation and guide. It is a natural way to get into difficult and dry material.

While the curiosity-driven study system may keep you motivated, it is not necessarily thorough or methodical. Therefore, you need some method to make sure that you eventually cover all of the aircraft systems, as well as establish a pattern for review, preferably on an annual or more frequent basis. You can accomplish this task in several ways. One is simply to "check off" areas in the table of contents as you master them. Some individuals prefer to design their own spreadsheet to monitor their study. Occasionally, your local training program or the manufacturer have examinations that can test your knowledge of various parts of the flight manual or technical order. Exercise caution with this approach, however, as mastering the examination questions may give you a false belief that you have mastered the systems themselves, which can lead to a dangerous self-delusion. Always keep in mind that the *real* test will come in the air, with little or no time for review.

Some aviators also keep a log of time spent studying, as well as the areas covered. This log accomplishes two things. First, it can give you trends on your own study habits. If you notice that your last study session was before Michael Jordan retired from baseball, you might get the hint that it's time to get back to the books. The log approach is also part and parcel to understanding yourself. If you are systematic in something as mundane as the study of aircraft knowledge, you may well be building a habit pattern that carries over into the air. Airmanship means being thorough. Airmanship means being prepared.

Keeping current

Aircraft are built to last. Because aircraft don't wear out in the normal sense, it is necessary to frequently update and modify various parts and systems as the airframes age. Additionally, as a particular type of aircraft matures, the experiences of those who fly them become incorporated into new procedures. This is especially true if the "experience" involves an accident or incident that might have resulted due to a lack of guidance in the current manuals.

Staying current is essential but often difficult. Unfortunately, changes to technical orders or flight manuals don't always coincide with our available study periods. Be that as it may, there is a systematic and very important way to handle changes. As a change occurs, quickly read it over to find out its essence. Is it a procedural change, a technical update, or simply correcting some data that has finally been flight-tested instead of computer-modeled? Once you are aware of the essence of the change, *post it!* Do not carry around loose changes stuck in the back of your checklist or flight manuals. Always keep your publications current. Make a note to yourself to look over the change more thoroughly as soon as you get the opportunity, and then close the loop by discussing the change with another aviator, to be sure you both got the same meaning from it. As before, if you have any unanswered questions, contact the experts at the source and pump them for information and explanations until you are satisfied with your grasp of the changed information.

Individual aircraft personalities

Aircraft are individual entities, even when they are of the same type. Yet we often treat all aircraft of a given type as identical, which can be a big mistake. Differences in manufacturing, maintenance, and flight stresses give each aircraft its own personality. Two aircraft that came off the production facility one behind the other could be as different as night and day, with one leading the fleet in hours flown, and the other a "hangar queen." Unless you are familiar with the aircraft as an individual, you might think you are taking off in Dr. Jeckyll but end up flying Mr. Hyde.

There are several ways of finding out about the individual quirks of a given aircraft. The first, most obvious method is through a detailed review of the maintenance records. Recurring discrepancies or "could not duplicate" (CND) corrective actions may give you your first hint of what to expect. Beyond the standard maintenance books, most organizations unofficially track trends on individual aircraft. These unofficial logs vary in depth and accuracy and are often just handwritten pilot inputs that would not normally appear in formal maintenance records. For example, you might see write-ups like "difficult to keep trimmed up" or "windscreen difficult to see through when landing into the sun." These clues can prove extremely valuable in gaining insights to individual aircraft personalities.

Perhaps the most reliable way to find out about an individual aircraft is to talk to others who have flown it or those who repair it. Ask other pilots general questions about how the aircraft "feels" or if there is anything special about it that you should know. Merely asking these questions immediately identifies you as having a high level of aircraft sophistication, and you will likely get honest and helpful information. Beware the maintainer who is hesitant to discuss a recent repair or who claims that he or she is "not the regular crew chief." Occasionally, maintainers are pushed to make an aircraft flyable and are likely to hedge on information that may cast doubt on the airworthiness of the aircraft. If you have any doubts, push for more details, or simply turn down the aircraft. Many obstacles may inhibit your quest for aircraft knowledge, but you have no good excuses for not acquiring it.

Inhibitors to knowing your team

Effective teamwork is part of the vocabulary and lexicon of nearly every flying organization today, from small flying clubs at a local airfield to major commercial and military operations. We have recently experienced a so-called "management revolution" that keys on the importance of teamwork to success, and yet there remains many obstacles to effective teamwork, including the tradition of the solo performer in aviation, egocentricism, poor communication, individual agendas, familiarity or unfamiliarity with team members, and poor credibility.

The solo pilot mentality

Perhaps the most difficult of these obstacles to overcome is the tradition of the lone-wolf aviator, with scarf and goggles, taking on the hostile environment armed with nothing but his or her steel nerve and wits. While this approach may have been necessary and appropriate during the early military and barnstorming eras, the modern cockpit and flight environment has given the pilot far greater resources than in earlier times. If we are to maximize our performance, both as individuals and team members, we must overcome the obstacles that prevent cohesion and synergy.

It has not been an easy task. Although the aviation community has adopted an effective team training concept called cockpit or crew resource management (see Chapter 6 for a complete discussion), there has been a great deal of debate and resistance to this proven approach. An early commercial airline anecdote had a senior captain talking to a new copilot, stating "Let me tell you all you need to know about cockpit resource management. This is the cockpit, you are the resource, and I am the management! Any questions?"

On a more theoretical note, NASA-Ames researcher Robert Helmreich found that some pilots were so set against this perceived threat to their authority that they actually came out of CRM training worse than when they came in. While the vast majority of participants in CRM training programs experience significant growth and change, these so-called "boomerangers" were negatively affected. Even more disturbing is that this group was characterized by a personality that was neither expressive nor instrumental, meaning that they typically were neither good communicators nor focused on the task (Helmreich and Wilhelm 1989). Lack of communication skills and task orientation coupled with a "boomerang effect" away from teamwork leaves these pilots with very little left to command an aircraft or crew.

Beyond traditional resistance to teamwork are other obstacles, such as poor communication, hidden agendas, and egocentricism. Consider the following example.

Poor communication and egocentricism

During a training flight on a large crewed aircraft, the instructor pilot decided that he would "test" the other pilot by "cutting out" the electric stabilizer trim without communicating his intentions to anyone on the crew. The other pilot, whose technique included trimming off pressure in the flare just before touchdown, landed nose gear first and began to porpoise violently. Although the crew managed to get the aircraft under control, there was considerable damage to the nose gear strut, and the mission had to be aborted.

This example highlights two more obstacles to effective teamwork: poor communication and a hidden personal agenda. The result illustrates what can occur in an environment that is designed for and demands teamwork, like the multiplace cockpit. In a similar military example of this same selfish mindset, the copilot got the last laugh, but this time it was a navigator who had the agenda.

Following a weapons release at a low-level bomb range, the navigator "sequenced" the automatic navigation system to the wrong waypoint for the racetrack pattern the bomber was required to fly for their next bomb run. The young copilot, who had flown this low-level route just two days before, asserted "the other day we were further north." The navigator, who felt his professionalism and pride being challenged by a mere copilot, replied sarcastically, "The other day you were wrong." Although the copilot could clearly see out his window that their current track would take them out of the confines of the low-level route, he kept further comments to himself. Within moments, the range controller directed the crew to abort the route because they had busted out of their protected corridor.

These two scenarios demonstrate that personal agendas and "the need to be right" are clear inhibitors of efficient teamwork and that both have negative safety and mission-effectiveness implications.

Familiarity with the team

Effective teams are not mere collections of talented individuals, and there is a growing body of research that suggests a range of familiarity optimizes team performance. Team members obviously should not be total strangers, and Chapter 6 highlights the need to understand the strengths, weaknesses, fears, capabilities, and limitations of teammates. However, some studies suggest that too much familiarity breeds complacency and creates an environment in which team members who may have become close friends are less willing to point out mistakes (Barker et al. 1996).

This complacency can be avoided through good team leadership, which constantly redefines the goals and creates new challenges for group improvement. This focus on leadership points to another possible inhibitor related to overfamiliarity with team members—credibility of both leader and follower.

Poor credibility

Effective team members are credible. That is to say, the rest of the team believes that each person can deliver when called upon to do so. A team environment establishes expectations, and a failure to perform as expected creates a credibility problem. If these events occur frequently to the same individual, the word is likely to spread to others within the organization. Once a credibility problem exists, a poor reputation is built up around the individual in question, making it extremely difficult to function in a team environment even when the individual's performance improves. In short, don't make a name for yourself as one who can't be counted on to pull his or her share of the weight. Conversely, don't be too quick to label others based on one or even two subpar performances. Other factors might be responsible for poor performance, and an individual with a credibility problem damages the entire organization.

Inhibitors to knowing your environment

Knowing your environment is a great deal like knowing your aircraft. The knowledge base is large, and the task is initially difficult. There is a great deal of material to get your arms around. It is not surprising, therefore, to discover that some of the same obstacles apply to both. The enormity of the challenge and the issue of currency apply to environmental knowledge just as they do with aircraft knowledge, so they are not covered in detail here. However, there are several differences and unique challenges posed by the three layers of environmental knowledge, beginning with a knowledge of the physical environment.

Physical environment

Knowledge of the physical environment, especially weather, will reap rewards for the duration of your flying career. Unlike aircraft or regulatory knowledge, few certainties are associated with this knowledge. Understanding weather patterns can help you predict and prepare, but never lose sight of the fact that localized weather patterns are notoriously unpredictable, and what you see on a forecast may not be what you get upon arrival. Mother Nature is a fickle lady, and those who feel that they have a complete understanding of her are often surprised by her sense of humor or terrified by her wrath.

Geographical differences

Learning about the physical environment can be hampered by a geocentric or single location-based point of view. Weather phenomena vary greatly from location to location. One example is the height of the tropopause, the layer of the atmosphere that separates the troposphere from the stratosphere, which varies in altitude depending on your latitude. This is not just "gee-whiz" atmospheric information. The "trop" has a significant effect on the tops and strength of thunderstorms, which is one weather phenomenon a pilot does not want to misjudge. A Midwest flyer who doesn't consider a thunderstorm to be a "big one" unless it tops 35,000 feet will be rudely awakened by the strength of a 20,000-foot TRW on the Florida panhandle or the British Isles. A variety of weather hazards are unique to particular areas, and those who venture out without adequate knowledge or respect for them do so at their own peril.

Case study: Unexpected encounter (NTSB 1991)

The pilot took off in his Piper PA-28-180 from Bullhead City, Arizona, en route to Rialto, California, at around 9:30 in the morning. The weather forecast was good—a few scattered stratus clouds en route—and the pilot was relatively unconcerned. After all, he had been taking some instrument training, 27 hours already, and was confident he could survive an encounter with a cloud if it came to that.

As he approached San Bernadino, he began to see some building and dissipating clouds near the mountain passes that he would have to get through up ahead. He was hoping that there would be a "hole" when he got there, so he made the decision to press ahead. He had a coworker waiting for him in Rialto, and he didn't want to leave him waiting. Besides, the highway he was tracking down went right through the center of the pass, so if he could just stay visual . . .

The aircraft wreckage was found on the side of a 10,000-foot mountain at an elevation of 3200 msl. Sheriff's patrol helicopters reported the common local phenomenon of rapidly forming and dissipating clouds in the mountain passes, which frequently obscured the hillside and floor of the pass. The sheriff's patrol knew better than to chance a crossing in such conditions, but, unfortunately, the cross-country pilot was not familiar with the local hazard.

Those who have never experienced valley, sea, or ice fog may not realize how quickly weather can form and that a landing field or mountain pass that is "clear and a million" can degenerate to zero-zero conditions in a matter of minutes. Many aviators who thought that they could squeeze in one more practice approach before the weather went down have found themselves shaking their heads in disbelief while diverting to alternate airfields.

One technique to avoid this problem is to call ahead to the location where you intend to travel and speak to a weather forecaster or fellow aviator about the local weather hazards. Another excellent source of information is the IFR and VFR Supplements. The bottom line is to seek information on the area to which you are intending to fly and not to rely too much on experience gained from a single location.

Overreliance on forecasters or automation

A second inhibitor to environmental knowledge is overreliance on automation or weather forecasters. Chapter 7 details the dangers of relying exclusively on weather forecasters for weather-avoidance procedures. Forecasting is imperfect, and many times the information you really need is not included in the forecast. The following example illustrates this point clearly.

Case study: Unbriefed hazard

The pilot and a single passenger received a weather briefing that indicated the possibility of icing conditions above 5000 msl for their route of flight from Hayward to El Monte, California. Although the pilot and aircraft were certified for instrument flight, the single-engine Beech was not equipped for flight in icing conditions, so the pilot listened carefully and then planned their route of flight appropriately. What they did not receive at their weather briefing was information on SIGMET LIMA 1, which cautioned pilots about the possibility of moderate to severe icing across a large portion of the local flying area. The reason that they did not get this critical piece of information was that their planned route of flight narrowly skirted around the SIGMET area. Once airborne, however, the pilot received an amended clearance that unknowingly took him through the valid area for SIGMET LIMA 1, and he was not advised of the hazard.

Shortly after leveling off at 11,000 feet, the ice began to form on the wings of the Beech 35-B33. The pilot knew he was in trouble and immediately radioed air traffic control to say that he was unable to maintain altitude. As the controller was coordinating for lower altitudes, the pilot reported in desperation that the aircraft was stalling—the final call made. The aircraft impacted the ground in a right-wing down, nose-low attitude and burned, killing both aboard.

Many aviators rely too much on automation for weather avoidance. The DC-9 crew that relied primarily on ground and airborne radar for the penetration of a line of thunderstorms (see case study, Chapter 7) paid the ultimate price for this misplaced trust by losing both engines due to ice and water ingestion. As far as possible, try not to make life-and-death decisions based solely on the input from a machine. When it comes to weather, what you don't know can kill you.

Regulatory environment

The regulatory environment is knowable, and there are few obstacles to knowledge acquisition beyond those we have already mentioned in other areas. There is a large amount of material that requires a solid plan of study and assessment. Currency of information applies as much to the regulatory environment as it does to other knowledge areas such as the aircraft and physical environment. The major inhibitor to regulatory knowledge is apathy, which can be overcome with disciplined study habits—one of the distinguishing marks of professional airmanship.

Organizational environment

The organizational environment is more challenging to understand than other environmental aspects for two primary reasons. First, there are seldom complete written guidelines on what is expected from individuals, and the *real* priorities of the organization might not be written or formally stated in any form. Secondly, company politics and pet ideas often interfere with good airmanship and decision making.

Unstated priorities

Organizational priorities naturally shift over time, depending on the nature of the larger environment and near-term challenges. For example, in many organizations it can be considered "bad form" or poor teamwork to openly compete with others within your organization. Yet nearly all ambitious supervisors realize that they are graded, at least in part, on what their part of the organization accomplishes relative to others. Unwritten and sometimes unspoken priorities emerge from this spirit of competition and can shape the organizational environment in which we operate. Individuals can get caught up in these unhealthy competitions and are often asked to "take sides" on issues that have little to do with the overall mission of the organization. Occasionally, these unstated priorities drift over the bounds of common sense and good airmanship. Further defined, these priorities are often referred to as "company politics."

Company politics and pet ideas

Company politics can muddy the water even further. Hidden agendas relative to personal advancement or pet ideas or projects can interfere with normal mission priorities and, therefore, good airmanship. An example of a "pet idea" that deterred good airmanship is told by one former B-52 crewmember who was part of a squadron whose commander had "a better idea." His commander had "discovered" a new way to achieve great bomb scores during the upcoming operational readiness inspection (ORI) by issuing binoculars to the lead crew in each "cell" or stream of bombers. The plan was to have a crewmember, who would ride in the jumpseat between the pilots, use his technologically enhanced vision to acquire the target much sooner than usual and "vector" the pilots to the visual release point. In theory, the follow-on bombers would be able to line up on the lead jet and drop better bombs as well. The end result would be much improved bomb scores, a victorious squadron, and, presumably, promotion for the commander. There are simply too many flaws in this theory to go into here, but suffice to say that this plan was unsound.

To test his idea, the commander rode in the jump seat of the lead bomber on a practice mission. Approximately 2 miles out from the target area, he spotted what he believed was a series of terrain features that would lead him into the designated target area and began to direct the lead bomber to turn left a few degrees to "get centered up early." As the terrain features became clearer in the binoculars, the vectors became more aggressive, and a final turn lined the bomber up on "the target." Although the pilots and navigators were less convinced than their vision-enhanced boss, they complied with his instructions and dropped their inert practice bombs along a county road approximately 400 yards outside the bombing range. The trail bombers, incidentally, made their normal radar-directed runs and dropped their bombs well within acceptable limits.

Although this is obviously an extreme example and in no way typical of the professional approach taken by most military commanders, it does highlight the problem that can occur with pet ideas or projects, especially when they are advocated by senior leadership.

In any case, our responsibility is to maintain high standards of airmanship, regardless of the organizational pressure to do otherwise.

Inhibitors to knowing risk

There are literally dozens of inhibitors and obstacles to knowing all of the sources of risk that are present in the aviation environment. For a detailed discussion of these factors, review Chapter 8 in its entirety. However, two primary challenges come to the surface when we begin to look at the nature of risk and hazard avoidance. The first is the lack of a systematic approach to hazard avoidance and risk management. The second is complacency.

Lack of a systematic view of risk sources

Hazards and pitfalls are involved with nearly every aspect of airborne operations, including ergonomic, psychological, physiological, and environmental hazards. At first glance, it would appear that perhaps only luck can save an aviator from disaster amongst this apparent chaos of operational hazards, but the airmanship model provides a starting point for a systematic approach to risk management. It gives the operator a means to break down the various parts of airmanship into more manageable pieces. Yet this is only a start. Chapter 8 outlines one method for taking a proactive individual approach, but any approach is only as effective as the individual's willingness to use it. The following example illustrates how even an identified risk can lead to disaster if ignored.

Case study: A hazard ignored (Hughes 1995)

An F-16 crashed while flying a preplanned airfield-attack mission in support of an operational readiness evaluation. Four aircraft on the ground and several buildings sustained damage. No ejection was attempted. The aircraft was destroyed, with two fatalities.

The mishap pilot took an unauthorized observer on an orientation flight. On top of this breech of discipline, he did not brief the passenger on a well-known potential hazard relating to inadvertent actuation of the side-stick flight controller. His unit was deployed to a temporary location, and the normal lines of supervision had broken down.

The requirement to perform an additional airfield overflight was passed to the pilot and his wingman the morning of the mishap after the flight prebrief. After the training portion of the mission, the mishap pilot brought the flight in to the airfield at approximately 300 feet agl and in excess of 450 KIAS in direct violation of command regulations (Tactical Air Command supplement to Air Force Regulation 60-16). As he began an aggressive loaded rolling turn to the right, apparently the passenger's G-suit inflated and pressed against the rear side-stick controller. That caused the aircraft to overbank to approximately 135 degrees in a nose-low attitude. Although the pilot tried full left stick and rudder, he was unable to recover the aircraft.

The F-16 B/D Dash One contains warnings about side-stick interference causing uncommanded right roll inputs. Numerous similar occurrences of side-stick interference had occurred prior to this mishap but were not reported. The pilot's failure to ensure that his backseat passenger was advised on a known hazard is indicative of an undisciplined approach to risk management. Aviators are often required to make changes to their plans or intentions, and risk management can get lost in the shuffle if a disciplined approach is not ingrained in airmanship habit patterns.

Complacency and overconfidence

One reason for a less-than-disciplined approach to risk is complacency. In a classic study, more than 1000 pilots were asked to define complacency, and the definition that emerged gives a clear indication of why it is the enemy of risk management. "Complacency is a mental state where the pilot acts, unaware of actual dangers or deficiencies. He still has the capability to act in a competent way—but for some reason or another this capacity is not activated. He has lost his guard without knowing it" (Fahlgren and Hagdal 1990). Using this definition, it is easy to see how complacency inhibits risk management, but how do we inhibit complacency?

Richard Jensen lists several recommendations for pilots to use in an effort to short-circuit this insidious element we call complacency. First and foremost, he recommends awareness of the phenomenon, using mental rehearsals, and reading accident and incident reports and other sources of information on the phenomenon. Beyond basic awareness, Jensen believes that using self-critiques, a training improvement focus that keeps your head in the game, and frequently asking yourself "what if?" type questions keeps you sharp in the cockpit (Jensen 1995). Try different approaches. Find out what works for you and your team.

Overconfidence is different than complacency. Overconfident flyers are frequently aware of the risks involved with a particular course of action but feel that they have the abilities to handle them. In a nutshell, overconfidence is when your ego writes checks that your skills can't cash. It masks the severity and relative risk of a known hazard to you, your aircraft, and your mission. Overconfidence can lead a pilot to select hazardous courses of action that are beyond actual capabilities and further defines the need for realistic and effective self-assessment.

Inhibitors to situational awareness and judgment

All elements of the airmanship model feed into the capstones of situational awareness and judgment. Therefore, *any* inhibitors to the foundation blocks (discipline, skill, and proficiency) or to the pillars of knowledge (self, aircraft, team, environment, and risk) naturally inhibit the capstones as well. A failure to fully develop each subcomponent of airmanship detracts from situational awareness and good judgment.

On a more fundamental level, the lack of personal motivation to understand and improve stands out as the single largest obstacle to improving situational awareness and judgment. Until the desire and discipline to systematically improve all areas of airmanship is instilled, developing these critical capstones to their maximum potential will remain beyond reach.

References

Barker, John M., Cathy C. Clothier, James R. Woody, Earl H. McKinney, and Jennifer L. Brown. 1996. Crew resource management: A simulator study comparing fixed versus formed aircrews. *Aviation, Space, and Environmental Medicine.* 67 (1): 3–7.

Fahlgren, G., and R. Hagdahl. 1990. Complacency. Proceedings of the 43d Annual International Air Safety Seminar, Rome, Italy. Flight Safety Foundation.

Helmreich, Robert L., and John A. Wilhelm. 1989. When training boomerangs: Negative outcomes associated with cockpit resource management programs. Proceedings of the 5th International Symposium on Aviation Psychology, ed.: R.S. Jensen. pp. 692–697. Columbus, Ohio.

Hughes Training Inc. 1995. Aircrew coordination workbook. Abilene, Tx.: CS-33.

————. 1995. Aircrew coordination workbook. Abilene, Tx.: CS-33, Computer disk.

Jensen, Richard S. 1995. *Pilot Judgment and Crew Resource Management*. Aldershot, UK: Avebury Aviation.

National Transportation Safety Board (NTSB). 1991. Report brief. NTSB File No. 1630, Accident ID #LAX89FAO78.

Yeager, Brig. Gen. Charles. 1996. Personal communication. January 4.

13

Instructing and evaluating airmanship

by J. D. Garvin and Tony Kern

The greatest challenges facing education are how to navigate the perilous course between adventure and discipline.
Robert Corrigan

A personal view from Al Mullen

I see us moving towards a fully integrated approach to flying — one that incorporates all aspects, from basic flying skills to complex human factors, into a single training approach. The organizational and regulatory cultures are beginning to support this shift. Let me give you an example of how we are moving in this direction.

Five years ago, if a check airman observed a pilot in command attempting to take off into unacceptably bad weather, the corrective action for the bust from the chief pilot would likely have been something along these lines: "Take him over to the simulator, give him a warm-up and another check. Give him a good stern talking to and tell him to use his head. If it all works out, get him back on the line as soon as you can."

Five years later, the same situation occurs, but this time the corrective actions are different. You are able to debrief the busted pilot using specific procedural guidance that has been incorporated into the regulations. This time the chief pilot says, "Take him over to the simulator and take him through the microburst profiles that are preloaded for windshear training to show him the hazards involved with this type of poor decision, and give him some training in case he inadvertently encounters one." You see the difference. Five years ago, most of us knew that taking off into severe weather was a dumb move, but for those few who didn't, we had no way to quantify or remediate the errors.

OK. Five years from now—same situation, same result, same remediation for the pilot in command. But now, the check airman swivels around and looks the first and second officers in the eye and says, "Didn't you observe all of the same things that I just observed? Why didn't you intervene with the appropriate assertiveness like your procedures dictate and the regulations require?" Now when we go over to the simulator, we videotape the remedial actions and preserve the words and actions so that we can approach the issue as a crew error instead of a single poor decision by an individual pilot. We have come light years in our approach to identification and remediation of cockpit error.

In the future, I see our flight instructors and evaluators so dialed into the integration of human factors and other aspects of flight that feedback on interpersonal aspects will come just as normally as feedback on what speed you retracted your flaps. When this happens, it will be difficult to even talk about "human factors" as a separate entity. It will be that well ingrained into the mindset of the individual aviators.

A former Navy Top Gun instructor, Al Mullen is an MD-11 captain and president of Crew Training International.

A flight instructor/evaluator is at once a resource provider, black hat evaluator, motivator, and evangelical enthusiast for airmanship. He or she must strike a delicate balance between advocacy and assessment, between teaching the student the adventure of learning and the discipline required to maintain a margin of safety while doing it. It is best accomplished in three distinct phases. First, the instructor must explain the big picture of airmanship and instill a motivation for achieving balanced and continuous improvement across all areas of airmanship. Second, the instructor must model and teach airmanship, stressing its multiple aspects during prebriefing, inflight instruction, and critiques. Finally, airmanship must be actively and aggressively evaluated, and the student assessed, objectively and subjectively, over all areas of airmanship. This final step is essential to facilitate individual reflective learning in between formal instruction or evaluation.

This chapter discusses how the airmanship model applies to the multiple facets of aviation instruction and evaluation. It is divided into five sections. First is a discussion identifying the vital importance of the instructor pilot in establishing and reinforcing airmanship. Second is a research overview defining the gap in current airmanship training practices. It underscores the need for a unified definition and new approach to airmanship education and training. The third section illustrates the significance of the airmanship model in developing and training instructor pilots for their unique role, showing how the model can be used to increase instructor training, specifically for common instructor vulnerabilities such as proficiency, intervention, and instructor/student relationships. The fourth section describes how the airmanship model can be applied to the instructional process, instrumentally weaving a comprehensive application of airmanship into daily training. The final section discusses a second hat that many instructors find themselves wearing, that of evaluator. How to measure airmanship with specific empirical instruments such as the LINE/LOS checklist, a crew resource management (CRM) metric, is discussed. This chapter focuses primarily on the instruction of new students with little or no previous experience, but aspects of good instruction certainly apply to advanced students as well, and some of the special considerations of instructing advanced students are covered as we move through the discussion.

A typical day?

Another day, another student. The instructor pilot reports to work distracted by a list of errands from home, only to be met with a stack of paperwork that has been building since last week. A few personal conversations, a change to the daily flying schedule, and suddenly she has a new student who has been waiting for half an hour. The airplane is double-booked this morning, so after a quick handshake, introductions, and a shallow prebrief, it's off to fly. Upon landing, the debrief takes place on the walk in from the aircraft, with perhaps another 5 to 10 minutes over a Coke in the instructor's office or debriefing room. It's not necessarily a typical day on the flight line, but it's certainly not unusual either. What has transpired in this example does not simply affect that morning's training, it also has a transcending effect on future airmanship. A serious, perhaps permanent, flaw has just been laid in the footing of an individual's airmanship. Students emulate their instructors, and what this student has just witnessed is an instructor with better things to do than to take a serious approach to airmanship instruction. Flight training and the role of the instructor pilot is paramount in establishing the attitudes and foundations for excellence.

On a more positive note, experienced flight instructors know that nothing in aviation is as satisfying as instruction well done. To watch a student begin to unlock the secrets of flight and know that you had a large part in the process is immensely gratifying. But the responsibilities of the flight instructor leave no room for half-measures or unprofessional approaches. As the student's primary source of habit patterns and attitudes, you must take this responsibility seriously. Airmanship, good or bad, is transferred from airman to airman, and the flight instructor is *the* key link in this chain.

The vital role of the instructor

Without doubt, the foundation for good airmanship begins with the flight instructor. Most students are new to aviation. They have little or no aircraft experience and therefore, their attitudes about airmanship are embryonic. New students are impressionable, and they are often looking for a role model, someone to show them how to act in this new environment. Research has determined that the type of experience the pilot has with the instructor and evaluator is a good predictor of that pilot's future performance, both technically and behaviorally (Connelly 1994). No other individual has as much impact on future airmanship as a student's first flight instructor. Instructors significantly develop student airmanship by introducing a flying culture, modeling aircrew behavior, and enforcing high standards and expectations.

The flying culture communicates expectations, standards, and biases. The flight instructor begins this process on day one. Through personal conduct, adherence or nonadherence to regulations, and attitudes

13-1 *The student/instructor relationship has always been critical. Here, Wilbur Wright explains the mysteries of flight to King Alphonso of Spain.* USAF Academy Library Special Collections

towards flying-skills development, an aviation culture is communicated. The very nature of flight training, in which instructors demonstrate maneuvers and students learn through duplication, distinguishes instructors as role models. This relationship is further amplified by the common use of "techniques" that instructors employ to better facilitate individual learning. Students quickly identify instructors as mentors or coaches and develop personal student/instructor relationships. Emulation and admiration often transcends beyond the cockpit, and students begin to imitate other airmanship behaviors. How to learn, study, and prepare for flying is established at this early stage of training and is fine-tuned by the flight instructor. Attitudes toward safety, regulations, and discipline are transmitted by the flight instructor. In fact, literally everything the instructor does in the presence of the student potentially touches that student's development. This influence occurs in the air, on the flight line, even downtown in social settings. The Air Force's Air Education Training Command, which has trained tens of thousands of pilots for combat duty, underscores the significance of holistic modeling for students, stating "your words and actions influence your student. Your influence has a BIG impact on air discipline (emphasis in the original text)" (USAF 1990).

As the new student's primary role model and mentor, the instructor pilot is often the first to introduce airmanship concepts to the student. Beginning on day one, philosophies and attitudes concerning how we do business in aviation are established. The influential relationship the instructor shares with a student must emphasize the significance and understanding of airmanship as a whole. An aviator's future commitment, understanding, and discipline begin with the flight instructor. The flight instructor is the guide, mentor, and enforcer for airmanship training.

The need to change instruction paradigms

There is a substantial void in current aviation training practices concerning airmanship education. The aviator's role has changed, but airmanship education has not evolved with it. Multiple, complex components of aviation now blend together to create an integrated system. New skills are required in areas such as automation, flight planning, crew management, and human factors. More than ever before in the history of aviation, the aircrew member is a systems manager as well as an operator. Education and training has not yet fully transitioned to the new systems paradigm. Many flight training programs remain entrenched in a simple skills-development approach that has failed to adapt airmanship education to a systems environment. Current educational practices in many flight training programs are either methodologically out of date or severely ignored. Although professional airline training, with increased emphasis on systematic instruction and evaluation of human factors, appears to be more in tune, it still seems to lack a unified, comprehensive systems approach that views airmanship as a whole.

A curriculum review of general and military aviation education highlights the need for a fresh look at training instructors. Many current programs are developed around an outdated, narrow, behaviorism philosophy towards education, focusing on operant conditioning techniques (motivating with rewards and punishment). Higher education over the past 20 years has abandoned behaviorism techniques in favor of individual learning strategies. Instructional methodology has evolved and is now directed at developing critical thinking skills through group learning and mental maps. Visual models (schema), knowledge presented as systems, and thinking skills development are the emphases throughout U.S. higher education. But aviation training is lagging and may have failed to evolve with these new, more effective instructional methods. The airmanship model represents a holistic approach that may help catalyze aviation education into the twenty-first century. The model presents airmanship as a way of thinking, reinforced with a visual model representing a system of integrated components. Instructors need to be trained (or train themselves) to communicate and reinforce the new approach.

The military method of training aviation instructors may also benefit from a more holistic approach. In the U. S. Air Force's Pilot Instructor Training (PIT) course, most of the focus in the three-month instructor upgrade program is aimed towards building demonstration skills in the aircraft. Only a few hours of the classroom program is dedicated to discussing student learning styles and student/instructor relationships (USAF 1990). Modern training programs must not stagnate in outdated instructional methods or neglect the impact of new technology on airmanship. Advances in human factors, CRM, and instructional techniques must be brought into the classroom and the cockpit. The instructor pilot graduate must be made proficient in a variety of pedagogical areas and develop beyond merely demonstrating maneuvers in the aircraft. If instructors are not brought "up to speed" on using a systems approach to airmanship instruction, the linchpin for cultural change towards better airmanship is missing.

Crew resource management and instruction

Driven by repeated pilot-error accident findings, crew resource management (CRM) has been developed by professional airline practitioners to address pilot judgment and decision-making concerns. Although revolutionary in itself, CRM is still incomplete in addressing the "whole" system of airmanship and decision making. As Pete Connelly from the NASA/UT/ FAA Aerospace Crew Research Project explained, "For the student pilot, the technical skills are spelled out in the FARs from Private through ATP. But, in terms of human factors/CRM skills, there is nothing spelled out in the FARs for the flight instructor to teach beyond his own personal style" (Connelly 1994). This deficiency is further underscored by the International Civil Aviation Organization (ICAO). "There are disparate understandings of human factors within the aviation community. The limitations in our current state of knowledge about the nature of human capabilities and limitations in aviation have resulted in a somewhat incoherent and incomplete approach to human factors training in the past" (ICAO 1991). Although by definition CRM includes the use of "all available resources," it has traditionally focused primarily on the "team" pillar of airmanship and lacks a comprehensive approach to other interconnected components, such as aircraft systems knowledge, skill, and proficiency. Even within CRM, there remains a lack of philosophy and understanding of the multiple, interrelated components to airmanship that are necessary for the development of a complete aviator.

The airmanship model fills a void in current aviation instructional practices. It offers a comprehensive definition of airmanship and provides a universal schema for all levels of aviation training. Perhaps more importantly, the airmanship model represents a long-overdue paradigm shift in aviation education. Individual learning strategies for continuous improvement beyond the formal setting complements current, and perhaps outdated, behaviorism educational approaches. A concrete illustration, the airmanship model, provides a comprehensive schema defining the formerly ambiguous components of airmanship. With this new model, airmanship instruction moves beyond skill training to developing a critical way of thinking, which in turn leads to a better integration of skills and knowledge in the cockpit.

Using the airmanship model to train instructors

Becoming an effective flight instructor requires specialized training and skill. Many challenges face the instructor, but three common instructor vulnerabilities—proficiency, intervention, and poorly developed instructor/student relationships—are worth a closer look. This section illustrates how the airmanship model can be applied to flight instructor training to increase instructor awareness and effectiveness for these common pitfalls.

When reviewing pilot-error accident reports involving instructor pilots, deficient instructor proficiency and late intervention are often cited. Staying on top of the situation, keeping one's own skills honed, and knowing where instruction ceases and intervention begins are the key instructor responsibilities, as the following case study reminds us.

Case study: Failed intervention—too little, too late (Hughes 1995)

A KC-135 aircraft was making a VFR touch-and-go landing with a simulated engine failure, takeoff continued emergency to be practiced during the takeoff phase. This training maneuver involves the instructor pulling one throttle to idle while the aircraft is on the runway to simulate the loss of an engine. The objective is for the student to maintain aircraft control, quickly analyze the problem, and get the jet safely airborne to work the necessary checklists for recovery. During this maneuver, the student failed the first step of this procedure and the aircraft departed the left side of the runway with approximately 5300 feet of runway remaining. The aircraft broke up, caught fire, and was destroyed. Five of the seven crewmembers were fatalities; two crewmembers were uninjured. The aircraft was airworthy for the entire duration of the flight up to the point of the mishap itself.

The aircraft had launched on a normal training mission and after an uneventful refueling and celestial navigation leg, the crew returned to base for transition training. The aircraft began its descent from the TACAN initial approach fix for a Hi-TACAN penetration and approach to a missed approach. The crew subsequently performed five additional approaches. One was followed by a missed approach, the last four by touch-and-go landings. Following the fourth touch-and-go landing, the crew requested and received a clearance from tower for a VFR closed pattern. When the aircraft was on downwind, the instructor pilot briefed the crew that this would be a VFR four-engine approach and touch-and-go landing and that they would be doing a simulated engine failure, takeoff continued, during the takeoff portion of the touch-and-go.

Training records indicate that this was the student pilot's first simulated engine failure, takeoff continued maneuver in 20 days and his first ever from the left seat. There is no indication that the instructor had demonstrated the maneuver in flight prior to this attempt. However, he had performed this maneuver several times in the cockpit procedural trainer (CPT), a low-fidelity simulator. The training received in the CPT teaches the student to look into the cockpit to determine which engine is failed. This response is contradictory to what is taught in the aircraft, which is to look outside. Visual cues should be the only method used to determine control inputs while on the ground. The student pilot had a tendency to undercontrol the rudder during the previous engine failure, takeoff continued maneuvers, which he had previously accomplished from the right seat. In addition, previous instructors stated that he tended to rotate before the aircraft was under control while performing this maneuver, an extremely dangerous trend.

The aircraft touched down in the center of the runway and bounced two or three times. Following the touchdown, the aircraft proceeded down the runway in a three-point attitude. The flaps were reset to 30 degrees and the trim to 3.0 nose-up in accordance with normal touch-and-go procedures. Power was advanced and the engines stabilized at touch-and-go engine pressure ratio (EPR). Prior to the 6000-feet-remaining marker, the instructor pilot simulated failure of the number-one (left outboard) engine by retarding the throttle to idle. This loss would cause the aircraft to yaw to the left and called for the smooth application of right rudder while remaining in a three-point attitude until the aircraft was at rotation speed.

At that moment, the student pilot rotated the aircraft sufficiently to lift the nose wheel off the runway and simultaneously applied full *left* rudder. The absence of nose wheel skid marks indicated the nose wheel never returned to the runway surface. One second after the incorrect control inputs were completed, the aircraft was 2 feet right of centerline, slightly nose up, right wing low, and beginning a rapidly increasing sideslip to the left.

At this point, the instructor pilot had approximately 2.25 seconds to react and accomplish full rudder reversal before the vertical fin stalls, nullifying the rudder input. If the instructor pilot does not initiate positive counteraction by the time the rudder reaches full left, recovery is virtually impossible because the yawing moment cannot be reversed. The aircraft continued left, and the right main gear crossed the centerline and lifted off the runway after 2.75 seconds from the student's incorrect control inputs. At 3.25 seconds of elapsed time, the left main gear began to leave the runway, and the aircraft heading increased from 16 to 25 degrees left of centerline. The sideslip angle increased from 8 to 16 degrees, and the vertical fin stalled. The violent left yaw and roll rate continued to increase until the outboard number-one engine nacelle contacted the runway. The number-one engine departed the aircraft prior to the aircraft leaving the runway. Elapsed time from incorrect control inputs was now six seconds.

The nose section impacted the ground approximately 70 feet from the edge of the runway. The initial impact was on the left side of the fuselage just forward of the crew entry hatch. The aircraft pivoted on its nose after initial impact as the tail continued in a counterclockwise rotation. This impact caused the cockpit compartment to separate from the main fuselage in the area of the cargo door. The tail of the aircraft continued to rotate counterclockwise and impacted the ground in the area of the boom pod after approximately 220 degrees of rotation. The aircraft slid backwards on its tail nearly 3500 feet to its final resting place after turning 285 degrees from runway centerline. The two crewmembers in the rear of the aircraft exited uninjured by jumping out of the aft escape hatch. The cockpit, fuselage, and other components of the wreckage continued to burn after coming to rest.

Although the student in this case did almost everything wrong at the worst possible moment, the instructor was still capable of saving the aircraft and crew if he had been on his toes. With less than 3 seconds to respond and correct the student error, the IP should have been "hard-wired" to the yoke, throttles, and rudder, and prepared to immediately intervene if a dangerous situation arose. He was not.

Several factors may have played into the instructor's complacency. First, there may have been a lack of cautiousness on the part of the instructor pilot since the student was older, previously experienced, and knowledgeable. This factor is often referred to as the "halo effect," in which high expectations can lead to slow intervention. Second, there was perceived pressure to get all the training required by the mission profile. Additionally, the instructor pilot may not have been fully conscious or aware of the full implication of the maneuver should controls be applied incorrectly. Flight tests had not identified the acute danger of misapplied control inputs during this maneuver, although common sense should have told him that limited time was available to correct the situation. Finally, the instructor was known as someone with a high self-image, and he may have been overconfident and felt that he could salvage the maneuver.

When the student pilot suddenly induced an unanticipated surprise factor into the equation, such as steering with ailerons, premature rotation, and improper rudder, the instructor was put behind the power curve, overcome by events, and failed to intervene in time to fulfill the sacred trust placed in him by his commanders, his crew, and his student.

The tragic end to this story makes it the exception rather than the rule. How often have you heard of a student putting the aircraft in a grave situation, only to have the instructor save the day with his or her skill and situational awareness? But the real question is not who saved the day but rather who *really* put the aircraft in a dangerous situation in the first place. It seems obvious that often instructors let students go too far and fail to intervene in a timely manner. In many of these cases, learning had probably stopped well before the incident, and intervention was overdue.

When should an instructor intervene?

Intervention should be based on three key factors: safety of the aircraft and crew, proficiency level of the instructor, and student learning. When any one of these are compromised, it is time for the instructor to take the aircraft and reestablish safe learning conditions. Instructor proficiency can be difficult to maintain, and it can be challenging for an instructor to achieve enough stick time to remain proficient with basic, let alone advanced, maneuvers. Even so, it is ludicrous to allow a student to put the instructor in a position that pushes the instructor's own envelope. Intervention, proficiency, safety, and learning are all inextricably linked to good instruction, so how does the airmanship model help? Let's take a look at a hypothetical case to see the various elements as they come into play.

Every instructor pilot has likely encountered a compromising situation while teaching landings to an inconsistent student. When coached, the student makes fair to good landings, but when simply observed, the student is erratic. After an hour or so in the pattern, the instructor begins to tire and decides that it's time to let the student demonstrate progression. This time, you think, I'll try to shut up and see what happens. The approach is unstable. First, the student's aim point constantly shifts, then the airspeed begins to fluctuate 5 to 10 knots. With several thousand feet of runway ahead, you're still thinking that the student might get it down. If he simply levels off over the runway, bleeds off airspeed, and nurses out the flare, he will have a landing. After all, that's what you would do. Instead, the fast, steep final ends up in a balloon resulting in a high flare with low airspeed. You take the aircraft, add power for the go-around, but unfortunately your 1½ second response exceeds the recovery envelope, the wing stalls and dips, and then . . .

What have you allowed to happen? Where were the critical decision points? First we must ask ourselves how much real airmanship was being taught by allowing the student to fly with erratic airspeed and a shifting aim point on final. Perhaps the unintended lesson of non-intervention is that any approach can be salvaged, and that airspeeds and glide paths are ballpark figures. Is this the lesson we wish to convey? If standards are consistently enforced, most students will rise to meet them. Those who do not should be removed from flying training, for their safety as well as the safety of others. If slop and salvage are accepted, the student will never develop good airmanship. Learning in this example stopped on short final, and intervention was overdue. Safety was exceeded during the roundout. A balloon should result in an automatic go-around in a normally functioning aircraft. Attempted landings from severe balloons can only result in a long landing or a stall. Intervention was called for twice in this scenario, once during short final (learning), and again during the landing phase (safety). Generally, under instructional settings, the call for intervention always rings twice — and in this order.

Proficiency can also be easily critiqued in this situation. When is the last time the instructor made either a go-around or landing attempt from a high flare? The instructor faces a unique dilemma. His pride says "grease every landing," but his real need may be to see the ugly stuff once in a while. If you as the instructor haven't been "there" lately, certainly don't let the student take you there.

The airmanship model can be used to better educate instructors for more disciplined interventions and systematic teaching techniques. By emphasizing a systems thinking process rather than simple skill development, an instructor should become more aware of ideal behaviors and establish more-disciplined boundaries and instruction. Consider first the bedrock principles of airmanship, *discipline, skill,* and *proficiency.* Tasked with developing a new aviator, the instructor's first challenge must be to build a disciplined aviator, which implies enforcing belief and practice of high standards and fostering an uncompromising attitude towards bending the rules. Too fast or too steep on final is not acceptable. Discipline considerations demand the student be taught to recognize the need for and execute a go-around — every time. Self-assessment of skill and proficiency may require the instructor to ask "Can I do this?" or, perhaps, "Have I done this recently?" But the instructor should seldom debate, "Should I do this, is this student's action acceptable?" These decisions should be set in stone long before they actually occur.

The bedrock principles of airmanship should prevent instructors from intentionally pushing the envelope in training. The model should shape ways of thinking in both instructor and student that constantly complement airmanship and flight discipline.

Instructor/student relationships

Instructor/student relationships require progressive development and an individual philosophy. The new student begins training in a tourist capacity, watching instructor demonstrations with naive amazement. The student's role quickly expands to participating in a part-task execution of maneuvers. Eventually, the student conducts most of the sortie with limited instructor guidance. This evolving student maturity of skills and responsibility requires a dynamic and complementary relationship with the instructor. Throughout the training process, the student transitions from a passenger to pilot in command. The instructor must complement this transition and evolve with the student. CRM skills incorporating students as crewmembers are now being introduced to instructor pilots, but without concrete illustrations, awareness and applications are minimized.

The student as a cockpit resource

Consider the following incident described by Gene Hudson in *SoCAL Aviation Review*. A presolo student and CFI are practicing maneuvers in a Cessna 152 when an actual engine failure occurs in the practice area. The instructor immediately commands, "I got it!" and the cockpit suddenly goes quiet at the same time the instructor's work load increases, eventually to the point of task saturation. Hudson describes the deteriorating student/instructor relationship. "Soon the instructor is completely consumed in the task of dealing with the emergency, and the student is reduced to the role of terrified spectator. The spectator-student, having but 5 hours, believes without question that Chuck Yeager's alter ego in the other seat must surely know about the dirt strip they just turned away from. The student is reluctant to express doubts about what the instructor is doing. There must be some compelling reason why the instructor rejected that option. The spectator-student considers asking about the decision, but decides to remain mute in order not to disturb the concentration of the instructor, who is now very busy and fully absorbed. The 152 comes to rest in a plowed field, not 2 miles from the dirt strip. There are no serious injuries, but the aircraft is substantially damaged" (Hudson 1992).

Why didn't the student speak up? Why didn't the instructor include the other set of eyes and hands in the cockpit? As a narrative, this case illustration is helpful in increasing both student and instructor awareness of using the student as another resource, and it underscores the need for CRM skills in the instructor/student relationship. But how can this relationship be emphasized preemptively? How do we know that we are fostering the correct relationship to maximize learning *and* safety during the process? The airmanship model can provide both a training tool in developing instructor and student awareness of this asset and a prebriefing tool to refresh awareness prior to flight instruction. An instructor who assesses the team pillar application of the model quickly realizes that the student's presence constitutes added strength to the team. Even if it is merely to fly straight and level for 30 seconds while you troubleshoot systems problems or communicate on the radio, the student can provide valuable and sometimes lifesaving assistance. A good instructor has the responsibility to use it.

Teaching airmanship: A model for daily operations

Ideally, specific aspects of airmanship are reinforced and discussed every time we fly. Used as a briefing tool, the airmanship model underscores awareness of the concrete tenants of airmanship and their interconnections. Many briefing and debriefing checklists are in use, some of which even have a separate "CRM considerations" or a "human factors" section. While the intention to include these valuable subjects is admirable, the approach is wrong. Human factors should not be singled out from the rest of the mission, but rather integrated into the human-machine-mission equation. The airmanship model accomplishes this difficult task by integrating human factors into "big-picture" airmanship. It additionally facilitates discussion and may be focused to specific training objectives.

Training objectives will most likely vary between students, depending on their experience and skill levels. Consider how a discussion of prior experience during the prebrief could have averted the following general aviation accident, which also points out the need for instructor vigilance during ground operations.

I was taxiing with a three hour student. He had problems taxiing due to past experience operating earth movers. In an earth mover when you step on the right pedal you turn left and vice versa, the opposite control inputs of taxiing a small aircraft. We practiced slow turns on the taxiway. We began to come within five feet of the left of the taxiway while steering left. He panicked and reverted back to his initial steering methods. He heavily applied left rudder while applying the brake and locked up his leg. In an effort to get away from the grass quickly he applied full throttle. I couldn't fight him on the rudder, and realized we were going off the hard surface taxiway (so) I pulled the mixture. The left wheel settled into a hole and the nose came down with the propeller striking the dirt twice . . . stopping due to the mixture" (NASA ASRS 1995a).

Skill development in taxi should have been a primary learning objective, and appropriate discussion should have been accomplished before they got in the aircraft. Instructor emphases such as "step on the centerline, and when in doubt stop straight ahead" should have been keenly developed in student's ground awareness. Agreement on acceptable deviations should have been established under discipline considerations. The airmanship model may not have developed actual taxi skills, but it would increase airmanship thinking and establish awareness of taxi procedure, learning emphasis, and standards.

There is also a hazard associated with understanding *too* much about your student, which can lead to a trap associated with expectations.

Inflight applications of the airmanship model

Inflight application of the airmanship model concentrates on developing systematic airmanship thinking, resulting in improved situational awareness and judgment. We already employ this practice with contingency training and malfunction analysis. Drills such as "fly the aircraft—analyze the situation—and land as soon as possible," develop a prioritized method of thinking. The airmanship model, however, provides a more integrated and systematic approach. The five pillars of knowledge, *self, team, aircraft, environment,* and *risk,* become integrated considerations. A new way of thinking emerges. A more complete, comprehensive understanding is developed. Take, for instance, the drone time to the practice area. Instructors often employ "what if" scenarios" "What if your engine failed, what if your radio went out?" We don't actually accomplish checklist or maneuvers during these mock exercises but rather discuss considerations and actions. These are perfect times to employ the airmanship model inflight.

"What if your engine seized? I would glide the aircraft (aircraft), I would assess the best landing zone (risk), I would incorporate you to run checklist, radios, and look for a better field (team), I would analyze winds for landing direction (environment), I would mentally rehearse forced landing procedures (self)." This drill goes beyond the typical fly-the-airplane routine and reemphasizes a more systematic, integrated thought process. With the airmanship model, concrete consideration is given to previously ambiguous airmanship items, such as team, self, and risk. Inflight employment of the airmanship model is critical for comprehensive instruction.

Avoiding distraction

In addition to being an excellent training tool, well-developed systems thinking can help avoid distraction—the instructor's constant and potentially deadly foe. Flight instructors are constantly forced to balance the attention required to instruct against that required to maintain positive control of the aircraft and mission. Even the finest instructors can fall victim to distraction, as the following two cases dramatically illustrate.

Near miss (NASA ASRS 1995b)

As my student and I were returning after a training flight . . . we reported downwind abeam and were cleared to land following the (aircraft) downwind ahead. At this point, I got heavily involved in talking my student through the steps to be followed during the approach, and after looking for the traffic and not seeing it, I wrongly assumed it was already on the ground . . . A couple of moments later I observed the other (aircraft) take evasive action . . . Contributing factors to this incident . . . are: my lack of concentration on looking and positively identifying our traffic before landing (as I routinely do) due to the heavy "question and answer" situation that my student involved me in. After this incident, I have made it a very clear point to all my students to minimize the pilot-to-pilot chat during operation in the traffic pattern.

A military example drives home the same points, but the results were far worse.

Case study: A good instructor overwhelmed (Hughes 1995)

The crew of a C-130 aircraft was scheduled to depart on a five-hour local sortie. The mission was to provide copilot upgrade training, consisting of instrument approaches and visual pattern work, for two students. Due to local pattern saturation, the mission was scheduled and planned to be completed at a regularly used satellite base. The mission proceeded normally until the last visual pattern prior to departing for return to home base. The student maneuvered the aircraft to a closed downwind for a simulated engine-out pattern to a touch-and-go. The student briefed the landing and directed the instructor pilot to lower the flaps to 50 percent. As the aircraft approached midfield downwind for landing, tower directed the aircraft to make a left 360-degree turn. After completing the turn, the aircraft reentered the VFR pattern on downwind. The tower controller then directed the mishap aircraft to turn to base leg earlier than the crew planned to establish spacing between them and other aircraft in the traffic pattern. The mishap aircraft configuration was flaps 50 percent and gear-up with the landing gear warning horn silenced. After completing the turn to final approach, the aircraft touched down 1030 feet from the approach end of the runway, 5 feet left of centerline, and slid for 3140 feet before coming to rest 18 feet left of centerline. The aircraft sustained substantial damage to the lower portion of the fuselage and ramp. Six crewmembers egressed the aircraft with no injuries.

A closer look into the events on the final traffic pattern leading up to the gear-up landing are illustrative. After demonstrating a flaps-up touch-and-go landing, the instructor pilot planned for the student to accomplish two simulated engine-out visual patterns to the runway and then return to base. The first approach and go went fine, and then the normal habit patterns of the instructor and crew began to be disrupted. At this time the real warning horn should have gone off—and I'm not talking about the landing gear horn, but rather the one inside the instructor's head. Note the number of distractors that were in play.

After turning onto the downwind leg, the student briefed the approach and landing and called for flaps to 50 percent. Tower cleared another aircraft for an opposite direction takeoff from runway 36 and instructed the mishap aircraft to accomplish a left 360-degree turn to provide spacing for the departure. Entering the turn interrupted the student's habit patterns, resulting in his failure to call for landing gear extension and checklist initiation. The landing gear warning horn was sounding during the beginning of this maneuver. Approximately 15 seconds after reentering the downwind leg and when approximately abeam the threshold of the runway, tower instructed the mishap crew to turn base leg. Tower's intent was to allow the mishap crew to complete their touch-and-go landing before other landing traffic. Due to the proximity of the mishap aircraft to the runway, the instructor pilot directed the student to extend downwind a few seconds and to reduce the power to minimum. As the student started his turn to base, the instructor pilot made the radio call, "(callsign), base, gear down, simulated engine out, (request) option." But the gear was not down, and as the student reduced the power, he did not pull it back far enough to activate the landing gear warning horn switches, because the three-engine approach required higher-than-normal power settings.

After completing the turn to final, the student, using all his concentration, established the aircraft on centerline, on glidepath, and on airspeed. The instructor pilot was concentrating on the approach, coaching the student, and mentally preparing for the departure back to base. The flight engineer was monitoring the runway, descent rate, and the engine instruments. As the aircraft crossed the threshold, the student was on centerline, on glidepath, and on airspeed. The tower controller checked the mishap aircraft to confirm its position in the traffic pattern but did not notice that the landing gear was up. The student flared the aircraft and touched down in an appropriate landing attitude. The unusual noise initially confused the entire crew. Both the instructor pilot and the flight engineer thought the aircraft had a blown tire. The student brought the throttles into flight idle, then ground idle, and used a very slight amount of reverse thrust. The landing gear warning horn began sounding upon retarding the throttles to flight idle; however, the crewmembers on the flight deck do not remember hearing it. The instructor pilot saw the landing gear handle in the UP position and the warning light in the handle illuminated and realized they had made a gear-up landing.

Several factors contributed to the fact that the mishap aircrew failed to maintain situational awareness. The crew did not receive an aural gear-up warning. The horn would be expected to sound during the approach and landing as the throttles were retarded to near the flight-idle position. However, the student never retarded the throttles to this range because of the higher power setting required during a simulated engine-out landing and his intention of keeping the engines out of a negative torque condition. If the warning horn had sounded during the final approach, it may have averted the mishap. But the warning horn was not the real culprit here.

It is highly probable that the landing gear warning light was illuminated for the previous three-engine approach as well as for the mishap approach. This condition resulted in the warning light's effectiveness as a warning being lowered due to habituation, the adaptation and subsequent inattention to an environmental cue after prolonged exposure to it. With the aural portion of the landing gear warning system not activating and the visual portion being reduced in effectiveness for the mishap landing, the following human factors developed.

The left seat copilot, an inexperienced "Phase I" student, was in the seat approximately 1 hour and 37 minutes prior to the mishap. He became task-saturated during the critical phases of his last pattern and approach, caused by conflicting traffic and the tower's multiple requests. This task saturation interrupted his prelanding habit pattern, and his attention was directed to returning the aircraft to the proper final approach glidepath and airspeed. It was especially significant in that the student had been having difficulty landing on centerline on this and previous missions. Due to his preoccupation with flying the aircraft and aligning it with the runway, the student became understandably overwhelmed.

The flight engineer was subject to the same anomalies of attention. Throughout the 360-degree turn and base leg, the flight engineer was occupied with clearing for other traffic in the pattern. When the left-seat copilot failed to initiate the checklist, the flight engineer was deprived of a cue that would have initiated his own routine associated with checking the landing-gear configuration. During final approach, the flight engineer was observing the runway to prepare for a hard landing, if necessary. While these considerations are less compelling than those facing the student copilot, the distractions were significant enough to make a weak case that the flight engineer could not be expected to catch the error.

But there is no forgiving the IP. Although the instructor pilot was occupied by clearing and monitoring the student's progress through the 360-degree turn and instructing this relatively difficult short approach, the safety of the aircraft and crew rested solely in his hands. The instructor pilot accomplished highly professional instruction during the mission, through critique, coaching, and demonstration, up to the last approach. Even though this approach necessitated more attention and instruction by the instructor pilot than had been required on previous approaches and despite the fact that the instructor pilot was occupied with opposite-direction takeoff traffic and multiple radio calls, he committed an unpardonable sin. By calling "gear down" without actually checking the physical position of the handle and the down-and-locked indications, the instructor gave away his last line of defense—the checklist. By paying "lip service" to words instead of actions, the instructor lost his battle with distraction. He allowed himself to be put into a box that required more attention than he was capable of giving.

Debriefing airmanship

The effective and well-presented debrief is where airmanship is truly developed. It is the time when the student and instructor reflect on what has just been practiced and learned. The airmanship model provides a visual schema to facilitate a comprehensive review in this process. You can identify and process a single component of the model or assess the components' interconnections as a system. The student experiences concrete considerations of airmanship as an integrated system. Airmanship begins to have a real definition, and a big picture emerges. The model further facilitates discussion and assessment of learning objectives. The debrief extends beyond simple skill development into an integrated systems method of thinking.

Let's use the last case study as an example of how the airmanship model can be used as the conceptual framework for a thorough debrief, starting with the foundations of airmanship. Flight discipline was compromised by the instructor pilot when he succumbed to the temptation to make a gear-down call without actually checking the position of the gear handle or the indicators. The flight engineer and the student can also be faulted here, as they failed to provide the necessary backup. Bottom line: No checklist discipline.

Moving up the model to skill and proficiency, we see that the student's recent difficulties with centerline landings preoccupied both the student and instructor, focusing their priorities on salvaging a nonoptimal situation into one more landing. Clearly proficiency played a role in this incident.

The five pillars of airmanship knowledge provide an opportunity to talk about the crew's understanding of the student's and instructor's maximum work load capabilities (self), the landing gear warning system (aircraft), the role of the flight engineer in pattern operations (team; what does "preparing for a hard landing" mean anyway?), the busy traffic pattern at the satellite base (environment), and the hazards associated with distractions and breaking habit patterns (risk).

The clear result of these failures led to predictable problems in the capstones of airmanship. Situational awareness was lost, as the crew did not realize that the landing gear had not been lowered, and a poor judgment was made perhaps in continuing the approach at all and certainly by calling gear down without confirming it was actually down and locked.

This short synopsis of a debrief combines all aspects of airmanship in a systematic way so that the events are not simply viewed as a linear cause-and-effect chain of events, but rather as an interactive system, kept in balance only by the airmanship skills and knowledge of the aviator.

Airmanship evaluation

Flight evaluators, or "black hats," as we are sometimes called, hold a special position in airmanship development. Face it, flyers are far more likely to take airmanship issues to heart if they are at risk of failing to meet standards. Typically, flight examiners evaluate specific physical flying skills to a quantitative standard, for example, ±100 feet of altitude or ±5 knots of airspeed. But big picture evaluation should encompass far more than mere physical performance. Evaluation should not simply be a periodic or end-of-training inspection but is instead a constant reflective process. It is meant to firmly establish performance baselines on which to judge and identify areas for improvement. Educational psychologists assert that real, long-term learning occurs during this reflection phase, and flight examiners, black hat in hand, can be the key to serious reflection. But self-assessment and evaluation takes place much more frequently and is central to any discussion on evaluation.

At a minimum, pilots should self-critique and evaluate their performance after every sortie and establish future training objectives and methods. Formal evaluations complement this self-assessment and are instrumental in providing validation of our individual picture of competence. Formal evaluations generally occur under two conditions: training progress or completion assessment and periodic inspections. This final section discusses the role of evaluation and how, by applying the airmanship model, a comprehensive reflection of performance can be accomplished as a base line for personal improvement.

The first, and most important, application of evaluation is self-assessment. Self-critique initiates personal reflection, which fosters a continuous self-improvement process. Postsortie critiques and discussion have been standard operating procedure in military aviation since its inception in the early twentieth century. General aviation, on the other hand, has never, apart from training environments, formally implemented a postsortie debrief requirement. Regardless of discipline, postsortie assessment is universally accepted as good practice and is, at a minimum, encouraged. However, is this practice optimized to gain the most development?

Education and learning experts would advocate that reflection of performance requires a standards comparison template. In other words, self-performance must be assessed in comparison to an ideal model. Until now, there has been no such ideal model for airmanship. The aviation sectors that do participate in formal debriefs generally have specialized debrief checklists, customized for their unique environments. These tools generally cover company-specific interests such as on-time takeoffs, fuel conservation, and wingman considerations. There is no universally accepted debrief template for aviator performance.

The airmanship model is the first to offer a universal and comprehensive debrief template. Any aviator, ranging from F-15 flight lead to DC-10 flight attendant, can apply the airmanship model for self-performance assessment. The general tenets of the model can apply to all professions and be customized for specific applications. The "team" pillar for the F-15 flight lead may include item specifics such as wingman utilization, formation coordination, and ground-observation employment. The DC-10 flight attendant's "team" pillar may include more conventional CRM metrics, such as flight deck communication updates, ground service coordination, and coworker utilization. The airmanship model uniquely provides a universal template for all aviation disciplines, which may be customized for specific job tasks.

Equally important, the airmanship model provides a comprehensive evaluation template that sees airmanship as a system of interrelated components. With this evaluation tool, the individual can be guided to consider all the factors of airmanship and their associated impact on the sortie. Item components such as "team" may not be obvious to a new general aviation pilot and subsequently ignored. Encouraged by airmanship-based evaluation, the new student may be forced to make a fit. He or she may begin to realize that tower and unicom are a part of the team, and he or she can begin to build skills and discipline to better use this potential team in the future. The airmanship model used as a postsortie debrief template highlights and focuses the aviator on aspects previously ignored or neglected. Administering the airmanship model in a continuous self-improvement environment of postsortie debriefs will ultimately begin to hone a comprehensive, systems way of thinking and create a safer, better-educated aviator.

Evaluation of airmanship occurs at two significant levels, the daily postsortie self-assessment and annual or periodic training progress/completion inspection. The daily self-critique is critical for development since it represents the reflection phase of learning. The periodic inspection is important since it sends the signal of accountability and enforcement. The airmanship model offers a unified and comprehensive definition of airmanship, allowing evaluation metrics to be developed. Through evaluation, the significance of airmanship is underscored.

There are some difficult challenges for the instructor/evaluator. Common pitfalls can include the trap associated with expectations and the dilemmas faced by the role of the evaluator as an "enforcer."

Expectations

An instructor must be wary about confusing past and future performance of a student. If you are acting as the instructor or evaluator, you are ultimately responsible for the safe handling of the aircraft and the safety of its crew. The evaluator's dilemma leaves little room for error. You must give the individual a chance to prove his or her skill (or lack thereof) and leave the student enough space to demonstrate competence, at the same time being prepared to immediately assume control of the situation to prevent damage to aircraft or injury to personnel. It often requires split-second intervention. The frustration of the following check airman is illustrative of this "razor's edge" requirement and how the situation can often turn on a split-second decision to intervene.

Case study: The check pilot's dilemma (NASA ASRS 1995c)

The pilot in command (PIC) was an experienced private pilot, owner of the aircraft, to whom I was administering a biennial flight review (BFR). I had flown with this pilot in the same aircraft on four prior occasions and had no reason to doubt his competence. A short cross country leg had gone well. The pilot had requested and received both an ARSA and a TCA clearance, and had complied well with them. We received VFR Flight Following until canceling same with destination in sight. The pilot overflew the field about 500' above pattern altitude, observed wind sock, noted runway 28 favored with a left crosswind component. (I estimate winds were approximately 240 degrees at 20 knots, well within the crosswind capability of both aircraft and pilot, in my opinion.) The pilot flew a normal right traffic pattern, turned final with full flaps and proper airspeed. I suggested "left aileron, right rudder," but the pilot replied he preferred to crab rather than slip on final, and proceeded to set up a satisfactory crab to track runway centerline. Kicking out the crab over the threshold, the pilot let left wing come up in the flare, the plane weather-vaned to left, and the nose gear strut collapsed and right wing tip scraped runway. The aircraft came to rest on runway, aligned 45 degrees to the left (directly into wind), resting on main gear and lower engine cowling. . .

At no time prior to the left wing coming up, did I see any pilot actions which suggested his inability to handle the prevailing conditions. He appeared both competent and confident, and I suppose I relaxed my guard a bit. The (check airman) walks a tightrope: He is supposed to let the qualified pilot in command fly the plane and demonstrate his proficiency in a BFR (flight check). He is not supposed to let the pilot in command damage the airplane. Short of commanding a go around in the flare (and I doubt there was sufficient time to execute one safely), the only thing I could have done to prevent this incident would have been to override pilot's control pressures and force more left stick and right rudder. This I failed to do because I thought the pilot fully in command of the situation (poor judgment on my part, coupled with pilot apparently momentarily relaxing his

*crosswind vigilance). In retrospect, obviously I should have
taken over control of the aircraft the instant the upwind wing
began to rise. But if a Certified Flight Instructor never lets the
candidate land the airplane, how can he certify the latter's
competence to act as pilot in command?*

This instructor/evaluator's experience underscores the need for
both proficiency and decisiveness on the part of evaluators, especially when giving flight examinations. Faced with a decision to
give evaluatees "enough rope to hang themselves" (or prove themselves) balanced against the safety considerations of exceeding
your own personal tolerances, the answer is clear. But these situations are not always clear-cut, and some flight examiners are reluctant to "bust" someone without a "clean kill," or obvious and blatant
deviation well outside of standards. This trend is unsafe and unhealthy. As an evaluator, you are tasked with making subjective, as
well as objective, evaluations of an individual's airworthiness. This
leads to the second evaluator dilemma: To bust or not to bust? That
is the question.

The bust

The evaluator is responsible to ensure the subject pilot meets standards. He or she is, quite literally, the enforcer. Often uncomfortable, and sometimes unpopular, the evaluator pilot must make the
hard calls. Failing a pilot on a check ride or flight review, the evaluator expresses more than just individual feedback. Tough enforcement of minimum standards is conveyed to all pilots. The standards
will be met or you will not fly. An F-16 evaluator pilot uses an analogy that is harsh, but perhaps applicable. "Look at your product [the
evaluatee] in terms of the way society views toxic waste — whoever
put it there is responsible for it from the cradle to the grave . . .
make sure that the product you send into the operational world can
consistently meet or exceed standards — not just have a lucky day
or one strong ride. Asking yourself "would I want this product on
my wing or in my right seat?" will often put things in perspective for
you (Del Toro 1995).

This role is clear to most evaluator pilots but perhaps is less defined
for instructors. An instructor builds a relationship with students and
often becomes biased, wanting the best for "my" student. Instructor
evaluation responsibilities during training are best underscored by a
recent Air Force instructor guide expectation: "If a product fails to

demonstrate proficiency at a required point in the training continuum, show nonprogression and complete the required remedial training. Don't ever think you're helping things along by slipping it on to the next block of training or overlooking one area due to other strong ones. If the standards aren't met, syllabus flow stops until they are" (Del Toro 1995). The instructor has the same responsibility, perhaps even more, as the evaluator in enforcing standards.

A significant difficulty in assessing airmanship in terms of "meeting standards" is the lack of a unified definition. How can we assess airmanship without concrete criteria? The airmanship model offers a starting point for measuring and evaluating airmanship. The tenets can be further expanded to better facilitate empirical measures. A popular and cutting-edge example of an emerging tool for evaluation is the LINE/LOS checklist (see Appendix). This tool is a CRM metric developed to measure crew performance. Unfortunately, as promising as the new instrument has proven to be, it has not been comprehensively integrated into formal inflight evaluations. In lieu of a formal tool for airmanship evaluation, there is a real need for each evaluator to look at how he or she plans to assess airmanship.

The main concern is not *how* we should assess airmanship, but rather that we must begin to assess it as an integrated whole. Discipline, skill, proficiency, knowledge of self, aircraft, team, environment, risk, situational awareness, and judgment—all must be evaluated to provide holistic feedback to the student or flyer being evaluated. Formal evaluations occur infrequently, and you owe it to the customer to give him or her the most complete evaluation possible.

Summary of airmanship instruction and evaluation

Above all, airmanship must be taught and evaluated as an integrated system. We began this chapter by saying that a flight instructor/evaluator is at once a resource provider, black hat evaluator, motivator, and evangelical enthusiast for airmanship, and that he or she must strike a delicate balance between advocacy and assessment. The following recommendations and techniques are suggested to help you achieve these goals. (Many of these suggestions are paraphrased and quoted from Del Toro 1995.)

1. *Change yourself first.* If you are not practicing complete airmanship, work on your own improvement simultaneous with or before attempting to teach it to others. This leads directly to suggestion number two.

2. *Keep credibility intact.* The instructor/evaluator *must* preserve credibility to remain effective. Students will see right through any attempt to gloss over a mistake. If you screwed up a demonstration, a maneuver, or a briefing item, admit it! If you talk the talk but don't walk the walk, you've lost effectiveness and credibility as an instructor.

3. *Be prepared to fly, and demand that your student is also.* Preparation sets the stage for learning airmanship. A failure to properly study and prebrief will likely cause the flight to degenerate into a limited, skill-building exercise.

4. *Be consistent with grades, praise, and critique items.* Training an airman can be equated to training a puppy. He or she will reach your expectations only if you are consistent in your praise and demands. Inconsistency only confuses and frustrates your student, and poor airmanship will be the likely end result.

5. *Remember, you are the training aid.* Far too many instructors use instructional sorties as an opportunity to show the student how good a flyer they themselves are. You don't have anything to prove; you have already completed your upgrade and instructor qualification. While a good demonstration now and then is a great tool, never lose track of your principle reason for existence — to train the student.

6. *Bust a student when necessary.* Aviation training is no place for grade inflation. Don't let worries about career implications or the fact that the evaluatee is a "great guy" color your decision. The best way to build airmanship, competence, and confidence is to establish and enforce standards. Never forget that you may well be the last line of defense to prevent a marginal aviator from buying the farm. Take your quality control role seriously.

7. *Remind all students that formal training is only a starting point, and real improvement only occurs through dedication and work on their part.* Challenge them to improve across all areas of airmanship, keep in touch with former students, and when you see them from time to time, ask them how they are doing.

The role of the instructor in airmanship development cannot be overemphasized. Habit patterns are easily established early in an aviator's career; they are far more difficult to change after we have become "hardened and experienced." Every time you read a poor-judgment fatal mishap report or review an incident of poor airmanship, ask yourself, "where did that attitude or mistake originate?" and then think carefully about your instructional role.

References

Connelly, Pete. 1994. CRM in general aviation: Who needs it? Unpublished technical paper. Austin, Tx.: NASA/UT/FAA Aerospace Crew Research Project.

Del Toro, J. 1995. The Mishap in the Mirror: Instructor Responsibilities. *Torch Magazine*, March, p. 6.

Hudson, G. 1992. Cockpit resource management for a flight instructor. *Flying Safety Magazine*, December, p. 25.

Hughes Training Inc. 1995a. Aircrew coordination workbook. Abilene, Tx.: Hughes Training Inc., CS-61.

International Civil Aviation Organization (ICAO). 1991. *Human Factors Digest No. 3: Training of Operational Personnel in Human Factors*. Montreal: International Civil Aviation Organization (ICAO).

NASA ASRS. 1995a. Accession Number: 231117. The Aeroknowledge, [CD-ROM].

————. 1995b. Accession Number: 124564. The Aeroknowledge, [CD-ROM].

————. 1995c. Accession Number: 92283. The Aeroknowledge, [CD-ROM].

USAF Air Training Command Study Guide. 1990. Instructor Development. Randolph AFB, Texas: USAF Air Training Command.

14

Understanding airmanship error

Eventually you are going to make a mistake, that's a given. The trick is not to make it a fatal mistake.
Stephen Coonts, *The Minotaur*

A personal view from Alan Diehl, Ph.D., ATP

Having investigated dozens of civilian and military accidents and flown privately for many years, I have come to a few interesting conclusions about human error in aviation. First, I am convinced that aviators don't take full advantage of their mistakes—which often can and should be viewed as positive events. Second, aviators do not realize their own vulnerability to the most dangerous types of errors—errors of judgment and decision making. Finally, the tendency to hide personal errors must be overcome if we, as a community of aviators, want to fly safer and smarter.

Experience may be the best teacher, but the price of tuition is too high, especially in aviation. When a flyer makes a mistake, the consequences can be lethal. However, most mistakes are not fatal, and we need to view these errors as positive learning opportunities, or we are wasting the "tuition" we are paying for the experience. Unfortunately, most of us like to forget our errors as soon as possible after we make them. Either we are embarrassed about them or it doesn't fit our self-image of perfection.

If we have some difficulty admitting mistakes to ourselves, it can be even harder bringing them up in front of other aviators.

However, research has revealed that the best aircrew commanders make a conscious effort to point out their own fallibility during preflight briefings. Statements such as "I'm not perfect, so please tell me when you see something that doesn't look right" or simply "I'm not above making mistakes" shift the crew into a watchful posture and facilitate effective communication in the cockpit. Oftentimes, it is the most junior crewmember who will pick up on incipient errors that may go unnoticed by more "experienced" crewmembers. Decision-making errors are the most common—and the most often fatal—errors made by flyers. Ironically, the very factors and personal flaws that cause these individual judgment problems also blind us from recognizing them when they occur. This is one of the primary reasons that we should share our fallibility with others.

No one likes to make mistakes, but we have an obligation to learn from them. Although our natural tendency is to forget—or even hide — our errors, to do so wastes a valuable learning tool. Experience may be the best teacher, but we simply don't have time to make all possible mistakes ourselves. Therefore, we need to learn from the mistakes of others — and help others to learn from ours.

Alan Diehl is a former accident investigator for the National Transportation Safety Board and the United States Air Force.

More than 2000 years ago, the great Roman philosopher Cicero declared that it was in the "nature of man to err." Twenty centuries of human evolution have not weakened the strength of the great man's observation; we still continue to make a variety and multitude of human errors. Cicero went on to say, however, that "only the fool perseveres in error." On this score, he may be on slightly weaker ground, because the nature of human error is complex, and the tasks required in modern aviation come at us so quickly that there is often little time to adequately identify, let alone analyze and correct, all of our errors. To improve airmanship, however, we must deal with our mistakes and move away from the mindset that human error is mystifying, unavoidable, and therefore something we must learn to live with.

One of the oldest clichés in aviation is that "you can never eliminate pilot error." This adage has been used to "low ball" the value
of human-factors research and training on both an organizational
and individual basis for decades. While it may be true that *all* aviation errors may never be eliminated, to use this as an excuse for
a failure to improve is ridiculous. If Hippocrates had taken this apathetic approach, medical science would still be attaching leeches
to our temples to cure migraines. It would be like the American
Medical Association saying, "There is always going to be illness, so
we're going to stop healing." It is time to take a different approach
to error and use our mistakes as signposts to improvement.

14-1 *Overcoming adversity and learning from error are key
elements of successful airmanship. Here, Orville Wright climbs down
from a successful learning experience.* USAF Academy Library Special Collections

Who is responsible for addressing the problem of error?

Now that we have determined that we must address the challenge
of error in aviation, we must ask ourselves, at what level should the
battle take place? Should this be a research issue, an organizational
or training issue, or an individual issue? What roles should be
played by the scientists, the organization, and the individual? The
magnitude of human error leaves room for all to contribute, but in
the end game, it must become an individual challenge.

The scientific community

The scientific community has done its part. In aviation, safety and human factors researchers have long been fascinated with error. Historian Maurice Matloff highlighted this point when he stated that we "are overly concerned with the pathology of the human condition . . . rather than its so-called normality" (Jessup and Coakley 1978). Literally hundreds, if not thousands, of formal research projects have looked at human error from nearly every angle imaginable, and yet errors continue relatively unabated. Certainly research alone is not the answer, but it is the first step in addressing the problem by providing the analysis necessary for organizations and individuals to begin to understand the complexity of error.

Flying organizations

Organizations are naturally concerned with minimizing error, for obvious reasons. The organization's bottom line is affected by errors that contribute to poor safety, lack of effectiveness, or lack of efficiency. Organizations address this issue differently, however. Some organizations see errors as a natural part of the maturation process and design training to help flyers work to improve. Others use punishment and fear in an attempt to quash pilot error. Still others do not address the error issue beyond the safety and flight evaluation requirements mandated by the regulations. Since the focus of this book is on individual and not organizational approaches to airmanship improvement, I will leave the suggested approach in the words of one of history's greatest organizational commanders and motivators, Field Marshall Erwin Rommel, the legendary Desert Fox of World War II.

> *It is always a bad sign . . . when scapegoats are habitually sought out and brought to sacrifice for every conceivable mistake. It usually shows something is wrong in the very highest command. It completely inhibits the willingness of junior commanders to make decisions, for they will always try to get chapter and verse for every thing they do, finishing up with a miserable piece (Rommel 1943).*

Rommel's point is simply that organizations seeking effective decision making should not create an environment of fear and apprehension in which every error must be paid for by someone. The optimum organizational role seems to be as a motivator, trainer, and resource provider.

Individual aviators

It should come as no surprise that error reduction is best accomplished on an individual and personal level. We hold the keys to unlocking our own potential, but to do so, we must meet three prerequisites. First, we must have a basic understanding of what our mistakes mean. This undestanding requires a generic understanding of error theory, which we will accomplish in this chapter. Second, we must be able to recognize our errors and place them in the proper perspective. For example, a pilot who continually makes the same type of error should look for a single and specific source of the mistake, whereas random errors may well indicate a generalized lack of skill, proficiency, or a loss of situational awareness. The last of the three-step process for error reduction requires the motivation and discipline to make the appropriate interventions in our daily flying practices and habit patterns to reduce or eliminate the errors. Continued self-analysis provides the feedback to continue the improvement process indefinitely.

The most difficult step of the three is the first one, understanding the nature of human error in general and aviation error specifically.

The scope of error

There seem to be as many ways to classify error as there are researchers to study it. According to James Reason, the author of *Human Error*, who is considered by many to be the modern guru on the subject, we have every right to be optimistic in our attempt to combat and learn from our mistakes.

> *On the face of it, the odds against an error-free performance seem overwhelmingly high. There is only one way of performing a task correctly, or at best, very few; but each step in a planned sequence . . . provides an opportunity to stray along a multitude of unintended or inappropriate pathways . . . Fortunately, the reality is different. Human error is neither as abundant nor as varied as its vast potential might suggest. Not only are errors much rarer than correct actions, they also tend to take a surprisingly limited number of forms (Reason 1990).*

In this chapter, human error is examined from general to specific. Included in this discussion are some brief results of research done in industry, maritime operations, and other related fields, all of which apply to us as aviators by showing us the various types and common causes of error. This discussion will produce a better understanding of the categories, causes, and results of human error in general. Armed with this basic background, a more *specific* discussion of airmanship error is accomplished. I have attempted to distill the volumes of research into a few useful models, or *taxonomies*, which will help you to focus your error-reduction efforts in the areas most appropriate for you as an individual.

Human error

Human error permeates every aspect of society. Because this chapter focuses on aircrew error, the discussion of error is limited to related fields, such as industry, using examples to illustrate various categories of error. This review demonstrates that while error is a much-studied phenomenon, there has emerged no single, accepted method for classifying error. It is left to the individual to choose among several models that are offered in the literature or to develop his or her own to use as a starting point for error identification and improvement.

Error defined

Let us begin by asking two simple questions: "What is human error?" and "How bad is the problem of human error in aviation?" Webster defines error as "an act or condition of often ignorant or imprudent deviation from a code of behavior" (*Webster's* 1990). Two key elements of this definition lead to a generalized discussion of the nature of human error. The first is the term "ignorant," which implies that the error-maker did not know or was incapable of knowing that the action he or she was taking would result in an error. This lack of knowledge roughly equates to the concept of a loss of situational awareness (SA) in aviation jargon. It may well be that the error *per se* occurred in the loss of SA and not in the subsequent action, but that "chicken or the egg" question is best answered by others. For our purposes, an error is an error.

The second part of the definition that needs clarification is the concept of *imprudence*. This concept implies that the error-maker knew and understood that his or her actions might result in an error but took the action or actions anyway. The important key to discriminating between these two most basic classifications of error is *intention*.

In the case of lost SA, there is no intentional deviation. An example can help illustrate this difference. Assume, for simplicity, that regulations prohibit a pilot from flying within 10 miles of a thunderstorm. Pilot Bob flies within 10 miles of the storm because he has not adequately "tuned" his radar and is therefore unaware of the danger. The error is in Bob's failure to realize he has insufficient information, but an error is committed nonetheless. Pilot Sue, on the other hand, is fully aware of the location of the thunderstorm but flies within 10 miles because she needs to get home to see her son's soccer game. Hers is clearly an intentional deviation and brings up the airmanship issues of discipline, knowledge, and judgment. But the question "What is error?" has only been partially answered. There are other subcategories within this basic definition of error that are important for the aviator to understand.

Errors of perception, execution, and intention

One of the most flexible and elegant explanations of human error is outlined by J. W. Sender, a prominent researcher in aviation psychology. He breaks down error into categories of *perception, execution,* and *intention.* Senders uses a traffic light analogy to outline the difference between each (Sender 1983).

Let's assume that Bob (after he lands and debriefs) is driving home and sees a red traffic light. He correctly decides to stop but accidentally steps on the accelerator instead of the brake. This is an error of *execution.* In aviation, these errors are often, but not always, tied to training, currency, and proficiency. If you discover that most of your errors are execution errors, perhaps the fix is simply more flying time and a better training focus.

On the other hand, if Bob had been looking into the sun and misperceived the red light as green or yellow, he would have committed a *perception* error. These errors can be associated with human-factors engineering and ergonomic issues, such as field of vision, accessibility of gauges and readouts, or software displays. They can also be the result of task saturation, which causes a pilot to get "behind the aircraft" and points to work load management as a possible solution. Perception errors usually have warning flags attached that can put an aviator on a more watchful posture. Fatigue, task saturation, and complacency all hold the potential for errors of perception.

If Bob was just in a hurry and correctly perceived the light as red and yet decided to go through it anyway, he would have committed an error of *intention*. In aviation, these are the most difficult to address. Intentional errors are not really "errors" at all but rather lapses of discipline that indicate the lack of a solid foundation of airmanship.

Internal and external sources of error

The next level of discretion is primarily concerned with the source of error. An error that occurs because of internal personal factors is called *endogenous*, or internally caused. If Sue's desire to get home to see her son's soccer game led to an unacceptable risk being taken, she would have committed an endogenous error. These errors are also the traditional targets of aviation human-factors training. *Exogenous* errors, on the other hand, arise from events outside of the person, for example, the sun in Bob's eyes as he approached the stoplight (Sender 1983).

Errors of omission, repetition, insertion, and substitution

One final taxonomy completes the discussion on the generalized nature of error and is important to our diagnosis of personal error. The nature of aviation error lends itself well to four basic categories. Errors of *omission* occur when someone fails to accomplish a required task, such as lowering the flaps or landing gear prior to landing. In aviation, errors of omission are often related to stress factors at both ends of the spectrum. At the low work load end of the stress scale, complacency and boredom may cause the error. At the high end, the culprit can be task saturation or poor communication.

An error of *repetition* is when something is done over once it has already been accomplished. One example of this can be seen in the case of the United Flight 173 crash at Portland, Oregon (see Chapter 9), in which a DC-8 captain required multiple checks of the landing gear position and cabin preparation, all while his aircraft was running out of fuel.

Closely related is an error of *insertion*, when something is done that should not have been done. These errors are associated with inexperience or overeagerness and are best overcome by efficient monitoring and crew coordination. An eager student pilot who wants to do the "Before Takeoff" checklist in the chocks is often the type to make errors of insertion, but surely none of us are immune.

Finally, if the wrong thing is done at the right time, an error of *substitution* has been made (Sender 1983). These errors are often the result of a phenomenon called negative transfer of training, which occurs when aviators take habits from one type of aircraft and incorrectly apply them to another. Most aviators will indeed fly more than one aircraft in their lifetimes, so we all need to be on the lookout for this problem. One student at a military training base fell victim to a substitution error in the following incident. He was guilty of mentally substituting a checklist step to "dump cabin pressure" (required in his old aircraft) into his new "After Landing" checklist. The "canopy jettison" handle was in the approximate same position as his old "pressure dump" switch. He inadvertently jettisoned the canopy as he taxied in from an early training flight, which no doubt impressed his new instructor to a great degree. While some elements of experience and airmanship transfer successfully from one aircraft to another, procedures and techniques most often do not, and errors of substitution should be carefully watched for during new checkout programs.

Human error and accidents

Although errors are deceptively easy to classify, they are extremely difficult to predict and often even more difficult to correct. Because of the situational nature of error, there can be literally infinite numbers of actors and variables in the error equation. This has caused many psychologists and training experts to begin to approach the problem of error identification and eradication as one of pattern recognition rather than an "if-then" cookbook (or checklist) approach to behavior modification. One of the biggest challenges we face in aviation is to get professionals to recognize the magnitude and potential consequences of the problem of human error. Resources tend to be expended on error reduction only following an accident or incident. Aviators often refer to this after-the-fact approach as "blood money." Sender addresses this issue:

> . . . *the study or error is inextricably mixed with the study of accidents. Accidents are not psychological events. They are mostly physical events. Not all accidents are the result of error; not all errors lead to accidents. The latter is most fortunate since if all errors did result in accidents, aircraft would be raining down from the skies like leaves in a fall wind (Sender 1983).*

These remarks ring true. But it also remains true that most evidentiary data on error come from accident investigations. An overview of the recent history of industrial safety provides valuable insights as to the breadth and scope of error in the workplace. It also serves to contextualize the study of aviation error and shows that aviation error is not fundamentally different than error found in other areas of industry. With the possible exception of the combat environment (stress level) and the accelerated dynamics of the flight regime, the theories of error previously discussed are as useful for aviators as for those studying error in other fields.

Learning from nonaviation error

Aviators are not alone in their search for answers to the problem of human error. The analysis of maritime safety data is illustrative. In 1992, the shipping industry experienced one of the worst safety records in recent history, resulting in the deaths of 1204 people through a rash of collisions, fires, and "foundering" (sinking) (Westlake 1992). Lloyds Register, considered a benchmark measuring stick for maritime safety, shows that human error was a causal factor in more than 80 percent of these accidents at sea (Westlake 1992). This number is nearly identical to those identified in aviation safety studies. In a separate longitudinal study, "a review of major insurance claims in the shipping industry revealed that the number of human error incidents outweighed claims related to equipment and machinery failure by (a ratio of) five to one" (*Global Trade* 1993).

Similar results are found in studies on vehicular accident rates on land in the transportation industry. "Highway accidents (caused by human error) are the leading cause of death in the industrial setting" (Messer 1992). This has led to a radical shift in training over the past two decades, focusing more on human-factors training and error recognition and prevention in all industrial fields, not just aviation.

One of the most alarming and illustrative incidents in industrial safety occurred in Japan's nuclear power industry. In early 1991, human error resulted in Japan's worst nuclear power accident at the Mihama Number 2 Nuclear Power Plant in Fukui Prefecture, Japan. Although it resulted in some radiation leakage, emergency cooling procedures were correctly employed to narrowly avert a complete meltdown and a disaster (Washio 1991). The resulting public outcry threatened to derail the entire nuclear power industry in Japan—a nation with an extremely low tolerance for errors of a nuclear nature. This type of high-profile accident or incident can have lasting and far-reaching implications (Washio 1991).

Parallels can be made between these incidents and commercial and military failures in aviation. A rash of commuter airline accidents caused the FAA to increase the level of regulation for this part of the industry. Several U.S. Air Force incidents received great publicity and resulted in similar disciplinary crackdowns. The Navy's problems with the F-14 fleet led to a "stand down" and a great deal of soul-searching in trying to find out why the aircraft and pilots were crashing with such frequency. Simple human error can result in accidents of such tragic proportions that it can place the entire aviation community under a microscope.

Perhaps the single most-watched human-error accident in history was the tragic disintegration of the space shuttle *Challenger* in front of millions around the world who watched in horror as the spectacle was shown over and over again on the network news. There were multiple human errors in this case, ranging from design problems to missed meetings by senior NASA officials. Although the director of the solid rocket motor project at Morton-Thiokol urged "close out" (ignoring) of a known defect problem with the booster's O-ring seals, many personnel share the blame for ignoring many red-flag indicators that might have averted the tragedy (Reason 1990). Additionally, the staff ignored or dismissed warnings about the temperature at the time of liftoff. The *Challenger* disaster has become synonymous with human failing in high-tech operations, and an overhaul of NASA procedures was accomplished in its wake.

In answering the question, "How bad is the problem of human error?" it has been shown that error has dramatic impacts on operations as simple as operating a forklift, to the pinnacle of high technology—space and nuclear operations. It should come as no surprise that human error in the aviation has profound, and often tragic, historical implications.

Aviation error

The basic error types in aviation closely parallel those from other fields, but a more detailed analysis reveals several trends and models that are useful for individuals who are seeking to improve their airmanship through error reduction or elimination. Before we look at these, however, it is essential to understand that aircrew error is far more complex than many of the accident investigations lead us to believe.

The problem with "pilot error" (Roscoe 1980)

There has been a tendency throughout history to write off a multitude of complex physical and psychological error types into a single category—pilot error. The problem with this approach is that it focuses on "fixing the blame" rather than identifying and finding solutions to the problem. In addition, it explains little of value to prevent future error occurrences. "Pilot error" may often be used to exonerate training, supervision, or maintenance from the stigma of association with a mishap. It does not insinuate a cover-up of any kind, but it does suggest an overly simplistic, nonbeneficial approach on the part of mishap and accident investigation boards. Recent studies from aircraft mishap investigations have continued to identify aircrew error by type, but often human factors experts must sift through the voluminous reports to isolate the actual cause-and-effect activities leading up to the mishap. A B-1B accident in the mountains of west Texas on November 30, 1992, was identified as "pilot error" in the Air Force 110-14 accident investigation report. A careful analysis of the flight profile hints strongly at lost situational awareness, misperceived cues from the aircraft, and spatial disorientation, none of which were listed as causal in the official report (Kern 1992).

The concept of blaming the cause of an accident on pilot error is of no greater value than if we said "the airplane broke" and did not identify the failed mechanism or part. The value of error identification is to provide up-to-date improvement tools, but we must get beyond the oversimplistic use of "pilot error" as a single categorical description of a wide variety of issues and challenges. The following breakdowns of error types are helpful in this endeavor.

Taxonomies of aircrew error

Reducing individual errors requires a method with which to categorize the mistakes we make to develop strategies for improvement. Research provides us with several taxonomies, or models, with which to accomplish this task. Each taxonomy is discussed with strategies for reducing the error types addressed within the discussion of the model itself.

Alan Diehl, an aviation psychologist, recommends dividing tactical aircrew errors into three categories: procedural activities, perceptual-motor activities, and decision-making activities, what he refers to as "slips, bungles, and mistakes" (Diehl 1989). These categories are useful for our purposes because they hint at areas associated to the airmanship model and, as such, can be very useful for analysis and improvement. For example, perceptual errors are linked closely to physiological concerns and situational awareness. A misperceived cue can cause a pilot to lose track of the current dynamic state of the mission. If you find yourself committing multiple perceptual errors, it would be wise to look to physiological or situational awareness areas for improvement.

Bungles, or motor errors, indicate the likelihood of skill or proficiency problems. You may want to seek improvement by increasing your flying time, seeking instruction, or sharpening the focus of your own personal approach to training.

Mistakes, or decision-making errors, may find their origin in any part of the airmanship model that feeds into the capstone of judgment. All critical components of the airmanship model should be looked at if you find yourself continuously making poor decisions.

A second taxonomy is reported in the U.S. Army's *Flightfax* safety publication. It lists eight areas as crew errors responsible for the majority of Army aviation accidents (Brooks 1994). While these items were identified through analysis of military operations, it is obvious that they represent common errors for all of us who fly.

1. *Scanning.* This error is the improper direction of visual attention or use of a scan pattern that is not thorough or systematic or with crew overlap. A potential fix comes in the form of crew coordination while you work on your scanning techniques.

2. *Crew coordination.* This type of errror is the failure of crewmembers to properly interact (communicate) and act (sequence and timing) in performance of flight tasks. Good CRM training is the best form of improvement on an organizational level. A thorough review of the leadership and followership lessons from Chapter 6 should also help improvement.

3. *Maintain/recover orientation.* This error type refers to the failure to properly execute procedures necessary to maintain

or recover orientation in the flight environment. Recovering from a loss of situational awareness necessitates recognition and recovery after the aircraft is removed from immediate danger. Chapter 9 discusses these procedures in detail.

4. *Preflight planning.* This error type is the failure to choose appropriate flight options for known conditions and contingencies and develop courses of action to maximize probability of mission accomplishment. Personal discipline means complete preflight planning every time. Proficiency at the planning processes may require some brushing up if you have neglected them.

5. *Inflight planning.* The improper modification of a flight plan in response to unanticipated events or conditions can cause a host of problems for the unprepared. Contingency planning and a systematic method for handling changes in flight are the first steps toward improving the odds of success against this airmanship challenge.

6. *Estimation errors.* Estimation errors occur when a crew member is inaccurate in the estimation of distance between objects, rate of closure with objects, or fuel or time calculations. Improving estimation errors means improving proficiency in the specific types of flying with which you are having the problem.

7. *Detection.* Detection errors involve a failure to identify hazardous conditions inside or outside the cockpit. These errors are best dealt with through improving knowledge of all associated risks so that you know what to look for, as well as keeping yourself physically and mentally sharp. Chapter 8 discusses risk identification and management in detail.

8. *Diagnose/respond to emergency.* The U.S. Army also noted a trend of improper identification or response to an actual, simulated, or perceived emergency in its analysis of accidents. Procedural, systems, and environmental knowledge are keys to solving this challenge when coupled with good situational awareness.

Attempting to improve by using error analysis is nothing new to aviation. A final taxonomy of aircrew error was found in a 1951 study from the office of the USAF Inspector General (IG), titled *Poor Teamwork as a Cause of Aircraft Accidents.* This study covered a period from January 1, 1948, to December 31, 1951, and analyzed 7518 major aircraft accidents (DAF 1951). This number alone illustrates how far we have come in preventing accidents. Although the study is somewhat dated, the human-factors errors identified track closely with those seen in more recent accident analyses. The study found that four general areas—pilot proficiency, crew discipline, crew proficiency, and teamwork—were illustrative of the majority of human-factors errors. But the study further developed these general categories to specific error types that included lack of alertness, insufficient briefing, failure to turn back, failure to take controls (intervention), wrong course of action, poor judgment, poor flight leadership or supervision, confusion over who had the controls, navigation errors, incorrect information, and poor teamwork (DAF 1951). Like many of today's accident analyses, the 1951 study recommended an organizational approach to address the problem of aircrew error. This approach did not go far enough to ensure individual improvement and accountability in 1951, and an organizational approach still does not go far enough today.

The individual is the answer

The nature of error is so diverse that even the finest organizational approach is doomed to marginal success. This, in turn, gives the human-factors critics the ammunition to spout their negative mantras, such as "You will never solve the problem of human error" or "No matter what you do, people will always make mistakes." In a way they are right. *We* may never solve the problem of human error, but *you* and *I* can certainly solve it on an individual level. Others have.

I have seen and flown with many pilots who seldom, if ever, make serious mistakes while flying. This is not to say that they are incapable of error, just that they have developed their airmanship to a personal level at which they do the right thing just about every time. These aviators are somewhat mechanical in their approach and have well-established habit patterns that guide them to automatic responses to most all situations. When new challenges do arise, these aviators have the breadth and depth of knowledge to work through them and the discipline to stay within the regulations and their own skill and proficiency limitations. These are the pros, and what we should learn from them is that error can be tackled successfully on an *individual level.*

This level of airmanship doesn't happen overnight. You won't wake up tomorrow and suddenly stop making errors. But you can wake up tomorrow and suddenly start analyzing errors and developing personal strategies for eliminating them. In addition to using the taxonomies and strategies already listed, a big-picture approach is an excellent starting point.

Error patterns: A good place to start

Another way of looking at errors is illustrated by the rifle shots on the targets in Fig. 14-2. This simple analysis technique, originally put forward by Chapanis in 1951, can be an ideal way to begin your error elimination plan (Chapanis 1951). The figures show two patterns of error, random and systematic.

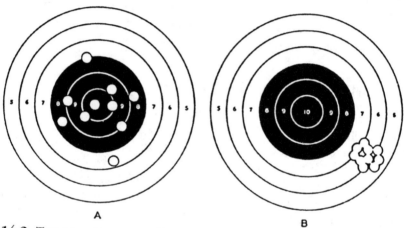

A   B

14-2 *Two target patterns illustrate random and systematic error. Identifying your own error patterns may allow you to zero in on small fixes that yield big results.* Chapanis 1951

The first target shows random error. It shows an individual who makes a variety of error types as well as various degrees of error. It may be indicative of a lack of skill or proficiency or perhaps just inattention or too large a personal tolerance for error, which is a motivational issue. In any case, it is clear that this individual is not an expert, and work may be required in multiple areas for real improvement.

The second target shows systematic error. The variability of the error is small, but all of the shots are consistently off the center mark. This individual shows considerable control and expertise but still has a single flaw. Error expert James Reason calls this phenomenon "poorly aligned sights" (Reason 1990). This is a significant insight for individuals and instructors, because if you can identify and fix the single source of the error, you have created an expert.

Sporadic, or occasional, error also exists—even among so-called experts. It is the most difficult to predict and analyze. Sometimes it is the result of an infrequent environmental factor. Other times it may reflect an emotional factor that only the individual is aware of. In any case, the occurrence of occasional sporadic error is reason enough for all aviators to maintain a watchful posture at all times.

Three targets for error reduction

Although aviators share many common error patterns, accident and incident analyses suggest that there are important systemic differences between commercial, military, and general aviation errors. In commercial accidents and incidents, poor communication seems to be the key element in many, if not most, of the accidents, and organizational and self-improvement efforts in this area may yield the greatest results. In military aviation, the problem of friendly-fire errors, or fratricide, is a tragic and preventable problem that must be addressed by all who fly with their finger on the trigger. Finally, in general aviation, the problem of inadvertent encounter with weather, or flying from VMC to IMC conditions when you are not prepared, must be addressed and, if corrected, would save hundreds of lives per year. If these three areas can be targeted for error reduction, the payoff will be measured in many lives saved.

Communications errors in commercial aviation

Perhaps because of the rapid pace and large scope of commercial operations, communications problems seem to occur more often here than in other aviation environments. In addition to the high density of commercial traffic, which creates greater potential for communications errors, we must also keep in mind that hazards on passenger-carrying flights can have disastrous consequences. The combination of the potential for error and the increased stakes involved in passenger operations makes communications error an ideal first target for self-improvement for commercial flyers.

NASA's aviation safety reporting system (ASRS) is inundated with reports of failed communications starting with confusion over predeparture clearances to final parking in the chocks. There appear to be four separate areas in which these errors of communication occur.

The first communications error type prevalent in commercial operations is between the flight crew and dispatchers. This arena is best left to the individual organizations to sort out, but it is highlighted in many ASRS reports as the first link of an error chain. The bottom line here is to ensure that no confusion exists at the beginning of a flight that might lead to greater problems later on.

Flight crew to air traffic control (ATC) communications errors

Miscommunications between the flight crew and air traffic control holds the single greatest potential for accidents and incidents within the realm of communications errors. Aviation researchers Billings and Cheaney looked at more than 28,000 ASRS reports over a five-year period (1976–1981) and discovered that the necessary information was usually available but not always transferred accurately. Their report explains:

> Close examination of the ASRS reports led to the finding that information transfer problems, as we have come to call them, did not ordinarily result from an unavailability of information nor because the information was incorrect at its source (although there are certain exceptions to this generalization). Instead, the most common findings showed that information was not transferred because (1) the person who had the information did not think it necessary to transfer it or (2) that the information was transferred, but inaccurately (Billings and Cheaney 1981).

These findings seem to indicate that a partial solution to communications error might be for all in this communications loop to provide *any and all* information that might be required and to consciously verify the accuracy of the translation of the information as it is passed along to those who require it. On the flip side of the coin, assertive questioning and reverification of information may be necessary by the user of the information, especially during operations in high-density traffic areas such as departure and arrival airspace.

Nonstandard phraseology

Another form of common miscommunication is nonstandard phraseology. A review of more than 250 ASRS incidents reported that "nearly half involved near midair collisions, loss of standard ATC separation, runway transgressions, or other conflicts with potentially serious safety consequences" (Matchette 1995). Many of these problems originate with an overfamiliarity with the system, which often leads to pilots who want to "spice up" their communications in a humorous way. Unfortunately, the results are not always so humorous, as the following captain found out:

> I called for clearance to St. Louis as follows: "Clearance delivery, company ident, ATIS info, federal aid to St. Louis." "Federal aid" was meant to mean FAA clearance in a joking fashion. The controller misinterpreted this to mean that we were being hijacked and called the FBI and airport police (Matchette 1995).

Most of us who deal with ATC on a regular basis know that nonstandard phraseology is common — more common than it should be, which is never. NASA Aircrew Safety Reporting System's Directline makes the following recommendations to combat this communications error (Matchette 1995).

1. Ask for clarification any time that the clearance or instruction is the least bit ambiguous. Continue to ask for clarification until there is no doubt in anyone's mind what is intended or required.

2. Communicate clearly and conscientiously, using only standard terms as identified in the *Aeronautical Information Manual* (AIM).

3. The pilot in command should discuss phraseology in the preflight briefing, emphasizing the importance of standard terms and precise communication.

4. Fight the urge to be cute or funny when communicating about clearances or expectations in the air. The potential results of miscommunication are nothing to laugh about.

Miscommunications within the flight crew

These observations on clarity and completeness of information also holds for the second area of common communications error, miscommunication between and among members of the flight crew. Often intracockpit errors are caused by stress or a break in the normal routine, as this ASRS report by an airline captain who landed without receiving landing clearance from tower clearly illustrates.

> *The weather and traffic were heavy. . . Just prior to the marker . . . we had a lightning strike which caused a momentary loss of navigation instruments. I believe that this caused the First Officer not to switch over to the Tower, and I forgot to verify that we had received a landing clearance. I landed and rolled out normally, and realized we were still on Approach Control frequency. When distracted or startled, even experienced pilots can make fundamental mistakes.*

This error points clearly to crew-coordination solutions. When a habit pattern is broken—as in the case of this crew, which was understandably distracted by the lightning strike—someone must take the lead to ensure that the standard procedures are followed. Ideally, this should take the form of a verbal statement to the crew such as, "OK, we just had a major distraction. Let's go back and check each other to make certain we have covered all of the required items. We've been set up for a mistake here; let's not make one." A statement such as this usually pulls the crew into a more watchful posture that allows the members to detect any latent errors that may have been caused by the distraction.

One type of communications error that may not be readily apparent to the pilots or crew is an error that occurs between air traffic control entities, such as ATC sector controllers and Approach control. Although aircrews cannot usually hear the communication that occurs between controllers, assuming that previous information has been passed along can be a dangerous error. If you are an emergency aircraft or have special handling needs, make it apparent to each controller after any frequency change.

More insidious communications errors can occur within the aircraft itself, and the results can be just as hazardous.

Cabin crew and flight deck interaction

Friction between cockpit and cabin crews can lead to situations in which safety is jeopardized. An ASRS report highlights the type of problem that can occur due to poor communication between the pilots and the flight attendants during a hectic night approach, when the flight attendant opened the door and flooded the cockpit with light at a critical moment.

The flight attendant had refused an earlier request by the captain to bring meals forward early in the flight, and the food arrived in the cockpit after the descent had begun. The captain wrote "the approach was unsafe" and described a serious breakdown in communication between the cockpit and cabin crew. "ATC became very busy. Constant airspeed changes, vector headings, and altitude changes. Suddenly approach control said, 'Stop descent immediately,' unidentified traffic at 12 o'clock.' (We) did not see any traffic, then looked out the left side of the aircraft and saw a light plane pass directly underneath us in the dark — might have been a near miss. During all the hectic action in the cockpit . . . the flight attendant opened the door."

The captain went on to say that "the captain is helpless to plan the approach anymore. The flight attendants ignore requests and directions from the captain. They work for the marketing department and don't hesitate to tell the pilots they don't have to listen to them (Chute and Wiener 1995).

Research by Chute and Wiener suggests several methods for improving this relationship to avoid communications errors that could put passengers at risk. They suggest that CRM training should include and address cockpit/cabin communications and include exercises to build solid habit patterns. They stress over and over the importance of good briefings to establish rapport as well as set guidelines and expectations for both halves of the crew. At a minimum, the crews should observe the courtesy of an introduction. The researchers point out that this concern was raised repeatedly and is a no-cost remedy. They also encouraged flight attendants to begin to take part in the ASRS self-report program, as they are a large and as yet untapped source of safety information. The bottom line is aircraft and passenger safety, and a group that is able to more effectively communicate produces a win-win situation.

Many other types of communications errors occur in commercial aviation, but perhaps the ultimate communications error is to lose it all together.

Lost communications

In an environment that relies almost entirely on effective two-way communications to provide separation between aircraft, it is surprising how many times lost communications actually occur in commercial aviation. The results can range from disaster to embarrassment. The following example highlights an example in which the pilots were attempting to eliminate "noise" coming from one of their two radios and inadvertently left their assigned approach control frequency.

After several minutes of radio silence, we noticed what had happened and switched approach (control) back on and called them. The controller was upset and announced we had delayed seven other aircraft due to our mistake.

This example is typical of the single largest reason for lost communications — a misset radio, which accounted for 52 percent of all lost communications events in a 1995 study of ASRS reports (NASA ASRS Directline 1995). But it is certainly not the only reason. Other reasons for lost communications include setting the aircraft radio too low, the controller assigning an incorrect frequency, "stuck" microphones, trying to communicate on the wrong radio, electrical failure, ATC facility failure, or occasionally a controller on the wrong frequency. Regardless of the reason, in commercial aviation, if you're not on frequency, you're in dangerous waters. The keys to fixing lost communications errors lie in awareness, prevention, recognition, and correction. Let's start with awareness.

Lost communications occur far more often than many aviators realize, and the average lost comm experience lasts 7.6 minutes (NASA ASRS Directline 1995) — an eternity in a high-density traffic area. Luckily, in air carrier operations, pilots are almost twice as likely to experience this problem in cruise. Unluckily, the general aviation pilots with whom we share the airspace are more likely to experience lost communications on approach. An even scarier statistic suggests that the lower the experience level, the greater the chances of experiencing lost communications — setting an inexperienced pilot up for failure in a high-density traffic area where he or she is likely to be unfamiliar with the emergency procedures for correctly handling the problem (NASA ASRS Directline 1995).

The ASRS Directline makes several recommendations related to preventing, recognizing, and recovering from lost communications errors:

1. The most common reason for a misset radio is inadvertent pilot action. The best solution to this problem is the old solution — proper attention to detail and good cockpit management and monitoring on the part of the flight crew.

2. Pilots should be aware that there is a significantly increased opportunity for a lost communications event when pilot experience in an aircraft type is low.

3. Pilots should write down assigned frequencies so that if lost communications occur on the new frequency, the pilot can easily return to the previous frequency.

4. During high-altitude flight, noting the location of ARTCC boundaries as marked on the charts alerts pilots to required hand-offs.

5. Pilots should monitor "guard" frequency (121.5/243.0) with adequate volume to ensure being able to hear a call in the event of lost comm on assigned frequency. Pilots should not hesitate to use guard as required to reestablish communications if lost communications are suspected. Better embarrassed than dead.

Speaking of embarrassment, the following ASRS report by an airline captain should caution all of us about the nature and potential of "stuck mike" events.

After three minutes of radio silence I had begun to wonder if I had lost communications with the radar controller. My instincts were right— we had lost radio contact . . . in the cockpit I had a few choice words to say about my aircraft and radios which should not have been said at any time.

Later, when asked to contact the facility supervisor on the phone . . . He [the supervisor] said "Now how are your blankblank radios doing? We have everything on tape, everything!

We had to go to a backup frequency because of your language. You apparently had a hot mike."

In summary, communications errors in commercial aviation represent an opportunity for individuals to make a major contribution to improving safety and effectiveness in their profession. Through awareness, prevention, recognition, and correction of individual communications errors, one of the larger problem areas in commercial aviation safety can be significantly improved. Better habit patterns, cockpit management practices, and individual awareness all hold great potential for reducing communications errors. It provides an excellent opportunity for individual airmanship to make a dent in an organizationwide challenge.

Military error: The challenge to eliminate fratricide

A historical analysis of human error in battle provides researchers with a long list of anecdotal evidence, ranging from Hannibal's lack of siege warfare apparatus to Lee's decision to send Pickett across open terrain. In spite of the relative abundance of historic case studies of human error, few of these analyses were of use for the purposes of this chapter. There are two reasons for this apparent contradiction. First, many historical analyses focus on command issues, often to the exclusion of discussing the tactical battlefield errors made by the individual soldier or airman, which is the focus of this book. The second, and more important reason, is that there has been little systematic historical recording of individual error, with one notable exception—fratricide, or "friendly fire" statistics. I have chosen to use these statistics as a benchmark of past military human error because it represents the closest approximation of a systematic identification of error types on the battlefield. In *Amicicide: The Problem of Friendly Fire in Modern War*, Lt. Col. Charles Shrader of the U.S. Army Combat Studies Institute looks at trends in fratricide statistics from air forces from WWI through Vietnam.

Aircraft fratricide

It comes as no great surprise that the statistics from aircraft fratricide show individual human error as the primary cause. Although the problem of friendly fire goes well beyond the realm of aviation, casualties from the air are far more damaging than those from other sources. Shrader explains, "In terms of the number of friendly casualties caused, air incidents clearly predominate, both as to total casualties and casualties per incident. In one WWII incident alone, 111 friendly troops lost their lives and 490 more were wounded" (Shrader 1982). A short historical overview illustrates the growing nature of the problem.

Fratricide incidents from aircraft were relatively few in WWI, as the aircraft had not yet matured in its air-to-ground mission. "Amid the tremendous casualties incurred by ground weapons, the deaths and wounds attributed to friendly air strikes went almost unnoticed" (Shrader 1982). In the next war, however, the role of the aircraft had grown.

WWII was filled with examples of fratricide from the air, ranging from the tragic to the almost humorous. One example tells of a German ship that was loaded with Allied POWs in Tunis on May 4, 1943. While at anchor for more than three days, the ship was continually strafed by "at least forty allied fighters and over 100 bombs were aimed at the target. Fortunately, the fighter pilots proved somewhat unskilled, only one of the bombs, a dud, hit the ship, and only one of the allied POWs was killed" (Shrader 1982). This example highlights the difficulty of positive target identification from the air, a primary difference between ground and air fratricide. With the advent of "smart" precision weapons, these marksmanship problems would all but disappear.

WWII also marked the beginnings of close air support (CAS) doctrine development in a concerted effort to avoid fratricide. Close air support is exactly what it sounds like: aircraft that provide firepower in close proximity to troops on the ground whom they are tasked to support. A "slow but steady" improvement in coordination between ground and air forces during CAS operations was seen. The one exception to this was an attempt to use heavy bombers in a CAS role, bombing from high altitude (Shrader 1982). Noncombatants also suffered the effects of human error from the air, as the following example illustrates:

> As a result of a gross error—due to poor navigation, poor headwork and misidentification of target, one group of medium bombers from the 9th Bombardment Division hit the Belgian town of Genck, twenty-eight miles west of the assigned target on the morning of 2 October 1944, killing thirty-four civilians and wounding forty-five" (Shrader 1982).

This example illustrates three error types that were prevalent in WWII and remain as common causes of fratricide across the historical spectrum: navigation, judgment, and target identification. The Pacific theater complicated the pilot's task with often heavy jungle foliage, a problem American flyers would face again a quarter of a century later in Vietnam.

Fratricide statistics from the Korean conflict did not significantly differ from WWII. It is noteworthy, however, that Korea saw the introduction of the helicopter and jet fighter-bomber to the battlefield, both of which offered unique human-factors challenges that were to surface in later conflicts.

Vietnam saw advancements in weapons technologies that made fratricide incidents far more damaging. Cluster bomb units (CBU) and napalm bombs had devastating effects. There were 23 casualties when a forward air controller (FAC) "failed to clear a target area properly and permitted an F-100 pilot to dump two CBU-2As in what the FAC presumed to be an authorized jettison area" (Shrader 1982). This incident points to the error of "the faulty assumption," which is a form of lost situational awareness.

Vietnam also showcased a new type of warfare—the air assault. Helicopter door gunners often found it impossible to control their lines of fire as the helicopter hit air pockets or changed altitude rapidly. On several occasions, these problems resulted in gunners killing troops that had just left the aircraft (Shrader 1982).

In general, the Vietnam conflict identified error types that are typical of those identified in safety mishap investigations in recent years. Lack of crew coordination, misidentified targets, disorientation, loss of situational awareness, navigation errors, and perception problems all are identified as leading to cases of fratricide (Shrader 1982).

The escalating stakes

The analysis of Desert Storm fratricide statistics show the same trends as previous wars, with individual pilot error as the single leading cause of the tragedies. But there are three significant differences that make the stakes of fratricidal error much greater than ever before. The first is the increased lethality of modern weapon systems. In the old days, if you fired at someone, they still had a better than average chance of survival. Precision weapons have changed that. If you are "locked up" by a modern targeting system, your chances of survival are almost nil.

The second difference lies in the nature of modern warfare. In the Gulf War, the percentage of friendly-fire deaths was much higher than in previous wars, due to the overall low coalition casualty rates from enemy fire. In addition, the perception of the problem is also greater, due to what I refer to as the "CNN syndrome," or the capability of modern news networks to broadcast breaking stories—such as a case of fratricide—within minutes of the occurrence. With modern satellite communications and broadcast technology, we no longer have to wait for tomorrow's headlines to hear about a case of fratricide; it will be on the next "30 minute update" and will be fully documented and analyzed by a group of "experts" by the next evening news.

The final reason that we must fight to avoid fratricide now more than ever is the nature of coalition warfare, in which multinational troops share the same battlefield. When a fratricide incident occurs between nations, and the tragedy is immediately broadcast around the world, the result is an automatic international incident. This was dramatically illustrated during two incidents during and after Desert Storm. In the first, two USAF A-10 pilots who were flying a close air support mission misidentified British Warrior armored vehicles as an Iraqi armored column. They fired Maverick missiles into the allied vehicles, killing 9 and wounding 11 British soldiers. A five-month British investigation into the incident attributed "no blame or responsibility to British forces." The British media splashed the incident across tabloid headlines for months afterward (Powell 1991).

During the enforcement of the no-fly zone following the war, two American F-15 pilots, under the control of an United States AWACs, misidentified, fired upon, and destroyed two friendly helicopters, resulting in an international incident. The wingman pilot lamented, "Human error did occur. . . It was a tragic and a fatal mistake which will never leave my thoughts, which will rob me of peace for time eternal. I can only pray the dead and the living find it in their hearts and their souls to forgive me." Further details are even more disturbing. Rules of engagement may not have been clearly understood, communicated, or followed (USAF 1994).

Fratricide is not limited to wartime operations, and deadly mishaps have occurred on and around practice gunnery ranges as well. In the past several years we have had several incidents in which pilots have misidentified range targets and dropped munitions on or near friendly forces or off-range civilian domiciles.

General aviation: Eliminating inadvertent IMC encounters

With the exception of loss of basic aircraft control, inadvertent or unauthorized flight from visual to instrument conditions accounts for more general aviation accidents than any other cause. If this judgmental error can be reduced, the impact on general aviation safety would be dramatic. We have already seen that judgment sits at the pinnacle of the airmanship model, so the root causes of these fatal errors may be varied and come from multiple sources. But a study conducted by the NTSB demonstrates the seriousness of the problem. From the NTSB accident files, 2026 fatal accidents were analyzed covering a nine-year span. In these accidents, a total of 4714 individuals died. During this same nine-year span, weather-related accidents accounted for more than one-third of all fatal general aviation accidents (Craig 1992).

The chances of a non-instrument-rated pilot surviving an encounter with the weather are not good. In 1955, the Aircraft Owners and Pilots Association (AOPA) sponsored a study by Bryan, Stonecipher, and Aron designed to determine the best way to train pilots how to get out of weather once they got into it. Using 20 non-instrument-rated pilots with between 31 and 1625 hours of flight time, they put these pilots under an instrument hood in an IFR-equipped aircraft and asked them to fly straight and level.

"Nineteen subjects placed the aircraft in a 'graveyard spiral' on the first attempt to fly instruments. The twentieth subject pulled the airplane into a whipstall attitude. Minimum time to reach the incipient dangerous attitude was 20 seconds; maximum time was eight minutes" (Bryan et al. 1955).

Based on extensive research, the NTSB has compiled a composite picture of the general aviation pilot who is likely to have a fatal weather-related accident. Do you recognize any of these traits or characteristics? (Craig 1992)

1. The pilot on the day of the accident received an adequate preflight weather briefing, most likely over the telephone. The information was "substantially correct" nearly three-quarters of the time. The weather conditions encountered were "considerably worse" than forecast only about five percent of the time.

2. The pilot was making a pleasure flight.

3. The pilot had a private pilot certificate, but was not instrument rated.

4. The pilot was relatively inexperienced, and typically had between 100 and 299 hours of flight time.

5. The pilot was typically not proficient, with less than 50 hours flown in the 90 days prior to the accident.

6. The accident occurred during daylight hours, with the pilot flying from visual into instrument conditions either intentionally or inadvertently.

7. The pilot had some instructional instrument time, typically between 1 and 19 hours under the hood with an instructor.

8. The pilot had logged no actual IFR time (at least until the fatal encounter).

9. The pilot did not file a flight plan.

10. The pilot was accompanied by at least one passenger.

This profile paints a picture of an inexperienced pilot with poor proficiency who intentionally flies into a cloud in broad daylight. In short, it profiles a pilot whose ego writes checks his or her skills can't cash. The role of the passenger may hold a key here. Most new pilots are justifiably proud of earning their wings and want to demonstrate their prowess to someone. This combination of ego and opportunity creates a condition that is conducive to bad judgment. When you throw weather into the equation, it is a volatile mixture indeed. A situation (weather) that should breed a healthy respect, if not downright fear, in the the new pilot somehow gets seen as another challenge to conquer when a friend is looking over your shoulder. This is a foolish and potentially fatal delusion. Even a proficient and experienced instrument-rated pilot has a healthy respect for flying in the weather, especially if the encounter is unexpected.

There are no easy answers to this challenge for general aviation pilots. On an individual level, perhaps fear is the key to good judgment. If you are not instrument-rated, or even if you are and are not proficient, you should run away from clouds like you would a red-hot poker.

Summary

The challenge of pilot error in aviation is conquerable, at least on a personal and individual level. We began by defining error as "an act or condition of often ignorant or imprudent deviation from a code of behavior." We then identified generalized human error types as those of perception, execution, and intention, pointing out that it is necessary to understand error to combat it. We also identified a few useful taxonomies, or models, of error, to assist aviators in classification and response to individual errors. By looking briefly at the nature of industrial error, we demonstrated that certain error types were seen across the spectrum of occupations.

Fratricide was highlighted as a specifically important challenge for military aviators, and we pointed out that even as the number of fratricide incidents decline, the increased lethality of the modern weapons of war make the odds of surviving a friendly fire attack almost nil.

Finally, and most importantly, we debunked the myth of pilot error as unassailable by pointing out that while organizational approaches may be limited, as individuals we are the masters of our own fate and error-free future.

References

Billings, C. E., and E. S. Cheaney. 1981. Information transfer problems in the aviation system. NASA Technical Paper 1875. Moffett Field, Calif.: NASA-Ames Research Center.

Brooks, Robert. 1994. Crew errors and the night environment. *Flightfax*. August 5.

Bryan, L. A., J. W. Stonecipher, and K. Aron. 1955. AOPA's 180 degree rating. AOPA Foundation, Inc. and Aircraft Owners and Pilots Association, Washington, D.C.

Chapanis, A. 1951. Theory and method for analyzing errors in man-machine systems. *Annals of the New York Academy of Science* 51:1179–1203.

Chute, R. D., and E. L. Wiener. 1994. Cockpit and cabin crews: Do conflicting mandates put them on a collision course? *Flight Safety Foundation Cabin Crew Safety*, 29(2):1.

Craig, P. 1992. *Be a Better Pilot: Making the Right Decisions*. Blue Ridge Summit, Pa.: TAB/McGraw-Hill.

Department of the Air Force (DAF). 1951. Poor teamwork as a cause of aircraft accidents. USAF IG Report.

Diehl, Alan. 1989. Understanding and preventing aircrew judgment and decision-making errors. HQ USAF Inspection and Safety Center. November 13.

Global Trade and Transportation. 1993. Study cites human error in maritime accidents. 113(7):12.

Jessup, J. E. II, and Robert W. Coakley. 1978. *A Guide to the Study and Use of Military History.* Washington D.C.: Center of Military History, United States Army.

Kern, Anthony. 1992. Personal testimony to USAF 110-14 accident investigation board, B-1B crash, Dyess AFB. November 30.

Matchette, R. 1995. Nonstandard phraseology incidents. ASRS Directline, Issue 7.

Messer, M. E. 1992. Transport vehicle accident rates drive operator training programs to change. *Occupational Health and Safety* 61(1):20–22.

NASA ASRS Directline. 1995. Lost com. Issue 6, p. 18.

NASA ASRS monthly update. 1995. Say the magic words. Number 196.

Powell, Stewart. 1991. Friendly fire. *Air Force Magazine.* December: 59.

Reason, James. 1990. *Human Error.* New York: Cambridge University Press.

Rommel, Erwin. 1953. The Rommel papers. *Dictionary of Military and Naval Quotations.* ed.: R. D. Heinl, Jr. (1966), p. 194. Annapolis, Maryland: Naval Institute Press.

Roscoe, Stanley. 1980. *Aviation Psychology.* Ames, Iowa: Iowa State University Press.

Sender, J. W. 1983. On the nature and source of human error. Proceedings of the Second Symposium on Aviation Psychology, Ohio State University, Columbus, Ohio.

Shrader, Charles R. 1982. Amicide: The problem of friendly fire in modern war. Combat Studies Institute, Fort Leavenworth, Kansas.

United States Air Force. 1994b. Report on Blackhawk shootdown. Vol 12: 13, July.

Washio, A. 1991. Nuclear "incident" report fails to calm public fears. *Japan Times Weekly International Edition.* 31(49):4.

Webster's Ninth New Collegiate Dictionary. 1990. Springfield, Mass.: Merriam-Webster, Inc.

Westlake, Michael. 1992. It was a bad year at sea. *Far East Economic Review.* November 19: 50–51.

15

The marks of an airman

The winds and the waves are always on the side of the ablest navigators.
Edward Gibbon

There is no cookie cutter for creating good airmen. Although this book has provided a model for airmanship development, it is doubtful that any two aviators who follow this path will end up with the same result. We are all unique and bring our own special talents, strengths, and weaknesses to flying. We should not expect—or seek—uniform results from the airmanship model. However, we should expect some benefits, beginning with a better understanding of what it means to be a good aviator.

The need for standards

Every profession or even hobby has established and accepted standards that are used as benchmarks for assessment, competition, evaluation, and self-improvement. Yet the flying game is curiously void of anything more than a small set of skill-based criteria for individuals to gauge their own personal development. These criteria generally come in the form of flight evaluation standards such as airspeed and altitude-deviation tolerances for flight training and periodic evaluations. The reasons for this peculiar lack of standards are debatable, but perhaps the origin of this apathy can be traced to the modes of advancement within aviation career fields.

In commercial aviation, advancement and promotion are tied primarily to seniority. This is certainly not to say that commercial aviators do not need to be skilled—they certainly do! But once a certain level of competence is reached, there is no incentive—beyond personal pride and professionalism—to strive for a higher level of expertise. Many are content to fly the schedule, pocket their paychecks, and go on about their lives. Military aviation is not significantly different in this respect.

Promotion for a military aviator is based on a variety of attributes, but it is certainly not necessary to maximize your airmanship potential to get promoted. Of course you must be competent, but if you meet qualification requirements and stay out of trouble, striving for personal airmanship excellence is not a requirement for promotion. In fact, many aviators believe that staying in the cockpit too long can actually hurt their chances for promotion. The so-called "whole person" concept requires that a military aviator spend much of his or her time on outside activities such as obtaining a master's degree, completing professional military education (PME—which has nothing to do with airmanship), and other career-broadening activities.

General aviation certification requirements also do little to inspire an aviator to seek personal achievement. Once a particular rating or certificate is obtained, the external motivation is gone unless one seeks a higher certification or rating.

All of these factors may account for the lack of a comprehensive set of airmanship standards. Perhaps none have been developed because we thought none were needed. A combination of seniority, experience, and basic competence has traditionally satisfied the requirements for climbing the career ladder. But we may have been missing something. While comprehensive airmanship standards may not have been necessary for organizations, they are absolutely essential for *individuals* who wish to improve themselves. Loss of efficiency, effectiveness, and compromised safety continue to result from poor airmanship and human-error mishaps. As an industry, we have searched for decades to find an answer for this problem. It may have been closer than we thought.

Personal improvement and accountability are the keys. By establishing a clear—albeit uphill—path to good individual airmanship, those who seek self-improvement can do so through systematic means. The ills that plague aviation as a whole can be tackled by any one of us who has the willpower and discipline to improve. The cultural change towards good airmanship occurs from within—flyer by flyer—until the judgment-error mishap becomes the exception rather than the rule. The day is coming and the means are here. There have been two primary inhibitors to personal improvement. The first was the lack of an accepted definition and model for improvement. Hopefully this book has begun to help remedy that situation. The second inhibitor is motivation for the individual improvement, and that is where organizations can still play a critical role.

The organization's role in improving airmanship

Even the best-intentioned individuals need motivation to improve. The aviation organization—be it governmental, commercial, military, or general aviation—can and should provide incentives designed to motivate individual airmanship improvement. These incentives could take on a variety of forms, from certification awarded for continuing education airmanship courses to actually linking promotion and advancement to demonstrated proficiency across the spectrum of airmanship. Airmanship awards could be designed that reward steady performance as opposed to miracle recoveries of stricken aircraft. In the military, airmanship badges such as command and senior pilot wings could be made to represent something more than longevity and flying time. They could be tied to completion of specific airmanship training across all areas of the airmanship model. Instructor upgrade in all areas of aviation should be based on demonstrated proficiency across the airmanship spectrum and not be viewed as merely another rite of passage that everyone qualifies for eventually. In short, the organization should be a cheerleader for airmanship and provide encouragement and concrete reinforcement for those who choose to improve themselves.

Second, the organization should be a resource provider. Airmanship education and training materials should be easily accessible and readily available to anyone who has a desire to improve. These materials can run the gamut from simple reading materials, such as pamphlets, to structured multimedia courses. It is important that material on all areas of airmanship are offered, because individuals will vary in their self-assessed needs. The temptation is to target recent problem areas or mishaps, but this is a mistake. Troubleshooting approaches are reactive in nature and defeats the entire purpose of self-assessment and individual improvement. If the organization dictates what training materials should be on the shelf, it nullifies the principle tenet of personal accountability for improvement based on individual needs.

Cost factors will certainly come up in any discussion about providing a new service or training, but they must be viewed in terms of future savings. What would you be willing to pay to make your organization 10 percent safer? How about 20 percent, or even 50 percent safer—how much is that worth? What about efficiency, effectiveness, job satisfaction, and retention? Improving airmanship makes sense to the bottom line.

Finally, the organization needs to remove obstacles to airmanship. These obstacles come in a variety of forms, but two of the most critical are undisciplined aviators and unnecessary taskings. A single undisciplined aviator who is allowed to continue to operate within an organization does immeasurable damage. Others see unchecked examples of poor airmanship as evidence of the organizational malaise and react accordingly. The organization should also seek to eliminate unnecessary taskings. Perhaps the single largest inhibitor to individual airmanship improvement will be time. By freeing up time and communicating the organization's reason for doing so (so that aviators can pursue personal improvement), you can send a clear message of sincerity and organizational commitment to the cause.

In a sense, this approach makes the organization a "servant leader" by empowering individuals to improve themselves through motivation, resources, and removing obstacles. Some individuals will use the opportunity wisely and productively; others undoubtedly will not. But by taking a proactive, rather than the traditional reactive troubleshooting approach, you will maximize all individuals' ability to improve their airmanship in a meaningful and personal way within the organizational setting.

The organization must also enforce standards of airmanship. The words of Richard Thornburgh, the former governor of Pennsylvania, echo like prophecy to those in aviation organizations:

> *Subordinates cannot be allowed to speculate as to the values of the organization. Top leadership must give forth clear and explicit signals, lest any confusion or uncertainty exist over what is and is not permissible conduct. To do otherwise allows informal, and potentially subversive, "codes of conduct" to be transmitted with a wink and a nod, and encourages an inferior ethical system based on "going along to get along" or the notion that "everybody's doing it" (Hughes et al. 1993).*

The following principles and standards of airmanship are suggested to help organizations and individuals accomplish this critical task.

Ten principles and standards of airmanship

Airmen need more than regulatory guidance, procedures, and word-of-mouth techniques to define — and become — experts. We need the same type of guiding principles that other fields enjoy to measure our progress on the road to airmanship excellence. The following principles are proposed as signposts and standards of airmanship development. Each principle of airmanship is followed by a standard by which to judge development. These standards are not quantifiable in the traditional sense (i.e., ±10 knots), but rather are qualitative measuring sticks for use in determining personal levels of airmanship. It is hoped that flyers who now understand the nature of good airmanship will use these principles as tools to take the next step — personal action.

Principle one: Airmanship must be viewed as a whole

All aspects of airmanship play upon each other. Failure to understand the interrelated nature of each part weakens the entire structure. Historically, aviators have tended to identify with single-trait flyers who perform great feats — the miracle recovery, the lowest pass, the tightest traffic pattern, the smoothest landings. Some see the systems expert or tactics aficionado as their role model. Both of these approaches are flawed unless they are accompanied by a holistic view of what complete airmanship means: discipline, skill, proficiency, knowledge of self, aircraft, team, environment, and risk. Airmanship means situational awareness and good judgment based on these attributes. Missing pieces of the airmanship structure signify either a lack of understanding or an apathetic attitude towards airmanship. It always signifies the potential for disaster.

The standard: Multidisciplinary competence

Airmanship encompasses physical, mental, and emotional skills, or, as the educational psychologists like to say, the psychomotor, cognitive, and affective domains. Airmanship means riding all three of these horses simultaneously and consistently well. Obviously this is not an easy task, and some will balk at the attempt to achieve such competence. But a golden-hands pilot who can't control his or her emotions in flight is not exhibiting good airmanship. Nor is the calm and cool systems expert who can't land in a crosswind. True airmanship requires physical, mental, and emotional competence. Specific benchmarks for these areas of airmanship are left for individuals or organizations to develop according to their aircraft and type of flying.

Principle two: Airmanship demands consistency

Expertise demands consistency. The demands of flying operations are constantly changing, like the lights on a stereo equalizer that move up and down depending on the nature of the music. One part of a flight may require concentrated risk analysis as you are faced with a hazardous weather front, while the next mission segment may require close teamwork or a decision based on personal capabilities and limitations. Because we seldom know where or when the next airborne challenge will arise, we must be consistently prepared in all areas. Our actions should be congruent with our personal assessment of our aircraft, team, and selves. This is not to say that we can't have a bad day. Even the best professionals on the pro bass tour don't catch fish every day on the lake. But an expert rides out the rough days with an expectation of success the next time out. This confidence is built on real skill and knowledge and leads to consistency of action and well-deserved success.

The standard: Predictability

"Surprise" is a bad word in aviation. Unfortunately, it can't always be avoided when it comes from an outside source, but we should never surprise ourselves. Consistency means avoiding surprises by approaching each situation with a confidence born of preparation. Given a common set of circumstances, your approach to the situation should be nearly the same each time. You should not be surprised when you succeed at something you are prepared for. Likewise, when you operate near the margins of your performance capabilities, you should not be surprised with less-than-perfect results. If you surprise yourself on a regular basis, you are likely lacking consistency or are unable to make accurate self-assessments.

Principle three: Airmanship requires balance

No single-focus flyer can approach airmanship excellence. A natural consequence of taking a holistic viewpoint towards airmanship is to ensure your airmanship structure is in balance. This means making a conscious effort to advance airmanship along two fronts — maintenance and development. We all have strengths and weaknesses, and our natural tendency is to gravitate our attention to areas of strength. We like to be good at things, so we do the things we are good at. However, this is only appropriate after we have achieved competence in *all* areas of airmanship, which means addressing our weak areas first, no matter how uncomfortable it makes us. While we are shaping our weaknesses into strengths, we must not completely neglect our strengths. This maintenance function is often a tricky proposition, because only you know when your proficiency is beginning to deteriorate. Try to give your strong areas enough attention so that they are maintained as strengths.

The standard: No weak areas

The ability to shift our education and training focus to areas of need, while simultaneously maintaining areas of strength and specialization, is one of the clear indicators of a mature aviator. The standard is competence across the board.

Principle four: Specialization occurs only after balance is achieved

All of us are eager to be the best at something — a recognized expert. Some of us are drawn to tactics, others to instrument procedures, aerodynamics, flight characteristics, or systems knowledge. But whatever our area of interest, specialization first requires balanced competence and readiness. A firm grasp of all areas of airmanship imparts credibility and relevance to the area of specialization. Remember, nothing in airmanship exists in isolation, so a solid grounding in all areas is required to fully appreciate and develop any selected area of specialization.

The standard: Broad-based competence

Much like the principle of balance, specialization demands readiness, and readiness means that our airmanship is fully up to speed. Seek first to be a competent airman — then to be a specialist.

Principle five: Airmanship is uncompromising flight discipline

This principle needs little further explanation. There is no room in good airmanship for intentional deviations from accepted regulations, procedures, or common sense. Violations of flight discipline create a slippery downhill path towards habitual noncompliance. Once you take that first step in this direction with a willing and intentional deviation, you are far more likely to do it again. Good airmanship is not compatible with flight discipline violations of any kind or of any magnitude.

The standard: Zero violations—zero tolerance

It is not enough to practice good flight discipline, you must also make it clear that you do not tolerate poor flight discipline in others with whom you fly. This may be initially difficult, because many people feel uncomfortable confronting others and value loyalty to friends above safety. Real loyalty speaks out against unsafe practices and makes it clear that poor flight discipline by anyone is unacceptable. Aviators share a moral obligation to each other to maintain safe operating conditions. Keep in mind, we all share the same sky.

Principle six: There is no substitute for flying skill

Good airmen fly well. They also understand that flying skills are perishable and that constant vigilance must be maintained if they are to be preserved. Unless we are in a formal training setting, this vigilance takes the form of mature self-assessment. An aviator must have or develop the kind of maturity that allows weaknesses to be recognized and the discipline to work on these areas, even though we would much rather be practicing areas of personal proficiency that we feel much better about. Airmanship goes beyond merely stick and rudder skills. It means honing and refining procedures and techniques to a personal level of excellence at which a missed checklist step or botched radio communication just doesn't happen anymore. Error-free flying—as well as "good hands"—is the mark of an airman.

The standard: Stick-and-rudder proficiency in all areas of flight and procedural perfection

Note the three parts of this standard. First, accurate and mature self-assessment must become part of your postflight routine. Only you know whether that crosswind caught you by surprise or if your stomach tightened up in knots when the controller changed the runway on you. No instructor can ever be as effective at pinpointing your weak areas as you are. Use this gift. Second, seek to achieve procedural perfection. This is one area where there are really no shades of gray. There are a finite number of checklist items and associated procedures. Learn them all and practice until you don't make omissions. It requires personal discipline and habituation, but it is well worth the effort. Finally, continuously hone your flying skills. Start with your weak areas, and when you get them up to speed, improve your strengths.

Principle seven: Airmanship requires multiple knowledge bases

Throughout history, superior airmen have drawn from deep pools of knowledge in several areas. We cannot hope to reach our potential without following the path they have established. Begin with the five pillars of knowledge identified by researchers as essential to good airmanship, and then add what you feel are relevant to your personal flying type. Expert airmen possess a knowledge of themselves, their aircraft, their team, the physical, regulatory, and organizational environments, and risk. Work these areas systematically until you reach a comfortable level in each, and then establish a procedure for periodic review.

The standard: Instant recall of critical items and sufficient knowledge of self, aircraft, team, environment, and risk to maximize performance

An old joke asks "what are the three things considered most useless to a pilot?" The standard answer is "runway left behind, altitude left above, and . . . (fill in the blank)." My blank filler would be "inert knowledge," that useless book learning that can be recalled at groundspeed zero for test purposes but is not known well enough to be recalled when your life is on the line. Throughout this book we have seen examples of aviators who were unable to recall important information when it was needed, who often paid the ultimate price for their lack of preparation. The only solution to inert knowledge is deeper study and drill so that critical knowledge recall becomes a subconscious event and leaps to the surface effortlessly when needed. The second half of this standard is knowing where to find other information if required—not only for use in flight, but to develop your knowledge across the breadth of airmanship. Topics like cockpit/crew resource management (CRM), situational awareness, weather, and others should all be readily available to deepen your understanding in these important airmanship topic areas.

Principle eight: Airmanship is maximizing situational awareness (SA)

No one maintains perfect SA at all times, but a consistently high state of SA is another mark of a superior airman. Situational awareness is directly correlated to an aviator's attention or lack thereof. Each of us only has a certain amount of attention to spread around all of our flight tasks, so development and expertise in lower parts of the airmanship model frees up more attention for situational awareness. For example, a disciplined, proficient, and knowledgeable pilot does not have to give much conscious thought to the procedures and skills required to fly an instrument approach; preparation makes it almost second nature. If a sudden distraction occurs, like a runway change or some unexpected weather, this pilot is usually quite capable of recognizing and reacting to the change in a safe manner. In contrast, a pilot who is less prepared and is struggling just to fly the approach is far less likely to handle the distraction and simultaneously complete a safe approach. The point is that each of us has an attention saturation point, beyond which we lose situational awareness. Airmanship is preparing ourselves, through discipline, skill, and knowledge, to have the maximum amount of "leftover" attention to handle the unexpected distraction. But since any of us can become overloaded, we must also be able to recognize the symptoms of lost SA and have the critical actions for recovery "hard-wired" to prevent disaster.

Three standards of situational awareness

The standards are

1. Understand components of preparation for maintaining SA.
2. Recognize lost SA in yourself and others when it occurs.
3. Know immediate action steps for recovery from lost SA.

Maintaining SA requires a solid undergirding of airmanship, with all that that necessitates. Recognition techniques are covered in detail in Chapter 9, and it is important to develop the ability to recognize lost SA in others, as well as yourself. Perhaps the most important aspect of understanding situational awareness—and the one that should be committed to memory first—are the steps to take to safely return home in the event of an episode of lost SA:

1. Get away from danger.
2. Stabilize conditions.
3. Give your mind a chance to get caught up.
4. Once on the ground, analyze the situation that led to the loss of SA, so that it doesn't happen again.

Principle nine: Solid airmanship leads to good judgment

Judgment has taken on an almost mystical quality among airmen, yet it is really quite simple. Once all of the prerequisites are in place, good judgment becomes a natural and automatic consequence of airmanship preparation. You show me an example of poor judgment, and I'll show you poor preparation. In nearly every case of poor judgment, you will find a problem with discipline, skill, or knowledge that existed prior to the episode of poor judgment.

There is an old adage in aviation that "you can't teach judgment." Like many dangerous misconceptions, this is partially true. Judgment cannot be taught as an independent objective, but it can certainly be accomplished by learning the fundamentals of airmanship. It is achievable and relatively uncomplicated. Yet the myths that judgment is either something you have or you don't or that it can only be obtained through experience are simply wrong. These myths have been accepted for decades by those who have not taken the time to understand airmanship. In fact, teaching judgment to ourselves is really quite uncomplicated, although certainly not effortless. All we must do is build a solid and complete airmanship structure, and good judgment will naturally flow from it. Nothing of value comes easy. You can't win judgment in the lottery or wake up with it one morning. You can't learn it from Chuck Yeager, your instructor, or from me. It is a personal journey through airmanship, based on individual strengths, weaknesses, and desires. The trip itself is enlightening and enjoyable, and the destination is well worth the price of the ticket.

The standard: Consistently sound decision making

Good judgment is the ultimate measuring stick of a superior airman. Nothing makes an aviator feel better than to have someone tell him or her that he or she exercised good judgment in a tough situation. But even poor airmen can make good decisions, and the true standard of judgment is consistency. Whoever coined the adage that "superior airmen use their superior judgment to stay out of situations where they must use their superior skills" was right on target. The inverse is also true. Superior skills, discipline, and knowledge create conditions (i.e. stability, SA, etc.) in which good judgment is easy to apply. These attributes of airmanship also create consistency in decision making—the mark of a superior airman.

Principle ten: Good airmanship is contagious

Airmanship excellence is self-sustaining and contagious. The pursuit of excellence is exciting, fun, and infectious. When others sense your enthusiasm with the journey, they too will begin to take a closer look at their own levels of and approaches to airmanship. Share your efforts with them. Peer review is one of the most effective and efficient forms of improvement known. Its utility is no secret in the business world or in the Israeli air force, where it is a formalized mandate for their combat pilots. Share your discoveries, resources, and insights. Find a partner or build a team.

The standard: Sharing what you've learned

Although the pursuit of airmanship excellence is by definition an individual project, there are great personal and organizational advantages to sharing your efforts with other aviators. First, it is likely that you share local airspace with your colleagues. Their predictability and airmanship directly benefits you, as well as all others who share the same sky. Second, it is always easier to stick with an improvement plan if you know that you are not alone in the effort. Finally, we have a moral obligation to share what works in a high-risk endeavor like flying. The little bit of information that you pass along may be what saves another's life—or yours—someday.

The ten principles of airmanship are not designed to be all inclusive or as a magic panacea for poor airmanship. They are offered in the hope that they will reinforce the material contained in the preceding chapters and remind us of the essentials as we pursue personal excellence. The traps of early specialization and gaps in knowledge are all too frequent in many of today's flyers, who then fail to understand why they occasionally get in over their heads and make poor decisions. Keeping the principles in mind forces us back to the work to be done—building and refining the entire airmanship package.

Jump-starting airmanship: Getting started with a six-month self-improvement plan

The longest journey begins with a single step, but that step is often a difficult one to take. With that in mind, the following structure is provided as a jumping-off point for individual improvement. It is based on a three-hours-per-week study plan and a regular schedule of flying, whatever that means for you. It provides 78 hours of self-instruction, which is more than many put into a graduate-degree course, so the depth of study should be adequate for an initial overview of airmanship topics. You need a planning calendar, a journal, and your logbook, along with a variety of study materials that we discuss as we describe the plan. The improvement plan requires self-assessment, self-instruction, and honesty. There is no one to impress or to fool but ourselves, and that is self-defeating.

The tools: Planning calendar, logbook, and journal

The jump-start plan is divided into months and steps for flying and nonflying activities, and you need a system to track your progress. The idea of the three tools is to allow you to establish priorities for your training and education, conduct the training, and then be able to look back reflectively at your performance to gain insights for further improvement.

The planning calendar

Success in any endeavor requires planning, and airmanship is no different. Your daily planning calendar is used to track both flight and ground activities. The purpose of this step is to add structure to your program through a scheduling function. In today's busy world, activities that are not scheduled are often the ones that are left undone. Prior to beginning your improvement plan, sit down with your calendar and decide when the optimum times for study are for you. Some prefer to put three hours in on a weekend, others prefer to spread it out across weeknights. Before you begin, however, you should have at least one month planned out in advance to help you prevent preemptive strikes against your plan from less-important distractions. There are definite advantages to scheduling self-improvement efforts. ("Honey, I can't mow the lawn this evening— I'm scheduled to study.") If a change in plans does occur, reschedule your study time immediately. To keep track of your program's fluidity, use ink on the calendar so that you can look back and see how many changes were required. It may help you to schedule smarter the next month.

The calendar should also be used to keep track of flight activities. Although this would seem to duplicate the function of your logbook, the calendar allows you to see how the two halves of your program (flying training and knowledge acquisition) are tied together. It also helps you to keep track of continuity and training priorities in your flying schedule.

The journal

The ever-present companion of the calendar should be your program journal. You should keep your journal in something small enough to bring with you when you fly; a small spiral notebook is ideal. Your journal is for reflections on various airmanship topics of interest, as well as to write down questions for which you wish to find answers. At the end of the six-month journey, your journal will help you to look back and refine your continuing improvement plan. It is a companion for both your logbook and study materials.

A typical journal entry is reflective, meaning that it should highlight what *you* think is important about what you just read or did. It is ideal for observations ("the crosswinds kicked my butt today") or reflections ("next time I'll try to put in cross-controls further out on final"). Although you should keep separate sections for flight and ground entries, you should attempt to tie the two together whenever possible. Knowledge and flying are inseparable in the real world, so we should try to bring them as close together as we can in our journal. The following is an example of a short journal entry that ties knowledge and action (or education and training) together.

Journal entry: 4 Feb.

I ran into a directional windshear in the traffic pattern today, but I didn't realize it until after I landed (only flew two patterns). Tower was calling the winds down the runway at 170/15G20 (winds from heading 170 degrees at 15 knots with gusts to 20); wind sock showed the same. I didn't put in any crab for directional control on downwind and got blown in way tight. I thought I just flew a bad heading, so next time I paid closer attention—same result. After engine shutdown, I saw some low scud clouds blowing due east—then I figured it out. Same cues were available in flight but I missed them. I bumped into a commercial pilot in the parking lot, who told me his INS (inertial navigation system) was showing pattern altitude winds of 260/30. It would have been nice to have a PIREP (pilot report)! Note: (added that evening) Reviewed windshear materials and found several visual cues that are available to "see" windshear, including blowing dust, dirt devils, bird flight patterns, low clouds, and observing the aircraft in front of you. Still, PIREPs are probably the best source of info. Next time I suspect w/s, I might ask tower if they have any pireps on pattern winds; it might prompt one of the high-tech guys to provide some verification.

This entry is typical of what you are attempting to do with your journal. Entries should be short and to the point and highlight lessons learned. There are no points taken off for grammar or spelling, so personal shorthand is acceptable. Review previous entries on both a weekly and monthly basis. This review provides critical reinforcement for your learning. This step cannot be overemphasized. In the final analysis, your journal is your record of a personal journey towards airmanship excellence. You may wish to explain your journey to others someday, so design your entries accordingly.

The logbook

The aviator's log completes the list of tools for accurate and reflective learning. The standard entries should be made but then cross-referenced in your journal by date or sortie number, for example, "4 February 1996/2nd sortie." This allows you to complete the picture of what occurred on a given day in your training program. You can look at your calendar for preparatory actions, view your log for specifics of the flight, and review your journal for lessons learned or questions regarding the flight.

Obtaining study materials

Throughout the six-month improvement plan, various readings are recommended from cutting-edge texts in the field. Although these books can be ordered and purchased through your local bookstore, some are relatively expensive, and they can also be obtained through interlibrary loan services at most local public libraries. If you do not wish to purchase the books, take a list of the texts to your local librarian and obtain them in this manner unless you want the books permanently for your personal collection.

Before you begin

There are a few entering guidelines and assumptions we need to establish before we begin our self-improvement program. We should first realize that we will reap what we sow and our returns will be greater the more dedicated we are to our improvement efforts. There are three key points to keep in mind:

1. Set moderate goals. Look for breadth of knowledge, not specialization in a subject area. This does not mean that you shouldn't look into areas that interest you, only that your study must move across many areas of knowledge in 26 weeks, so don't get bogged down.

2. Use existing instruction and evaluation tools whenever possible. A great time to start this program is right after a flight evaluation critique session. If you are in between check rides, talk to a local instructor and explain what you are trying to do. Get a thorough flight and ground evaluation and a no-holds-barred debrief. Armed with a professional assessment, you are better equipped to personalize your training and education plan.

3. Personalize the plan. The guidelines below are only a template. Insert, delete, and modify as you see fit. Only you know where you are and where you want to go. If you are already satisfied with your aircraft knowledge but are weak on the regulations, adjust the plan as required. Review the section in Chapter 3 titled "A layman's course on personal program development." It establishes techniques and guidance for valid and useful program development.

OK, let's get started.

The first month

The first month is designed to establish a starting point. Your objectives should be to gain an understanding of airmanship as a whole and determine your current status relative to the ideal. It is perhaps the most crucial month in the entire plan, because it will establish the seriousness and intensity of your effort. Your second objective is to begin the process of perfect flight discipline. Both are easily achievable goals for your first 30 days.

Step 1: Gain an appreciation and understanding of airmanship as a whole (four weeks/12 hours)

1a. Read this book cover to cover. Take time to write down any questions you have in the margins or in your journal. Discuss them with another aviator, an instructor, or write or call the author to obtain clarification. Don't leave any areas unless you have a handle on their meaning.

1b. Reflect on any personal experiences that you might have had that relate to the subject areas covered in the airmanship model. Write them down in your journal.

1c. Using a 1-to-10 scale, rate yourself in each of the airmanship areas, with 10 representing perfection and 1 representing the total absence of skill or knowledge. Questions are provided to help your personal assessment. Be honest with yourself.

Airmanship foundations

Flight discipline. How often do you knowingly bend or break the rules? Do you feel a secret admiration for a pilot who successfully "gets in" when the weather is below minimums? Are there times when the rules get in the way? Are your role models disciplined or undisciplined aviators?

How do you assess your current flight discipline?

1——2——3——4——5——6——7——8——9——10

Flying skills. What are your strengths and weaknesses? Are you comfortable flying the aircraft to its flight manual limits? Are there circumstances in which you doubt your abilities? If so, what are they and how often do they occur?

What is your current assessment of your flying skills?

1——2——3——4——5——6——7——8——9——10

Proficiency. How often do you fly? What is your normal interval between flights? Which maneuvers do you need to practice the most frequently? Which maneuvers *do* you practice most frequently? After a long layoff, what scares you?

How proficient are you today?

1——2——3——4——5——6——7——8——9——10

Pillars of knowledge

Knowledge of self. How well do you know your capabilities and limitations in the aircraft? When was the last time you really "pushed" yourself? How does your personality affect your flying? What hazardous attitudes do you exhibit in flight? Do you have any "personal minimums?" How will you know when you are in over your head?

How well do you know yourself?

1——2——3——4——5——6——7——8——9——10

Knowledge of the aircraft. Can you recite all of the critical action procedures from memory at any time? Have you read your flight manual cover to cover? Do you know all of the operations limits by heart? What percentage of the cautions and warnings do you know? How much do you know about the technical aspects of your aircraft systems?

How well do you know your aircraft?

1——2——3——4——5——6——7——8——9——10

Knowledge of team. How well do you know the capabilities and limitations of those whom you fly with? How about their proficiency, health, stress factors, fears, and attitudes? Do you trust them with your life? Should you?

How well do you know your team?

1——2——3——4——5——6——7——8——9——10

Knowledge of environment

Physical environment. How well do you understand weather phenomena? Are you completely familiar with the airfields and airspace you operate in? How much do you know about the effects of terrain? Do you fully understand the implications of other physical factors, such as altitude, darkness, birds, and glare? Where is the special use airspace in your flying area? Where are the towers and other obstructions relative to your home airfield?

How well do you know your physical environment?

1——2——3——4——5——6——7——8——9——10

Regulatory environment. How well do you know your FARs and other regulatory guidelines for you and your aircraft? Can you name and physically locate *all* of the regulations and procedures that guide your operations? What regulation is the source of your authority as a pilot in command?

How well do you know your regulatory environment?

1——2——3——4——5——6——7——8——9——10

Organizational environment. What are the priorities of your organization? Do any of these conflict with safety or your personal views? Who are the organizational heroes and villains? What did they do to achieve that status? Can you describe the "ideal" aviator in your organization—from management's point of view? Does he or she look like you?

How well do you know your organizational environment?

1——2——3——4——5——6——7——8——9——10

Knowledge of risk. Where will the next accident occur in your flying organization? What will cause it? If an accident occurred to you, what would be its likely cause? How many risk factors can you list off the top of your head? Can you name your local safety representative?

How well do you know your risk?

1——2——3——4——5——6——7——8——9——10

Capstones

Situational awareness. How often do you feel like you've "missed something" while flying? How often are you surprised by another aircraft, a weather phenomenon, or a controller's instruction? When was the last time you felt behind the aircraft? Do you dwell on errors made while flying? How well do you keep track of multiple aircraft in a traffic pattern?

What is your assessment of your situational awareness?

1——2——3——4——5——6——7——8——9——10

Judgment. How confident are you in your decisions? Do you often find yourself wishing you had selected another course of action? Do you find yourself having to fly out of situations you could have avoided? Do others question your decisions? How much confidence do you inspire in those who fly with you?

How would you assess your judgment?

1——2——3——4——5——6——7——8——9——10

This self-assessment, when combined with a thorough flight and ground evaluation from a qualified instructor, should help you determine where to place your improvement emphasis. If you wish, you can add your scores together and divide by 12 to give you an overall airmanship average. In a small-scale sample study using this instrument, the average score for midcareer military pilots was 6.5. However, these scores are not for comparison between flyers, but rather for your own personal insights on perceived areas of strength and weakness. Ironically, if you use this instrument to check your growth over the six-month program, your scores may actually go down. This is because as you begin serious self-analysis and get into the books, you begin to realize how much you *don't* know. It's been said that ignorance is bliss—but it's certainly not airmanship!

1d. Focus on flight discipline. Prepare for this step by reviewing Chapter 2 in its entirety. The first step towards establishing a strong foundation for airmanship is to make a personal vow to never willingly violate flight discipline—never. This includes following all regulations and policy guidance to the letter, even when no one is watching. Unless emergency circumstances dictate, stay between the lines. This should be the standard for most of us already. When this vow is made, you have taken your first major step towards airmanship excellence.

Month two

The second month focuses on basic skills and knowledge of self and aircraft. Review Chapter 3, with particular emphasis on the four levels of skill development, and the first half of Chapter 4, medical airworthiness. Our flying goal this month is to ensure that we meet safety levels in all authorized basic aircraft maneuvers. It may require some review to find out what all of these authorized maneuvers are, but they typically include slow flight, steep turns, stalls or approach to stalls, and others, depending on aircraft type. Since we will be practicing aircraft-handling characteristics, we will review the flight manual sections on flight characteristics. Our second educational goal this month is to gain insights on our physical self through a look at aerospace physiology.

2a. Flight objectives. Safe operation in all of the basic airwork. Challenge yourself safely, and find any limits you might have in these areas. If you need instruction, schedule it and attack the weak areas with gusto. Keep track of what you do well and what needs more practice. The goal here is safety in all maneuvers, so don't be frustrated by a lack of expertise in areas you may not have tried for a while.

2b. Knowledge objectives. Begin with an overview of the flight characteristics section or chapter in your aircraft's technical order or flight manual. Pay close attention to information on slow-speed handling characteristics and stall recovery. Memorize any relevant warnings or operating limitations. For the second educational objective, review the first half of Chapter 4 in this book, and I highly recommend reading *Fit to Fly: A Pilot's Guide to Health and Safety*, by Richard Reinhart, M.D. (TAB/McGraw-Hill 1993) and Chapters 4 and 10 in *Human Factors in Aviation*, by Earl Wiener and David Nagel (Academic Press 1988). If these cannot be located, any other text on aerospace physiology should provide a solid foundation of knowledge in this important area. Place particular emphasis on understanding the importance of nutrition and the hazards of fatigue, as well as the role of physical and emotional stress.

Month three

This month's objectives are traffic pattern operations and associated procedures, techniques, and other relevant systems information in the aircraft publications. Environmental knowledge is also stressed through a study of airspace considerations and regulations relating to airport traffic area operations.

3a. Flight objectives. Maximize proficiency in ground operations, takeoffs, visual traffic patterns, and landings. This should include a thorough study of local area procedures. Focus on precision in the traffic pattern, and narrow your acceptable airspeed, heading, and altitude deviations to as small a window as possible. For example, if you currently give yourself ±50 feet at traffic pattern altitudes, try narrowing that to ±20 feet, then ±10. Pretty soon you will be unconsciously making small corrections towards the desired parameters, and the precision of your pattern operations will have increased greatly. Try to find an opportunity to practice crosswind takeoffs and landings, and don't forget about pattern corrections in strong wind conditions. Record your efforts, self-critiques, and lessons learned in your journal.

3b. Knowledge objectives. This month's study focuses on aircraft, environment, and self as they relate to traffic pattern operations. Begin with a thorough review of normal and emergency traffic pattern procedures and techniques. These are found in your flight manual. Next, study your aircraft's flap, landing gear, and instrumentation systems (i.e., airspeed, AOA, heading, and attitude reference systems) that you rely on for pattern operations. Locate information on windshear operations for your aircraft. If none exists, read general material that explains how to calculate and use a reference groundspeed when windshear conditions might be expected. Review FAR Section 91 information on general flight rules (91.101 to 91.144), and visual flight rules (91.145 to 91.159), looking specifically at right-of-way rules and other materials relevant to traffic patterns and airport procedures. Since you plan to fly multiple traffic patterns, it might be helpful to review information on complacency found in Chapters 4 and 9.

Month four

The fourth month's objectives are related to navigation and weather. The flight objectives highlight filing procedures and inflight changes. Knowledge objectives focus on the physical environment and instrument flight rules.

4a. Flight objectives. Maximize your capability to use the airspace effectively, navigating safely and precisely by a variety of methods. Depending on the level and nature of your flying, you may be accustomed to navigating by primarily visual or instrument means. Break out of your mold. Be meticulous in your filing procedures, whether they are VFR or IFR. Obviously, this effort must be guided by your company policies and personal certification, but to the maximum extent possible, seek to understand and experience what the "other guy" must do to navigate and complete a mission. This aids greatly in improving situational awareness and teamwork between yourself, other aircraft, and air traffic control, as you gain a greater appreciation for the overall structure of the airspace and the various modes of operation within it.

Look carefully at the methods and hazards related to inflight course or destination changes. When you deviate from a planned flight profile, what safeguards to you employ to ensure terrain avoidance, adequate fuel reserves, and navigational capability? Remember, change is the mother of distraction, distraction often leads to lost situational awareness, and lost SA leads to . . . . So, study your habit patterns for dealing with inflight changes carefully. Record lessons learned and personal reflections in your journal, and see how the study materials tie in to your flight activities.

Your second flight objective is to look closely at all weather-related phenomena, noticing their effect on your operations, any hazards posed, or advantages to be gained. Study and discuss all options available related to weather hazards, beyond mere avoidance, which is usually the best, but not often practicable, solution.

4b. Knowledge objectives. This month's knowledge objectives focus on aircraft, environment, and team. Review Chapter 7 as well as FAR Part 91, Sections 91.161 to 91.193, which cover instrument flight rules. I also recommend *Weather Flying*, by Robert N. Buck (Macmillan Publishers 1978), especially Chapters 2, 8, 11, and 17. These chapters cover philosophies and techniques for weather flying that are useful for all types of flyers. Additionally, study your aircraft systems that are used for dealing with weather carefully, including radar, anti-icing equipment, and especially any cautions or warnings associated with weather operations. Review any and all associated company or military guidelines, policies, or regulations.

Month five

The fifth month is designed to review all emergency procedures in the air and in the books. Flight objectives should include all simulated emergencies from engine start through shutdown. If you have a simulator available, it is ideal for practicing critical emergencies that are too risky to attempt in flight. Knowledge objectives focus on risk, self, team, and aircraft, including the basic three-step procedure for all emergencies and all applicable procedures and regulations in the flight manual and company regulations.

5a. Flight objectives. After reviewing all applicable knowledge objectives for this month, schedule a minimum of two flights with an instructor to practice as many emergency procedures (EPs) as possible. If a simulator is available, use it for EPs that are too hazardous to attempt in the aircraft. If a simulator is not available, chair-fly these procedures. Don't forget the ground emergencies, such as engine fire, departing a prepared surface, and various abort scenarios. Debrief extensively to find your weak areas, record your reflections in your journal, and establish a plan to shore up any weak areas.

5b. Knowledge objectives. Review the basic three-step process for handling any emergency:

1. Fly the aircraft.

2. Analyze the situation.

3. Take appropriate action and land as soon as conditions permit.

Although these steps sound uncomplicated, myriad items fall into each category. Study your flight manual emergency procedure pages in detail to discover what all of these might be. If this starts to seem too much like an academic exercise, I recommend reading a few case studies from either of two excellent books, Norbert Slepyan's *Crisis in the Cockpit* (Macmillan Publishing 1986) or Robert L. Cohn's *They Called It Pilot Error* (TAB/McGraw-Hill 1994). Use the airmanship model to analyze the nature of the accidents and incidents, and mentally project similar scenarios into your own flight environment for lessons learned.

The second knowledge objective is to identify all available resources at your disposal to handle emergency situations. Using a team approach (i.e., other crewmembers, peers, instructors, etc.) see how many resources you can list that could potentially aid you in an emergency. Don't forget to look outside as well as inside the cockpit. For additional information on using all available resources and cockpit/crew resource management, I recommend Chapter 6 of *Pilot Judgment and Crew Resource Management*, by Richard S. Jensen (Avebury Aviation 1995) and *Cockpit Resource Management*, by Earl Weiner, Barbara Kanki, and Robert Helmreich (Academic Press 1993). Both of these books are excellent research-based overviews of relevant material on maximizing resources and decision making.

Month six

The final month in the jump-start program focuses on the capstones of airmanship, situational awareness, and judgment. The only specific flight objectives for this month are to monitor and analyze your inflight decision-making process and situational awareness.

6a. Flight objectives. Reflect after each mission, asking yourself and crew (if applicable) why you did what you did at each decision-making crossroad in the mission. Although most of us believe we maintain complete situational awareness throughout all flights, if we take a few moments after each sortie to ask ourselves where we were least comfortable or most distracted, we can identify personal high-risk areas for loss of SA. If you actually encounter momentary loss of SA, analyze the episode in depth, looking for causes and cures. Also be certain to assess your symptoms. How did you determine you were losing SA? Be sure to look closely at your reactions to the episode. What snapped you out of it? Were you lucky or good? Ensure you log these reflections and reactions in your airmanship journal.

6b. Knowledge objectives. The knowledge objectives for this month are to gain basic insights and understanding on the complex topics of situational awareness and judgment. Entire college courses are taught on these subjects, so avoid getting in too deep to the research. Review Chapters 9 and 10 to ensure an understanding of the role of SA in decision making, as well as various inputs to situational awareness. It is important to realize that there are both positive and negative "feeders" to our SA, and the sources of distraction are also critical pieces of information to integrate into our situational-awareness equation. It is also important to gain an understanding of the physiological aspects of SA and judgment. Once again, *Fit to Fly,* by Richard Reinhart, is an operationally oriented text that points out how various physical factors can dramatically affect our perceptions and performance in this area.

At the end of this six-month program, you will have touched base with every aspect of airmanship. There will still be work to be done, as improvement never stops, and it takes years to develop airmanship skills to a fine edge. But it is likely that you already possess many of the skills and much of the knowledge that makes a top-notch airman, so your journey may be one of mending, filling, and patching as opposed to one of complete construction. Hopefully, the jump-start program will open doors to a variety of new information and perhaps suggest a more systematic way to approach improvement in a very demanding, complex, and often hostile environment.

A final perspective

Aviators come in all sizes, shapes, religions, sexes, races, creeds, and degrees of airmanship proficiency. We fly different types of aircraft in different locations, from a grass strip in Mississippi to an aircraft carrier in the Indian Ocean. We fly for different reasons. Some fly for money, some for prestige, and some just for fun. But regardless of what background we bring to aviation, what aircraft type we fly, or what our motivations are for doing it, we all have a moral responsibility to each other to practice sound fundamental airmanship. This obligation also extends to the public at large, over whom we fly and who often become the innocent victims of our ineptitude. How we address this obligation is the subject of this book. The airmanship model is offered as a means for individual aviators to take a stand against poor airmanship at the only point at which we have total control — ourselves.

The cure for the rash of human-error accidents and incidents lies at our fingertips. Through self-improvement, we, as aviators, can effect a cultural change in aviation. We can make undisciplined, unskilled, or unknowledgeable aviators a thing of the past. Before we can expect changes in others, we must make certain that our own camp is in order. The standards suggested by the airmanship model are by no means the final word on this matter, but it does represent characteristics and traits of successful airmen since the dawn of flight.

The essence of what it means to be an airman cannot fall by the wayside. We *need* a shared sense of "who we are and what we stand for," as General Shaud so astutely pointed out in the foreword. The common structure and language suggested in this book may be the first step in this direction. The next step is yours.

Reference

Hughes, R. L., R. C. Ginnett, and G. J. Curphy. 1988. *Leadership: Enhancing the Lessons of Experience.* Homewood, Ill.: Irwin Press.

Appendix

LINE/LOS checklist

NASA/UT LINE/LOS Checklist

Observer is to complete one form for each flight segment

Date (Mo. Yr.)	
Observer ID	
Route	
A/C Type & Series	
Hrs Observed	

LOFT	
Scenario ID	
Line Obs.	

Demographics

	Capt.	1st Off	2nd Off
Domicile			
Years of experience - all airlines			
Yr. in Position this A/C			

Leg # of Legs Observed for this crew	of
Pilot Flying	

Check One Box

First leg flown together		More than one day flown together
First day together		

CREW PERFORMANCE RATING BY PHASE OF FLIGHT

1
Poor - Observed performance is significantly below expectations. This includes instances where necessary behavior was not present, and examples of inappropriate behavior that was detrimental to mission effectiveness.

2
Minimum Expectations - Observed performance meets minimum requirements but there is ample room for improvement. This level of performance is less than desired for effective crew operations.

3
Standard - The demonstrated behavior promotes and maintains crew effectiveness. This is the level of performance that should be normally occurring in flight operations.

4
Outstanding - Performance represents exceptional skill in the application of specific behaviors, and serves as a model for teamwork - truly noteworthy and effective.

The following performance markers are specific behaviors that serve as indicators of how effectively resource management is being practiced. They are not intended to be exhaustive lists of behaviors that should be seen, but rather as exemplars of behaviors associated with more and less effective crew resource management. When performance is rated either as (4) or (1), please describe the causes for the rating in the COMMENTS section.

Team Management & Crew Communications	Pre - Depart	T/O & Climb	Cruise	Des/Appr Landing	COMMENTS
1. Team concept and environment for open communications established and/or maintained (e.g., crewmembers listen with patience, do not interrupt or "talk over", do not rush through the briefing, make eye contact as appropriate).					
2. Briefings are operationally thorough, interesting, and address crew coordination and planning for potential problems. Expectations are set for how deviations from normal operations are to be handled					
3. Cabin crew is included as part of team in briefings, as appropriate, and guidelines are established for coordination between flight deck and cabin. Passengers are briefed and updated as needed, i.e. delays, weather, etc.					

431

Team Management & Crew Communications (Cont.)

	Pre-depart	T/O & Climb	Cruise	Desc/Appr Landing	COMMENTS
4. Group climate is appropriate to operational situation (e.g. presence of social conversation). Crew ensures that non-operational factors such as social interaction do not interfere with necessary tasks.					
5. Crewmembers ask questions regarding crew actions and decisions. i.e. effective inquiry					
6. Crewmembers speak up, and state their information with appropriate persistence, until there is some clear resolution and decision. i.e. effective advocacy & assertion.					
7. Operational decisions are clearly stated to other crewmembers and acknowledged, and include cabin crew and others when appropriate.					
8. Captain coordinates flightdeck activities to establish proper balance between command authority and crew member participation, and acts decisively when the situation requires					

Situational Awareness & Decision Making

9. Workload and task distribution is clearly communicated and acknowledged by crew members. Adequate time is provided for completion of tasks.					
10. Secondary operational tasks (e.g. dealing with passenger needs, company communications) are prioritized so as to allow sufficient resources for dealing effectively with primary flight duties.					
11. Crewmembers check in with each other during times of high & low workload to maintain situational awareness & alertness.					
12. Crew prepares for expected or contingency situations including approaches, weather, etc.(e.g. stays "ahead of curve").					

Automation Management

13. Guidelines are established for the operation of automated systems (i.e. when systems will be disabled, programming actions that must be verbalized and acknowledged).					
14. PF and PNF duties and responsibilities with regard to automated systems are outlined. (e.g. FMS entry and cross-checking)					
15. Crewmembers periodically review and verify the status of aircraft automated systems.					
16. Crewmembers verbalize and acknowledge entries and changes to automated systems parameters.					
17. Crew plans for sufficient time prior to maneuvers for programming of Flight Management Computer.					
18. Automated systems are used at optimal levels. (i.e. when programming demands could reduce situational awareness and create work overloads, the level of automation is reduced or disengaged)					

Special Situations

	Pre-Depart	T/O & Climb	Cruise	Des/Appr Landing	COMMENTS
19. Positive and negative performance feedback is given at appropriate times and is made a positive learning experience for the whole crew – feedback is specific, objective, based on observable behavior, and given constructively.					
20. Performance feedback is accepted objectively and non defensively.					
21. When conflicts arise, the crew remains focused on the problem or situation at hand. Crewmembers listen actively to ideas and opinions and admit mistakes when wrong. i.e. was the conflict resolved?					
22. During long duty periods, crewmembers are pro-active in maintaining high levels of crew alertness and taking fatigue countermeasures.					
23. Crew actions avoid the creation of self-imposed workload and stress. e.g. late descent due to lack of situational awareness/planning					
24. Crewmembers recognize and report work overloads in self and others.					
25. When appropriate, crewmembers take the initiative and time to share operational knowledge and experience i.e. new: crewmembers, routing, airports, situations					

Overall Crew Rating

Overall Observations

		COMMENTS
28. Overall Technical Proficiency		
29. Overall Crew Effectiveness		

Operational Considerations

		COMMENTS
30. Assess the severity of abnormals and other systems events that occur during flight. This item is rated 1=low to 4=high		
31. Assess the complexity of operating environment (e.g. WX, ATC, Traffic, MEL's, XCM's), rated 1=low to 4=high) - Comment on conditions affecting flight.		

In those cases where the actions of a particular crewmember may be particularly significant to the outcome of the observed behavior enter the relevant item number from above and the individual rating.

Item #	Crew position	Rating	Comments

Additional Comments on flight

Glossary

AAA	Anti-aircraft artillery
ABCCC	Airborne Command and Control Center
ACC	Air Combat Command
AFR	Air Force Regulation
agl	Above gound level
AIM	*Aeronautical Information Manual*
AMRAAM	Advanced medium-range air-to-air missile
AOA	Angle of attack
AOPA	Aircraft Owners and Pilots Association
armament	Weapons carried on an aircraft
ARSA	Airport radar service area
ARTCC	Air route traffic control center
ASRS	Aviation safety reporting system
ATC	Air traffic control
ATIS	Automatic terminal information system, which provides current, routine information to arriving and departing aircraft by means of continuous repetitive broadcasts
ATP	Airline transport pilot
BDA	Bomb damage assessment
BFR	Biennial flight review
BOLDFACE	Critical action emergency procedures that must be memorized

435

CAS	Close air support
CBU	Cluster bomb unit
CEVG	Combat evaluation group
CFI	Certified flight instructor
CFIT	Controlled flight into terrain
CG	Center of gravity
CLRM	Command leadership resource management
CND	Could not duplicate
CNN	Cable News Network
CPT	Cockpit procedural trainer
CRM	Cockpit/crew resource management
CVR	Cockpit voice recorder
DCS	Decompression sickness
DME	Distance-measuring equipment
DO	Deputy commander for operations
EMS	Emergency medical service
EP	Emergency procedure
EPR	Engine pressure ratio
EPT	Effective performance time
FAA	Federal Aviation Administration
FAC	Forward air controller
FAR	Federal Aviation Regulation
field grade	Military ranks between major and colonel
finis flight	Last flight in a given type of aircraft
flyby	Airshow overflight
FO	First officer
GA	General aviation
GPS	Global positioning system

GPWS	Ground proximity warning system
groupthink	A phenomenon in which the group prefers harmony over finding the best solution
IAS	Indicated airspeed, or what the airspeed dial normally reads
IAW	In accordance with
ICAO	International Civil Aviation Organization
IFR	Instrument flight rules
IG	Inspector general
ILS	Instrument landing system, which allows appropriately equipped aircraft to land in bad weather
IMC	Instrument meteorological conditions, or conditions below those required for VFR flight
INS	Inertial navagation system
IP	Instructor pilot
IP/AC	Instructor pilot/aircraft commander
ISD	Instructional system development
KIAS	Knots indicated airspeed
LOFT	Line-oriented flight training
MAC	Military airlift command
MAW	Medical airworthiness
MEA	Minimum en route altitude
MOA	Military operating area
MP	Mishap pilot
msl	Mean sea level, or altitude above the ocean
NASA	National Aeronautics and Space Administration
NDB	Nondirectional radio beacon
NOTAM	Notices to Airmen, which are regular updates on rules and regulations

NTSB	National Transportation Safety Board
OI	Operating instructions
ORI	Operational readiness inspection
OTC	Over-the-counter
OUE	Operational utility evaluation
PGM	Precision-guided munitions
PIC	Pilot in command
PIREP	Pilot report
PIT	Pilot instructor training
PME	Professional military education
QFE	The altimeter setting that gives you height above the airfield on your altimeter; the default landing datum in the United Kingdom (U.K. only)
QNH	The altimeter setting that gives height above sea level. It may be specific to an airfield (airfield QNH) or a regional setting (regional QNH).
RAF	Royal Air Force
RCO	Range control office
ROE	Rules of engagement
RN	Radar navigator
RVR	Runway visual range, or the range over which the pilot of an aircraft on the centerline of a runway can expect to see the runway markings or lights
SAM	Surface-to-air missile
SAR	Search and rescue
SD	Spatial disorientation
SID	Standard instrument departures
SIGMET	Significant information concerning en route weather phenomena that may affect the safety of aircraft operations

SOLL	Special operations, low level
SOP	Standard operating procedure
sortie	A combat mission made by one aircraft
SPINS	Special mission instructions
Stan-eval	Standardization and evaluation
TA	Transition areas
TACAN	Tactical air navigation
TCA	Terminal control radar areas
TCAS	Terminal collision avoidance system
TFR	Terrain-following radar
TRW	Thunderstorm
TUC	Time of useful consciousness
USAF	United States Air Force
UHF	Ultrahigh frequency
VFR	Visual flight rules
VHF	Very high frequency (30–300 MHz)
VMC	Visual meteorological conditions, which are at least as good as the minimums required for VFR flight
VMCA	Velocity minimum control when airborne
VOR	VHF omnidirectional radio range
VVI	Vertical velocity indicator
WSO	Weapon systems officer

Bibliography

Adams, Richard, and Jack Thompson, eds. 1986 Instruction in civil pilot training. *Aeronautical Decision Making for Helicopter Pilots.* Federal Aviation Administration: DOT/FAA PM-86-45. p. 3.

Air Force Historical Research Agency. 1937–1938. Bombardment aviation. 284.101–9.

————. 1943a. Report of psychiatric study of successful aircrews. 141.28J. October 11.

————. 1943b. Study of 150 successful airmen. 141.28J. Record Group 112, 730, Box 1328, National Archives. October. Office of the Surgeon General of the Army. Memorandum for the Director, Training Division.

————. 1943–1944. Aircraft accidents. 520.742–7.

————. 1943–1944. Physical standards for aviators. 520.7411–9.

————. 1944a. What fighter pilots say about their precombat training. 141.28. July.

————. 1944b. Survey of fighter pilots in Eighth Air Force: A comparison with heavy bomber pilots. Box 91, Spaatz MSS. Library of Congress. July.

————. 1944c. The reclassification of personal failures in the Eighth Air Force. 520.742–4. October 16.

AIR. 1944. The selection, classification, and initial training of air crew. March 31. 15/53.

Allen, Arthur H. 1983. Ramblings of a bomber pilot. *Stories of the Eighth: An Anthology of the Eighth Air Force in World War Two.* ed.: John Woolnough. Hollywood, Fla.: The 8th Air Force News.

Anonymous. 1979. In memorium. *The Talon.* November.

Armstrong Laboratory. 1993. Collection of critical incidents (#00144) from Operation Desert Shield/Storm. Brooks AFB (HRMA), Texas.

Arnold, Henry H., and Ira C. Eaker. 1943. *This Flying Game.* New York: Funk and Wagnalls Co.

Aviation Safety Reporting System (ASRS). 1993a. Heads up, somebody! *NASA CALLBACK.* Publication 169. Moffett Field, Calif.: NASA-Ames Research Center.

————. 1993b. A well-planned response. *NASA CALLBACK*. Publication 173. Moffett Field, Calif.: NASA-Ames Research Center.

————. 1994a. If misfortune knocks . . . don't answer. *NASA CALLBACK*. Publication 178. Moffett Field, Calif.: NASA-Ames Research Center.

————. 1994b. The pattern serves a purpose. *NASA CALLBACK*. Publication 187. Moffett Field, Calif.: NASA-Ames Research Center. December.

Aymar, Brandt. 1990. *Men in the Air: The Best Flight Stories of All Time from Greek Mythology to the Space Age.* New York: Crown Publishers.

Balwantz, CWO Chad. 1994. Personal interview. May.

Barker, John M., Cathy C. Clothier, James R. Woody, Earl H. McKinney, and Jennifer L. Brown. 1996. Crew resource management: A simulator study comparing fixed versus formed aircrews. *Aviation, Space, and Environmental Medicine.* 67 (1): 3–7.

Barker, Ralph. 1981. *The RAF at War.* Series, The Epic of Flight. Alexandria: Time-Life Books.

Billings, C. E., and E. S. Cheaney. 1981. Information transfer problems in the aviation system. NASA Technical Paper 1875. Moffett Field, Calif.: NASA-Ames Research Center.

Bond, Douglas D. 1945. A study of successful airmen with particular respect to their motivation for combat flying and resistance to combat stress. 520.7411-1. Air Force Historical Research Agency. January 27.

————. 1952. *The Love and Fear of Flying.* New York: International Universities Press.

Brooks, Robert. 1994. Crew errors and the night environment. *Flightfax.* August 5.

Bryan, L. A., J. W. Stonecipher, and K. Aron. 1955. AOPA's 180 degree rating. AOPA Foundation, Inc. and Aircraft Owners and Pilots Association, Washington, D.C.

Buck, Robert N. 1978. *Weather Flying: A Practical Book on Flying in All Kinds of Weather By a Pilot.* New York: Macmillan.

Bulfinch, Thomas. 1934. *Bulfinch's Mythology.* New York: The Modern Library.

CAE-Link Corporation. 1995. Aircrew coordination workbook. Abilene, Tx.: CS-33.

Caine, Philip D. 1991. *Eagles of the RAF.* Washington, D.C.: NDU Press.

Casey, Steven M. 1993. *Set Phasers on Stun: And Other True Tales of Design, Technology, and Human Error.* Santa Barbara: Aegean Publishing Company.

Chapanis, A. 1951. Theory and method for analyzing errors in man-machine systems. *Annals of the New York Academy of Science.* 51:1179–1203.

Childs, J. M., W. D. Spears, and W. W. Prophet. 1983. Private pilot flight skill retention 8, 16, and 24 months following certification. Atlantic City Technical Center: New Jersey Department of Transportation/FAA.

Chute, R. D., and E. L. Wiener. 1994. Cockpit and cabin crews: Do conflicting mandates put them on a collision course? *Flight Safety Foundation Cabin Crew Safety,* 29(2):1.

Cohn, Robert L. 1994. *They Called It Pilot Error: True Stories Behind Aviation Accidents.* Blue Ridge Summit, Pa.: TAB/McGraw-Hill.

Collins, Richard L. 1981. *Flying Safety.* 2d. ed. New York: Delacorte Press.

Connelly, Pete. 1994. CRM in general aviation: Who needs it? Unpublished technical paper. Austin, Tx.: NASA/UT/FAA Aerospace Crew Research Project.

Craig, P. 1992. *Be a Better Pilot: Making the Right Decisions.* Blue Ridge Summit, Pa.: TAB/McGraw-Hill.

Crew Training International (CTI). 1995. CRM course workbook (F-15). Germantown, Tenn.

Cross, Christopher. 1994. Telephone interview with tower controller. September 8.

Crouch, Tom D. 1989. *The Bishop's Boys: A Life of Wilbur and Orville Wright.* New York: W. W. Norton & Co.

Curtiss, Glenn, and Augustus Post. 1912. *The Curtiss Aviation Book.* New York: Frederick A. Stokes.

————. 1990. The first international aeroplane contest. *Men in the Air.* ed.: Brandt Aymar. pp. 220–225. New York: Crown Publishers.

D'Orlandi, Renato (translator). 1961. *The Origin of Air Warfare.* 2d. ed. Rome: Historical Office of the Italian Air Force.

Davis, Frederick B. (ed.) 1947. The AAF qualifying examination. Army Air Forces aviation psychology program. Research reports No. 6. Washington, D.C.: Government Printing Office.

Del Toro, J. 1995. The Mishap in the Mirror: Instructor Responsibilities. *Torch Magazine,* March. p. 6.

Department of the Air Force (DAF). 1951. Poor teamwork as a cause of aircraft accidents. USAF IG Report.

DeSeversky, Alexander P. 1942. *Victory Through Air Power.* New York: Simon and Schuster.

Diehl, Alan. 1989. Understanding and preventing aircrew judgment and decision-making errors. HQ USAF Inspection and Safety Center. November 13.

Diehl, Col. Billy. 1994. Taken from videotaped interview. May.

Doolittle, Gen. James H. 1979. Interview by Professor Ronald Schaffer. 239.0512-1206. Air Force Historical Research Agency. August 24.

Dorfler, Maj. Joseph F. 1988. *The Branch Point Study: Specialized Undergraduate Pilot Training.* Maxwell AFB: Air University Press.

Douhet, Giulio. 1941. *Command of the Air.* Translated by Dino Ferrari. New York: Coward-McMann. Reprinted 1983, Washington, D.C.: Office of Air Force History.

Eaker, Lt. Gen. Ira C. 1918–1960. Personal papers. Washington, D.C.: Library of Congress.

————. 1976. The military professional. *Air University Review,* January–February: 2–12.

Eighth Air Force Headquarters. 1943. Report of psychiatric study of successful aircrew. 00114382.

Endsley, Mica. 1989. Pilot situation awareness: The challenge for the training community. Paper presented at the Interservice/Industry Training Systems Conference, Fort Worth, Tx. November.

Ericsson, K. Anders, and Jacqui Smith. 1991. *Toward a General Theory of Expertise: Prospects and Limits.* New York: Cambridge University Press.

European Theater of Operations Headquarters Research Branch. 1994. Attitudes of fighter pilots toward combat flying. 141.28-21. Air Force Historical Research Agency. July.

Fahlgren, G., and R. Hagdahl. 1990. Complacency. Proceedings of the 43rd Annual International Air Safety Seminar, Rome, Italy. Flight Safety Foundation.

Federal Aviation Administration (FAA). 1987. *Introduction to Pilot Judgment.* FAA Accident Prevention Branch. FAA Pamphlet P-8740-53.

First Central Medical Establishment Headquarters (U.S. Army). 1944. A study of 100 successful airmen with particular respect to their motivation and resistance to combat stress. 00224740.

Flanagan, Lt. Col. J. C., and Major P. M. Fitts. 1944. Psychological testing program for the selection and classification of aircrew officers. *The Air Surgeon's Bulletin.* June 1–5.

Flying Magazine. 1993. *I Learned about Flying from That.* Blue Ridge Summit, Pa.: TAB/McGraw-Hill.

Foushee, H. C. 1985. Realistic training for effective crew performance. Proceedings, 4th Aerospace Behavioral Engineering Technology Conference. 177–181.

Foushee, H.C., and B. Kanki. 1989. Communications as a group process mediator of aircrew performance. *Aviation, Space and Environmental Medicine.* 60: 56–60.

Gabreski, Col. Francis S. 1991. *Gabby: A Fighter Pilot's Life.* New York: Orion Books.

Gajeski, Antone E. 1988. Combat aircrew experiences during the Vietnam conflict: An exploratory study. Master's thesis. Air Force Institute of Technology.

Galland, Adolf. 1953. Defeat of the Luftwaffe: Fundamental causes. *Air University Review.* Spring: 16–36.

Galland, Adolf. 1955. *The First and the Last*. London: Methuen.

Ginnett, R. C. 1987. First encounters of the close kind: The formation process of airline flight crews. Unpublished doctoral dissertation, Yale University, New Haven, Conn.

————. 1995. Groups and leadership. *Cockpit Resource Management*. eds.: E. L. Weiner, B. G. Kanki, and R. L. Helmreich. San Diego: Academic Press.

Global Trade and Transportation. 1993. Study cites human error in maritime accidents. 113(7):12.

Goodson, James A. 1983. *Tumult in the Clouds*. London: William Kimber.

Hallion, Richard, and Bildstein, Roger. 1978. *The Wright Brothers: Heirs of Prometheus*. Washington, DC: Smithsonian Press.

Hammel, Eric M. 1992. *Aces Against Japan: The American Aces Speak*. Novato, Calif.: Presidio Press.

Hawkins, Ian (ed.). 1990. *B-17s Over Berlin: Personal Stories from the 95th Bomb Group*. Washington: Brassey's.

Heinl, Robert Debs, Jr. 1966. *Dictionary of Military and Naval Quotations*. Annapolis: United States Naval Institute.

Helmreich, Robert L., and John A. Wilhelm. 1989. When training boomerangs: Negative outcomes associated with cockpit resource management programs. Proceedings of the 5th International Symposium on Aviation Psychology, ed. R. S. Jensen. pp. 692–697. Columbus, Ohio.

Helmreich, Robert. L., Roy E. Butler, William R. Taggart, and John A. Wilhelm. 1995. Behavioral markers in accidents and incidents: Reference list. NASA/UT FAA Aerospace Crew Research Project Technical Report, 95-1.

Hodgson, David. 1985. *Letters from a Bomber Pilot*. London: Thames Methuen.

Hudson, G. 1992. Cockpit resource management for a flight instructor. *Flying Safety Magazine*. December: 25.

Hughes Training, Inc. 1995a. Aircrew coordination workbook. Abilene, Tx.: Hughes Training Inc., CS-61.

————. 1995b. Aircrew coordination training (ACT) workbook, draft case studies. Provided on disk by Dave Wilson, Hughes Training Inc., Abilene, Tx.

Hughes, R. L., R. C. Ginnett, and G. J. Curphy. 1988. *Leadership: Enhancing the Lessons of Experience*. Homewood, Ill.: Irwin Press.

International Civil Aviation Organization. 1991. *Human Factors Digest No. 3: Training of Operational Personnel in Human Factors*. Montreal: International Civil Aviation Organization (ICAO).

Jensen, Richard S. 1995. *Pilot Judgment and Crew Resource Management*. Aldershot, UK: Avebury Aviation.

Jessup, J. E. II, and Robert W. Coakley. 1978. *A Guide to the Study and Use of Military History*. Washington D.C.: Center of Military History, United States Army.

Kelly, R. E. 1988. A two-dimensional model of follower behavior. *Leadership: Enhancing the Lessons of Experience.* eds.: Richard Hughes, Robert Ginnett, and Gordon Curphy. p. 229. Homewood, Ill.: Irwin Press.

Kern, Anthony. 1992. Personal testimony to USAF 110-14 accident investigation board, B-1B crash, Dyess AFB. November 30.

———. 1994a. Ignoring the pinch. *Torch Magazine.* Air Education and Training Command (AETC). Randolph AFB, Texas. September.

———. 1994b. A historical analysis of tactical aircrew error in Operations Desert Shield/Storm. U.S. Army Command and General Staff College (CGSC). Monograph. June 2.

———. 1995. What is airmanship? A survey of military aviators. Unpublished research in progress. United States Air Force Academy.

Klein, G. A. 1993. A recognition-primed decision (RPD) model of rapid decision making. *Decision Making in Action: Models and Methods.* eds. G. Klein, J. Orasanu, R. Caulderwood, and C. Zsambok. Norwood, New Jersey: Ablex.

Knoke, Heinz. 1953. *I Flew for the Führer: The Story of a German Fighter Pilot.* New York: Holt.

Lauber, John. 1996. Interview with the vice president for safety and compliance, Delta Airlines. January 21.

Longacre, R. F. 1929. Personality study. *The Journal of Aviation Medicine.* 33–50.

Magnuson, Kent. 1995a. Human factors in USAF mishaps, 1 Oct 89 to 1 Mar 95. USAF Safety Agency Life Sciences Division, Albuquerque, N.M.

———. 1995b. Human factors associated with mishaps database. Kirtland AFB, N.M.: Air Force Safety Center.

Marshall, S. L. A. 1947. *Men Against Fire.* New York. Reprinted 1978, Gloucester, Mass.: Peter Smith.

Matchette, R. 1995. Nonstandard phraseology incidents. ASRS Directline, Issue 7.

Maurer, Maurer. 1987. *Aviation in the U.S. Army, 1919–1939.* Washington, D.C.: Office of Air Force History.

McConnell, Col. Michael G. 1994. Executive summary. United States Air Force Regulation 110-14. USAF Accident Investigation Board B-52 Crash at Fairchild AFB. June 24.

McDonald, Major Ralph E. 1994. *Cohesion: The Key to Special Operations Teamwork.* AU-ARI-94-2. Maxwell AFB, Ala.: Air University Press.

McElhatton, J., and C. Drew. 1993. Time pressure as a causal factor in aviation safety incidents: The hurry-up syndrome. Proceedings of the Seventh International Symposium on Aviation Psychology, Ohio State University, Columbus, Ohio.

Messer, M. E. 1992. Transport vehicle accident rates drive operator training programs to change. *Occupational Health and Safety* 61(1):20–22.

Morrison, R., K. Etem, and B. Hicks. 1993. General aviation landing incidents and accidents: A review of ASRS and AOPA research findings. Proceedings of the Seventh International Symposium on Aviation Psychology, 975–980.

Nance, John J. 1986. *Blind Trust: How Deregulation Has Jeopardized Airline Safety and What You Can Do About It.* New York: William Morrow and Company.

NASA ASRS. 1995. Accession Number: 134927. The Aeroknowledge, ASRS [CD-ROM].

———. 1995. Accession Number: 245988. The Aeroknowledge, ASRS [CD-ROM].

———. 1995. Accession Number: 261766. The Aeroknowledge, ASRS [CD-ROM].

———. 1995. Accession Number: 297539. The Aeroknowledge, ASRS [CD-ROM].

———. 1995. Accession Number: 231117. The Aeroknowledge, [CD-ROM].

———. 1995. Accession Number: 92283. The Aeroknowledge, [CD-ROM].

———. 1995. Accession Number: 124564. The Aeroknowledge, [CD-ROM].

NASA ASRS Directline. 1995. Lost Com. Issue 6, p. 18.

NASA ASRS monthly update. 1995. Say the magic words. Number 196.

National Transportation Safety Board. 1978. Report NTSB-AAR-78-3, January 26.

———. 1980. Aircraft accident report, NTSB-AAR-80-2.

———. 1982. Report NTSB-AAR-82-10. August 24.

———. 1985a. Accident ID #CHI89FA039.

———. 1985b. Accident/incident summary report: Cockeysville, Md.: April 28, 1984. NTSB-AAR-85-01-SUM.

———. 1991. Report brief. NTSB File No. 1630, Accident ID #LAX89FAO78.

———. 1993. Report brief. Accident ID #LAX94FA005, NTSB File #1730.

———. 1994a. Report brief. Accident ID #LAX92FA202, NTSB File #2764

———. 1994b. Report brief. Accident ID #LAX94LA003, NTSB File #1740.

———. 1977. Aircraft accident report. L&J Company, Convair 240, N55VM, Gillsburg, Mississippi, October 20, 1977. Report number NTSB-AAR-78-6.

———. 1980. Report brief. File 140, #LAX89DUJO4.

———. 1989. Report brief. File 2280, #LAX89FA111.

————. 1989. Report brief. Accident ID #FTW89MA047, File #944.

Northwest Airlines instructor seminar facilitator guide. 1993. p. 4. Reprinted by permission.

Notes for instructors on the recognition of nervousness in pilots. 1943. AIR 2/6252.

NTSB ASRS. 1994. Accession number: 133773. Aeroknowledge [CD-ROM].

NTSB ASRS. 1994. Accession number: 180477. Aeroknowledge [CD-ROM].

Orasanu, Judith. 1993. Decision making in the cockpit. *Cockpit Resource Management.* eds.: E. Wiener, B. Kanki, and R. Helmreich. San Diego: Academic Press.

————. 1995. Situation awareness: Its role in flight crew decision making. NASA Ames research paper. In nasa.gov/publications/OSU_Orasanu. Internet.

Powell, Stewart. 1991. Friendly fire. *Air Force Magazine.* December: 59.

Prime, J. A. 1995. 1994 BAFB mission's near collision revealed. *The Shreveport Times.* December 21.

Provenmire, H. 1989. Cockpit resource management: An annotated bibliography. October.

Rasmussen, J. 1993. Deciding and doing: Decision making in natural context. *Decision making in action: Models and methods.* eds.: G. Klein, Judith Orasanu, R. Caulderwood, and C. Zsambok. Norwood, N.J.: Ablex.

Reason, James. 1990. *Human Error.* New York: Cambridge University Press.

Reinhart, Richard O. 1993. *Fit to Fly: A Pilot's Guide to Health and Safety.* Blue Ridge Summit, Pa: TAB/McGraw-Hill.

Rippon, T. S. 1918. Report on the essential characteristics of successful and unsuccessful aviators with special reference to temperament. *Aeronautics.* October 9.

Rippon, T. S., and E. G. Mannell. 1918. Report of the essential characteristics of successful and unsuccessful aviators. *The Lancet.* September 28: 411–15.

Rommel, Erwin. 1953. The Rommel papers. *Dictionary of Military and Naval Quotations.* ed.: R. D. Heinl, Jr. (1966). pp. 194. Annapolis, Maryland: Naval Institute Press.

Roscoe, Stanley. 1980. *Aviation Psychology.* Ames, Iowa: Iowa State University Press.

Sarter, Nadine B., and David D. Woods. 1994. *Pilot Interaction With Cockpit Automation II: An Experimental Study of Pilots' Model and Awareness of the Flight Management System.* 4(1):1–28.

Scharr, Adela Riek. 1986. *Sisters in the Sky.* 2d. ed. St. Louis: Patrice Press.

Schwartz, Doug. 1990. Reducing the human error contribution to mishaps through identification of sequential error chains. *Safetyliner.* 13–19.

———. 1992. Training for situational awareness. Presented at the Ohio State University Fifth International Symposium on Aviation Psychology, Columbus, Ohio.

Sender, J. W. 1983. On the nature and source of human error. Proceedings of the Second Symposium on Aviation Psychology, Ohio State University, Columbus, Ohio.

Senge, Peter M. 1990. *The Fifth Discipline: The Art & Practice of The Learning Organization.* New York: Doubleday.

Shrader, Charles R. 1982. Amicide: The problem of friendly fire in modern war. Combat Studies Institute, Fort Leavenworth, Kansas.

Sikorsky, Igor. 1990. The first flights of Igor Sikorsky. In *Men in the Air.* ed.: Brandt Aymar. pp. 161–162. New York: Crown Publishers.

Slepyan, Norbert, ed. 1986a. *Crisis in the Cockpit: Other Pilots' Emergencies and What You Can Learn from Them.* New York: Macmillan Publishing Company.

———. 1986b. *Defensive Flying.* New York: Macmillan Publishing Company.

Smith, Perry. 1986. *Taking Charge: A Practical Guide for Leaders.* Washington D.C.: National Defense University Press.

Spaatz, Gen. Carl. 1915–1953. Personal papers. Washington, D.C.. Library of Congress.

Spick, Mike. 1988. *The Ace Factor: Air Combat and the Role of Situational Awareness.* Annapolis: Naval Institute Press.

Stewart, J. 1995. A little bit of all of us. *Torch Magazine.* Air Education and Training Command (AETC). Randolph AFB, Texas.

Stokes, Alan F., and Kirsten Kite. 1994. *Flight Stress: Stress, Fatigue, and Performance in Aviation.* Aldershot, UK: Avebury Aviation.

Sventek, Jeff. 1994. Aircrew awareness and attention management briefing. San Antonio: Trinity University.

Trollip, Stanley R., and Richard S. Jensen. 1991. *Human Factors for General Aviation.* Englewood, Colo.: Jeppesen Sanderson, Inc.

U. S. Army Air Forces. 1945. Letter 35-8. 520.7411-3. Air Force Historical Research Agency. April 12.

U.S. Department of Transportation/FAA. 1991. Flight Instructor for Airplane, Single Engine. Practical Test Standards. Washington, D.C: U.S. Department of Transportation and Federal Aviation Administration.

United States Air Force. 1991. Air Force Regulation (AFR) 110-14. 1991. Report of aircraft accident investigation of B-52G 4300 Provisional Bombardment Wing (SAC) in the Indian Ocean. 59-2593. February 3.

———. 1992a. Aircraft Report, C-130E, SN 63-7881, Berkeley Springs, West Virginia. October 7.

————. 1992b. Accident investigation report, B-1B mishap, Dyess AFB, Texas. November 22.

————. 1994a. Report on B-52 mishap. June 24.

————. 1994b. Report on Blackhawk shootdown. Vol 12: 13. July.

United States Air Force (USAF). 1991. Accident Investigation Technical order 1B-52G-1-11.

USAF Air Training Command Study Guide. 1990. Instructor Development. Randolph AFB, Texas: USAF Air Training Command.

VEDA, Inc. 1988. AMRAAM OUE Tactics Analysis Methodology Briefing. As quoted in Stiffler, Don: Graduate level situational awareness, *USAF Fighter Weapons Review.* Summer.

Wagg, Wayne L. 1993. Preliminary report from the situational awareness integration team (SAINT). Briefing slides, Armstrong Lab, Mesa, Ariz.

Wagner, B., and Alan Diehl. 1994. Your wingman is your copilot. *Flying Safety Magazine.* Albuquerque: USAF Safety Agency.

Waller, Lt. Gen. Calvin. 1995 U.S. Army Command and General Staff College (CGSC) lecture slides.

Washio, A. 1991. Nuclear "incident" report fails to calm public fears. *Japan Times Weekly International Edition.* 31(49):4.

Webster's Ninth New Collegiate Dictionary. 1990. Springfield, Mass.: Merriam-Webster, Inc.

Wells, Mark K. 1995. *Courage and Air Warfare.* London: Cass Publishing.

Westlake, Michael. 1992. It was a bad year at sea. *Far East Economic Review.* November 19: 50–51.

White, Gen. Thomas. 1973. Air Force Fact Sheet 73-7, AF Aces. Secretary of the Air Force Office of Information.

Wiener, Earl L., and David C. Nagel (eds.). 1988. *Human Factors in Aviation.* San Diego: Academic Press, Inc.

Williams, Maj. Christopher, and Maj. John Auten. 1994. Situational awareness in tactical fighter operations. Unpublished instructional materials, aerospace medicine course. Brooks AFB, Texas.

Wills, Kelly, Jr. 1968. Feldfliegerabteilung 25—The Eyes of Kronprinz, from the recollections of Dr. Wilhelm Hubener, The Cross and Cockade 9(1): 22–32.

Wilson, W. B. 1973. The effect of prolonged nonflying periods on pilot skill in performance of simulated carrier landing task. Masters thesis. Monterey, Calif.: Naval Postgraduate School.

Woolnough, John H. (ed.) 1983. *Stories of the Eighth.* Hollywood, Fla.: The 8th Air Force News.

Yacovone, D. W., M. S. Borosky, R. Bason, and R. A. Alcov. 1992. Flight experience and the likelihood of U.S. Navy mishaps. *Aviation, Space, and Environmental Medicine.* 63:72–74.

Yeager, Brig. Gen. Charles E. 1980. Interviewed for the Albert F. Simpson
    Historical Research Center. 239.0512-1204. Air Force Historical Re-
    search Branch. April 28–May 1.
————. 1985. *Yeager: An Autobiography.* New York: Bantam Books.
————. 1990. Quoted in *Military Airpower: The CADRE Digest of Air-
    power Quotations and Thoughts.* ed.: C. M. Westenhoff. Maxwell Air
    Force Base, Ala.: Air University Press. p. 23.

Index

About the Author

Dr. Tony Kern is a Major in th U.S. Air Force and has flown extensively over his 16-year military career, including serving as an aircrew commander, instructor pilot, and flight examiner on the supersonic B-1B bomber. He has served in a variety of operational and training capacities, specializing in aircrew training for the last decade. Kern was the Chief of Cockpit Resource Management (CRM) Plans and Programs at the USAF Air Education and Training Command, where he designed and supervised a comprehensive, career-spanning training system, which was adopted by the Air Force for use in human factors training.

Kern holds master's degrees in public administration and military history, and a doctorate in higher education. He is presently an Assistant Professor of History at the U.S. Air Force Academy in Colorado Springs, and a T-3A "Firefly" instructor pilot for the USAF flight screening program. He is currently working on a second book on the critical aspects of flight discipline.